STUDENT'S SOLUTIONS MANUAL

SALVATORE SCIANDRA

FINITE MATHEMATICS & ITS APPLICATIONS
ELEVENTH EDITION

Larry J. Goldstein
Goldstein Educational Technologies

David I. Schneider
University of Maryland

Martha Siegel
Towson University

Boston Columbus Indianapolis New York San Francisco Upper Saddle River
Amsterdam Cape Town Dubai London Madrid Milan Munich Paris Montreal Toronto
Delhi Mexico City São Paulo Sydney Hong Kong Seoul Singapore Taipei Tokyo

The author and publisher of this book have used their best efforts in preparing this book. These efforts include the development, research, and testing of the theories and programs to determine their effectiveness. The author and publisher make no warranty of any kind, expressed or implied, with regard to these programs or the documentation contained in this book. The author and publisher shall not be liable in any event for incidental or consequential damages in connection with, or arising out of, the furnishing, performance, or use of these programs.

Reproduced by Pearson from electronic files supplied by the author.

Copyright © 2014, 2010, 2007 Pearson Education, Inc.
Publishing as Pearson, 75 Arlington Street, Boston, MA 02116.

All rights reserved. No part of this publication may be reproduced, stored in a retrieval system, or transmitted, in any form or by any means, electronic, mechanical, photocopying, recording, or otherwise, without the prior written permission of the publisher. Printed in the United States of America.

ISBN-13: 978-0-321-87831-1
ISBN-10: 0-321-87831-0

1 2 3 4 5 6 EBM 17 16 15 14 13

www.pearsonhighered.com

PEARSON

Contents

Chapter 1: Linear Equations and Straight Lines	1 – 1
Chapter 2: Matrices	2 – 1
Chapter 3: Linear Programming, A Geometric Approach	3 – 1
Chapter 4: The Simplex Method	4 – 1
Chapter 5: Sets and Counting	5 – 1
Chapter 6: Probability	6 – 1
Chapter 7: Probability and Statistics	7 – 1
Chapter 8: Markov Processes	8 – 1
Chapter 9: The Theory of Games	9 – 1
Chapter 10: The Mathematics of Finance	10 – 1
Chapter 11: Difference Equations and Mathematical Models	11 – 1
Chapter 12: Logic	12 – 1

Chapter 1

Exercises 1.1

1. Right 2, up 3
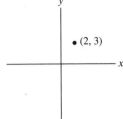

3. Down 2
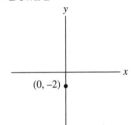

5. Left 2, up 1
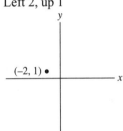

7. Left 20, up 40
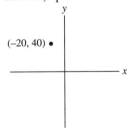

9. e

11. $-2(1) + \frac{1}{3}(3) = -2 + 1 = -1$ so the point is on the line.

13. $-2x + \frac{1}{3}y = -1$ Substitute the x and y coordinates of the point into the equation:
$\left(\frac{1}{2}, 3\right) \to -2\left(\frac{1}{2}\right) + \frac{1}{3}(3) = -1 \to -1 + 1 = -1$ is a false statement. So the point is not on the line.

15. $m = 5, b = 8$

17. $y = 0x + 3; m = 0, b = 3$

19. $14x + 7y = 21$
$7y = -14x + 21$
$y = -2x + 3$

21. $3x = 5$
$x = \frac{5}{3}$

23. $0 = -4x + 8$
$4x = 8$
$x = 2$
x-intercept: (2, 0)
$y = -4(0) + 8$
$y = 8$
y-intercept: (0, 8)

25. When $y = 0$, $x = 7$
x-intercept: (7, 0)
$0 = 7$
no solution
y-intercept: none

27. $0 = \frac{1}{3}x - 1$
$x = 3$
x-intercept: (3, 0)
$y = \frac{1}{3}(0) - 1$
$y = -1$
y-intercept: (0, -1)
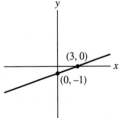

29. $0 = \dfrac{5}{2}$

no solution

x-intercept: none

When $x = 0$, $y = \dfrac{5}{2}$

y-intercept: $\left(0, \dfrac{5}{2}\right)$

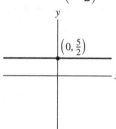

31. $3x + 4(0) = 24$
$$x = 8$$
x-intercept: (8, 0)
$$3(0) + 4y = 24$$
$$y = 6$$
y-intercept: (0, 6)

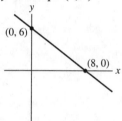

33. $x = -\dfrac{5}{2}$

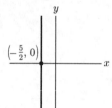

35. $2x + 3y = 6$
$$3y = -2x + 6$$
$$y = -\dfrac{2}{3}x + 2$$

a. $4x + 6y = 12$
$$6y = -4x + 12$$
$$y = -\dfrac{2}{3}x + 2$$
Yes

b. Yes

c. $x = 3 - \dfrac{3}{2}y$
$$\dfrac{3}{2}y = -x + 3$$
$$y = -\dfrac{2}{3}x + 2$$
$$y = -\dfrac{2}{3}x + 2$$
Yes

d. $6 - 2x - y = 0$
$$y = 6 - 2x = -2x + 6$$
No

e. $y = 2 - \dfrac{2}{3}x = -\dfrac{2}{3}x + 2$
Yes

f. $x + y = 1$
$$y = -x + 1$$
No

37. a. $x + y = 3$
$$y = -x + 3$$
$$m = -1, b = 3$$
L_3

b. $2x - y = -2$
$$-y = -2x - 2$$
$$y = 2x + 2$$
$$m = 2, b = 2$$
L_1

c. $x = 3y + 3$
$3y = x - 3$
$y = \frac{1}{3}x - 1$
$m = \frac{1}{3}, \ b = -1$
L_2

39. $y = 30x + 72$

   a. When x = 0, y = 72. This is the temperature of the water at time = 0 before the kettle is turned on.

   b. $y = 30(3) + 72$
   $y = 162° F$

   c. Water boils when y = 212 so we have $212 = 30x + 72$. Solving for x gives $x = 4\frac{2}{3}$ minutes or 4 minutes 40 seconds.

41. a. x-intercept: $\left(-33\frac{1}{3}, 0\right)$
   y-intercept: (0, 2.5)

   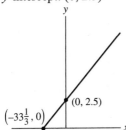

   b. In 1960, 2.5 trillion cigarettes were sold.

   c. $4 = .075x + 2.5$
   $x = 20$
   $1960 + 20 = 1980$

   d. $2020 - 1960 = 60$
   $y = .075(60) + 2.5$
   $y = 7$
   7 trillion

43. a. x-intercept: (−11.3, 0)
   y-intercept: (0, 678)

   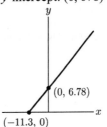

   b. In 1997 the car insurance rate for a small car was $678.

   c. $2000 - 1997 = 3$
   $y = 60(3) + 678$
   $y = 858$
   $858

   d. $1578 = 60x + 678$
   $x = 15$
   $1997 + 15 = 2012$
   The year 2012

45. a. In 2000, 3.9% of entering college freshmen intended to major in biology.

   b. $2011 - 2000 = 11$
   $y = 0.2(11) + 3.9$
   $y = 6.1$
   6.1% of college freshmen in 2011 intended to major in biology. The actual value is close to the predicted value.

   c. $5.3 = 0.2x + 3.9$
   $x = 7$
   $2000 + 7 = 2007$
   In 2007, the percent of college freshmen that intended to major in biology was 5.3.

47. a. $2008 - 2000 = 8$
   $y = 787(8) + 10600$
   $y = 16896$
   $16,896 will be the approximate average tuition in 2008.

b. $24000 = 787x + 10600$
$x \approx 17$
$2000 + 17 = 2017$
In 2017, the approximate average cost of tuition will be $24,000.

49. $y = mx + b$
$8 = m(0) + b$
$b = 8$
$0 = m(16) + 8$
$m = -\dfrac{1}{2}$
$y = -\dfrac{1}{2}x + 8$

51. $y = mx + b$
$5 = m(0) + b$
$b = 5$
$0 = m(4) + 5$
$m = -\dfrac{5}{4}$
$y = -\dfrac{5}{4}x + 5$

53. On the x-axis, y = 0.

55. The equation of a line parallel to the y axis will be in the form x = a.

57. $2x - y = -3$

59. $1 \cdot x + 0 \cdot y = -3$

61. $\dfrac{2}{3}x + y = -5$
$2x + 3y = -15$

63. Since (a,0) and (0,b) are points on the line the slope of the line is (b-0)/(0-a) = -b/a. Since the y intercept is (0,b), the equation of the line is $y = -(b/a)x + b$ or $ay = -bx + ab$. In general form, the equation is bx + ay = ab.

65. One possible equation is $y = x - 9$.

67. One possible equation is $y = x + 7$.

69. One possible equation is $y = x + 2$.

71. One possible equation is $y = x + 9$.

73. a. $y = -3x + 6$

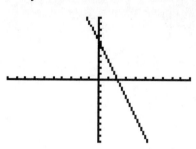
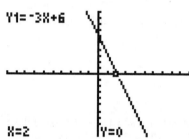

b. The intercepts are at the points (2, 0) and (0, 6)
c. When x = 2, y = 0

75. a. $3y - 2x = 9$
$3y = 2x + 9$
$y = \dfrac{2}{3}x + 3$

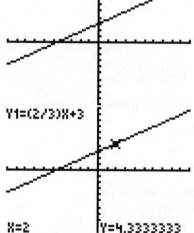

b. The intercepts are at the points (–4.5, 0) and (0, 3).
c. When x = 2, y = 4.33 or 13 / 3.

SSM: Finite Math   **Chapter 1:** Linear Equations and Straight Lines

77. $2y + x = 100$. When $y = 0$, $x = 100$, and when $x = 0$, $y = 50$. An appropriate window might be $[-10, 110]$ and $[-10, 60]$. Other answers are possible.

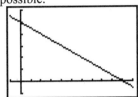

Exercises 1.2

1. False

3. True

5. $2x - 5 \geq 3$
$2x \geq 8$
$x \geq 4$

7. $-5x + 13 \leq -2$
$-5x \leq -15$
$x \geq 3$

9. $2x + y \leq 5$
$y \leq -2x + 5$

11. $5x - \frac{1}{3}y \leq 6$
$-\frac{1}{3}y \leq -5x + 6$
$y \geq 15x - 18$

13. $4x \geq -3$
$x \geq -\frac{3}{4}$

15. $3(2) + 5(1) \leq 12$
$6 + 5 \leq 12$
$11 \leq 12$
Yes

17. $0 \geq -2(3) + 7$
$0 \geq -6 + 7$
$0 \geq 1$
No

19. $5 \leq 3(3) - 4$
$5 \leq 9 - 4$
$5 \leq 5$
Yes

21. $7 \geq 5$
Yes

23.

25.

27.

29.

1-5

31. $x + 4y \geq 12$

$y \geq -\dfrac{1}{4}x + 3$

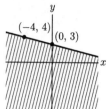

33. $4x - 5y + 25 \geq 0$

$y \leq \dfrac{4}{5}x + 5$

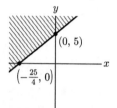

35. $\dfrac{1}{2}x - \dfrac{1}{3}y \leq 1$

$y \geq \dfrac{3}{2}x - 3$

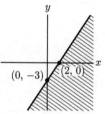

37. $.5x + .4y \leq 2$

$y \leq -1.25x + 5$

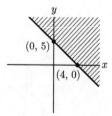

39. $\begin{cases} y \leq 2x - 4 \\ y \geq 0 \end{cases}$

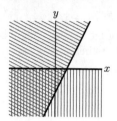

41. $\begin{cases} x + 2y \geq 2 \\ 3x - y \geq 3 \end{cases}$

$\begin{cases} y \geq -\dfrac{1}{2}x + 1 \\ y \leq 3x - 3 \end{cases}$

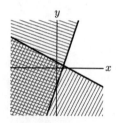

43. $\begin{cases} x + 5y \leq 10 \\ x + y \leq 3 \\ x \geq 0, y \geq 0 \end{cases}$

$\begin{cases} y \leq -\dfrac{1}{5}x + 2 \\ y \leq -x + 3 \\ x \geq 0, y \geq 0 \end{cases}$

45. $\begin{cases} 6(8)+3(7) \le 96 \\ 8+7 \le 18 \\ 2(8)+6(7) \le 72 \\ 8 \ge 0, 7 \ge 0 \end{cases}$

$\begin{cases} 69 \le 96 \\ 15 \le 18 \\ 58 \le 72 \\ 8 \ge 0, 7 \ge 0 \end{cases}$

Yes

47. $\begin{cases} 6(9)+3(10) \le 96 \\ 9+10 \le 18 \\ 2(9)+6(10) \le 72 \\ 9 \ge 0, 10 \ge 0 \end{cases}$

$\begin{cases} 84 \le 96 \\ 19 \le 18 \\ 78 \le 72 \\ 9 \ge 0, 10 \ge 0 \end{cases}$

No

49. For $x = 3$, $y = 2(3) + 5 = 11$.
So $(3, 9)$ is below.

51. $7 - 4x + 5y = 0$

$y = \frac{4}{5}x - \frac{7}{5}$

For $x = 0$, $y = \frac{4}{5}(0) - \frac{7}{5} = -\frac{7}{5}$.

So $(0, 0)$ is above.

53. $8x - 4y = 4$
$y = 2x - 1$
$8x - 4y = 0$
$y = 2x$

$\begin{cases} y \ge 2x - 1 \\ y \le 2x \end{cases}$

55. d

57. $4x - 2y = 7$

$y = 2x - \frac{7}{2}$

a.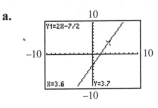

$(3.6, 3.7)$

b. Below, because $(3.6, 3.7)$ is on the line.

59.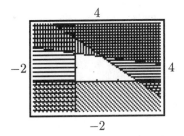

Exercises 1.3

1. $4x - 5 = -2x + 7$
$6x = 12$
$x = 2$
$y = 4(2) - 5 = 3$
$(2, 3)$

3. $x = 4y - 2$
$x = -2y + 4$
$4y - 2 = -2y + 4$
$6y = 6$
$y = 1$
$x = 4(1) - 2 = 2$
$(2, 1)$

5. $y = \frac{1}{3}(12) - 1 = 3$
$(12, 3)$

7. $\begin{cases} 6-3(4) = -6 \\ 3(6)-2(4) = 10 \end{cases}$
$\begin{cases} -6 = -6 \\ 10 = 10 \end{cases}$
Yes

9. $\begin{cases} y = -2x+7 \\ y = x-3 \end{cases}$
$-2x+7 = x-3$
$-3x = -10$
$x = \dfrac{10}{3}$
$y = \dfrac{10}{3} - 3 = \dfrac{1}{3}$
$x = \dfrac{10}{3},\ y = \dfrac{1}{3}$

11. $\begin{cases} y = \dfrac{5}{2}x - \dfrac{1}{2} \\ y = -2x - 4 \end{cases}$
$\dfrac{5}{2}x - \dfrac{1}{2} = -2x - 4$
$\dfrac{9}{2}x = -\dfrac{7}{2}$
$x = -\dfrac{7}{9}$
$y = -2\left(-\dfrac{7}{9}\right) - 4 = -\dfrac{22}{9}$
$x = -\dfrac{7}{9},\ y = -\dfrac{22}{9}$

13. $\begin{cases} x = 3 \\ 2x + 3y = 18 \end{cases}$
$y = -\dfrac{2}{3}x + 6 = -\dfrac{2}{3}(3) + 6 = 4$
$A = (3, 4)$
$\begin{cases} y = 2 \\ 2x + 3y = 18 \end{cases}$
$x = -\dfrac{3}{2}y + 9 = -\dfrac{3}{2}(2) + 9 = 6$
$B = (6, 2)$

15. $A = (0, 0)$
$\begin{cases} y = 2x \\ y = \dfrac{1}{2}x + 3 \end{cases}$
$2x = \dfrac{1}{2}x + 3$
$x = 2$
$y = 2(2) = 4$
$B = (2, 4)$
$\begin{cases} y = \dfrac{1}{2}x + 3 \\ x = 5 \end{cases}$
$y = \dfrac{1}{2}(5) + 3 = \dfrac{11}{2}$
$C = \left(5, \dfrac{11}{2}\right)$
$D = (5, 0)$

17. $\begin{cases} 2y - x \le 6 \\ x + 2y \ge 10 \\ x \le 6 \end{cases}$
$\begin{cases} y \le \dfrac{1}{2}x + 3 \\ y \ge -\dfrac{1}{2}x + 5 \\ x \le 6 \end{cases}$
$\begin{cases} y = \dfrac{1}{2}x + 3 \\ y = -\dfrac{1}{2}x + 5 \end{cases} \Rightarrow (2, 4)$
$\begin{cases} y = -\dfrac{1}{2}x + 5 \\ x = 6 \end{cases} \Rightarrow (6, 2)$

$\begin{cases} y = \frac{1}{2}x + 3 \\ x = 6 \end{cases} \Rightarrow (6, 6)$

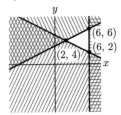

19. $\begin{cases} x + 3y \le 18 \\ 2x + y \le 16 \\ x \ge 0, y \ge 0 \end{cases}$

$\begin{cases} y \le -\frac{1}{3}x + 6 \\ y \le -2x + 16 \\ x \ge 0, y \ge 0 \end{cases}$

$\begin{cases} y = -\frac{1}{3}x + 6 \\ y = -2x + 16 \end{cases} \Rightarrow (6, 4)$

$\begin{cases} y = -\frac{1}{3}x + 6 \\ x = 0 \end{cases} \Rightarrow (0, 6)$

$\begin{cases} y = -2x + 16 \\ y = 0 \end{cases} \Rightarrow (8, 0)$

$\begin{cases} x = 0 \\ y = 0 \end{cases} \Rightarrow (0, 0)$

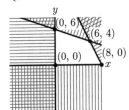

21. $\begin{cases} 4x + y \ge 8 \\ x + y \ge 5 \\ x + 3y \ge 9 \\ x \ge 0, y \ge 0 \end{cases}$

$\begin{cases} y \ge -4x + 8 \\ y \ge -x + 5 \\ y \ge -\frac{1}{3}x + 3 \\ x \ge 0, y \ge 0 \end{cases}$

$\begin{cases} y = -4x + 8 \\ y = -x + 5 \end{cases} \Rightarrow (1, 4)$

$\begin{cases} y = -x + 5 \\ y = -\frac{1}{3}x + 3 \end{cases} \Rightarrow (3, 2)$

$\begin{cases} y = -\frac{1}{3}x + 3 \\ y = 0 \end{cases} \Rightarrow (9, 0)$

$\begin{cases} y = -4x + 8 \\ x = 0 \end{cases} \Rightarrow (0, 8)$

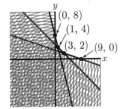

23. **a.** $p = .0001(19{,}500) + .05$
 $= \$2.00$

 b. $p = .0001(0) + .05$
 $= \$.05$
 No units will be supplied for $.05 or less.

25. $\begin{cases} p = .0001q + .05 \\ p = -.001q + 32.5 \end{cases}$

$.0001q + 0.05 = -.001q + 32.5$
$.0011q = 32.45$
$q = 29{,}500 \text{ units}$

$p = .0001(29{,}500) + .05$
$p = \$3.00$

27. a.
$$p = -.15q + 6.925$$
$$5.80 = -.15q + 6.925$$
$$-1.125 = -.15q$$
$$7.5 = q$$

$$p = .2q + 3.6$$
$$5.80 = .2q + 3.6$$
$$2.2 = .2q$$
$$11 = q$$

Demand will be 7.5 billion bushels and supply will be 11 billion bushels

b. The equilibrium point occurs when supply is the same as demand. Therefore,
$$-.15q + 6.925 = .2q + 3.6$$
$$-.35q = -3.325$$
$$q = 9.5$$

To find the equilibrium price, substitute the value into either equation.
$$p = -.15(9.5) + 6.925$$
$$p = -1.425 + 6.925$$
$$p = 5.5$$

Equilibrium occurs when 9.5 billion bushels are produced and sold for $5.50 per bushel.

29. Let $C = F$, then
$$C = \frac{5}{9}(F - 32)$$
$$F = \frac{5}{9}(F - 32)$$
$$\frac{9F}{5} = F - 32$$
$$\frac{4F}{5} = -32$$
$$F = -40$$

Therefore, when the temperature is $-40°$, it will be the same on both temperature scales.

31. Let x = numbers of shirts and y = cost of manufacture.
$$\begin{cases} y = 1200 + 30x \\ y = 500 + 35x \end{cases}$$
$$1200 + 30x = 500 + 35x$$
$$-5x = -700$$
$$x = 140$$

$$y = 1200 + 30x$$
$$y = 1200 + 30(140)$$
$$y = 1200 + 4200$$
$$y = 5400$$

The manufactures will charge the same $5400 if they produce 140 shirts.

33. Method A: $y = .45 + .01x$
Method B: $y = .035x$
Intersection point:
$$.45 + .01x = .035x$$
$$.45 = .025x$$
$$18 = x$$
For a call lasting 18 minutes, the costs for either method will be the same, $y = .035(18) = 63$. The cost will be 63 cents.

35. $$\begin{cases} 3x - y = 3 \\ x + y = 5 \\ y = 0 \end{cases}$$
$$\begin{cases} y = 3x - 3 \\ y = -x + 5 \\ y = 0 \end{cases}$$

$\begin{cases} y = 3x-3 \\ y = -x+5 \end{cases} \Rightarrow (2, 3)$

$\begin{cases} y = -x+5 \\ y = 0 \end{cases} \Rightarrow (5, 0)$

$\begin{cases} y = 3x-3 \\ y = 0 \end{cases} \Rightarrow (1, 0)$

Based on the above points of intersection, the base of the triangle is $5-1=4$ and the height is 3. Therefore the area of the triangle, in square units, is:

$A = \dfrac{1}{2}bh$

$A = \dfrac{1}{2}(4)(3)$

$A = 6$

37. Let x = weight of first contestant
y = weight of second contestant

$\begin{cases} x+y = 700 \\ 2x = 275+y \end{cases}$

$\begin{cases} y = 700-x \\ y = 2x-275 \end{cases}$

$700 - x = 2x - 275$

$975 = 3x$

$x = 325$ pounds

Answer (c) is correct.

39.

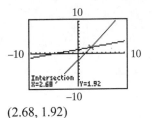

(3.73, 2.23)

41. $\begin{cases} x-4y = -5 \\ 3x-2y = 4.2 \end{cases}$

$\begin{cases} y = \dfrac{1}{4}x + \dfrac{5}{4} \\ y = \dfrac{3}{2}x - 2.1 \end{cases}$

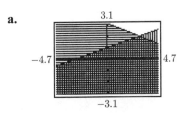

(2.68, 1.92)

43. $\begin{cases} -x+3y \geq 3 \\ .4x+y \geq 3.2 \end{cases}$

$\begin{cases} y \geq \dfrac{1}{3}x + 1 \\ y \geq -.4x + 3.2 \end{cases}$

a.

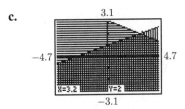

b. (3, 2)

c.

$\begin{cases} -(3.2)+3(2) \geq 3 \\ .4(3.2)+2 \geq 3.2 \end{cases}$

d.

$\begin{cases} 2.8 \geq 3 \\ 3.28 \geq 3.2 \end{cases}$

No

Exercises 1.4

1. $m = \dfrac{2}{3}$

3. $y - 3 = 5(x+4)$
$y = 5x + 23$
$m = 5$

5.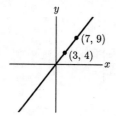

$$m = \frac{9-4}{7-3} = \frac{5}{4}$$

7.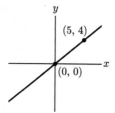

$$m = \frac{4-0}{5-0} = \frac{4}{5}$$

9. The slope of a vertical line is undefined.

11.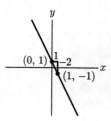

13.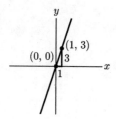

15. $m = \frac{-2}{1} = -2$
$y - 3 = -2(x - 2)$
$y = -2x + 7$

17. $m = \frac{0-2}{2-1} = -2$
$y - 0 = -2(x - 2)$
$y = -2x + 4$

19. $m = -\frac{1}{-4} = \frac{1}{4}$
$y - 2 = \frac{1}{4}(x - 2)$
$y = \frac{1}{4}x + \frac{3}{2}$

21. $m = -1$
$y - 0 = -1(x - 0)$
$y = -x$

23. $m = 0$
$y - 3 = 0(x - 2)$
$y = 3$

25. $y - 6 = \frac{3}{5}(x - 5)$
$y = \frac{3}{5}x + 3$
y-intercept: $(0, 3)$

27. Let y = cost in dollars.
$y = 4x + 2000$

29. a. Let x = altitude and y = boiling point.
$m = \frac{212 - 202.8}{0 - 5000} = -.00184$
$y - 212 = -.00184(x - 0)$
$y = -.00184x + 212$

   b. $y \approx -.00184x + 212$
$y \approx -.00184(29029) + 212$
$y \approx 158.6°F$

31. a. Let x = quantity and y = cost.
$m = \frac{9500 - 6800}{50 - 20} = 90$
$y - 6800 = 90(x - 20)$
$y = 90x + 5000$

b. $5000

c. $90

d.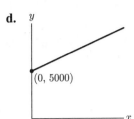

33. a. 100(300) = $30,000

   b. $6000 = 100x$
   $x = 60$ coats

   c. $y = 100(0) = 0$
   (0, 0); if no coats are sold, there is no revenue.

   d. 100; each additional coat yields an additional $100 in revenue.

35. a.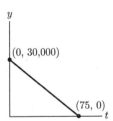

   b. On February 1, 31 days have elapsed since January 1. The amount of oil y = 30,000 – 400(31) = 17,600 gallons.

   c. On February 15, 45 days have elapsed since January 1. Therefore, the amount of oil would be y = 30,000 - 400(45) = 12,000 gallons.

   d. The significance of the y-intercept is that amount of oil present initially on January 1. This amount is 30,000 gallons.

   e. The t-intercept is (75,0) and corresponds to the number of days at which the oil will be depleted.

37. a. $y = 0.10x + 220$

   b. $y = 0.10(2000) + 220$
   $y = 420$

   c. $540 = 0.10x + 220$
   $x = \$3200$

39. $m = -\dfrac{1}{2}$, $b = 0$
   $y = -\dfrac{1}{2}x$

41. $m = -\dfrac{1}{3}$
   $y - (-2) = -\dfrac{1}{3}(x - 6)$
   $y = -\dfrac{1}{3}x$

43. $m = \dfrac{1}{2}$
   $y - (-3) = \dfrac{1}{2}(x - 2)$
   $y = \dfrac{1}{2}x - 4$

45. $m = -\dfrac{2}{5}$
   $y - 5 = -\dfrac{2}{5}(x - 0)$
   $y = -\dfrac{2}{5}x + 5$

47. $m = \dfrac{3 - (-3)}{-1 - 5} = -1$
   $y - 3 = -1[x - (-1)]$
   $y = -x + 2$

49. $m = \dfrac{-1 - (-1)}{3 - 2} = 0$
   $y - (-1) = 0(x - 2)$
   $y = -1$

51. $\begin{cases} 4x + 5y = 7 \\ 2x - 3y = 9 \end{cases} \Rightarrow (3, -1)$

$$y-(-1)=2(x-3)$$
$$y+1=2x-6$$
$$y=2x-7$$

53. Changes in x-coordinate: 1, –1, –2
Changes in y-coordinate are *m* times that or 2, –2, –4: new y values are 5, 1, –1

55. The slope is $\dfrac{-1}{4}$ Changes in x coordinates are 1, 2, –1. Changes in y coordinates are m times the x coordinate changes. New y coordinates are $\dfrac{-5}{4}, \dfrac{-3}{2}, \dfrac{-3}{4}$

57. a. $x+y=1$
       $y=-x+1$
       (C)

    b. $x-y=1$
        $y=x-1$
        (B)

    c. $x+y=-1$
        $y=-x-1$
        (D)

    d. $x-y=-1$
        $y=x+1$
        (A)

59. One possible equation is $y=x+1$.

61. One possible equation is $y=5$.

63. One possible equation is $y=-\dfrac{2}{3}x$.

65. $m = \dfrac{212-32}{100-0} = \dfrac{9}{5}$

    $F - 32 = \dfrac{9}{5}(C-0)$

    $F = \dfrac{9}{5}C + 32$

67. Let 1995 correspond to x = 0. So in 2009, x = 14. When x = 0, tuition is 2848. When x = 14, tuition is 6695. Using (0,2848) and (14,6695) as ordered pairs, find the slope of the line containing these points: $\dfrac{6695-2848}{14-0} = \dfrac{3847}{14}$. Since the y-intercept is 2848, the equation becomes $y = \dfrac{3847}{14}x + 2848$. Therefore, in 2002 when x =7, the tuition should approximately be $y = \dfrac{3847}{14}(7) + 2848 = 4771.50$.

69. Let x = number of pounds tires are under inflated. When x = 0, the miles per gallon (y) is 25. When x = 1, mpg decreases to 24.5. The equation is $y = -\dfrac{1}{2}x + 25$. Thus, when x = 8 pounds the miles per gallon will be $y = -\dfrac{1}{2}(8) + 25 = 21$ mpg.

71. Let 1991 correspond to x = 0 and 2009 correspond to x = 18. Then, the two ordered pairs are on the line: (0, 249,165) and (18, 347,985). The slope of the line is $\dfrac{347,985 - 249,165}{18-0} = 5490$ The equation of the line is therefore $y = 5490x + 249,165$. In the year 2014, x = 23, so the number of Bachelor's degrees awarded can be estimated as $y = 5490(23) + 249,165 = 375,435$.

73. Let 2010 correspond to x = 0 and 2012 correspond to x = 2. Then, the two ordered pairs are on the line: (0, 2.5) and (2, 3.5). The slope of the line is $\dfrac{3.5-2.5}{2-0} = .5$. The equation of the line is therefore $y = .5x + 2.5$. In the year 2011, x = 1, so the cost of a 30-second advertising slot (in millions) can be estimated as $y = .5(1) + 2.5 = \$3$ million.

75. The slope is $\dfrac{3.4-3}{6-5} = .4$

    $p - p_1 = m(q - q_1)$
    $p - 3 = .4(q - 5)$
    $p - 3 = .4q - 2$
    $p = .4q + 1$

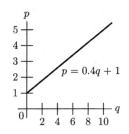

77. $m = \dfrac{9-5}{4-2} = 2$

$y - 5 \le 2(x-2)$

$y \le 2x + 1$

79. $m_1 = \dfrac{8-5}{2-(-2)} = \dfrac{3}{4}$

$y - 8 = \dfrac{3}{4}(x-2)$

$y = \dfrac{3}{4}x + \dfrac{13}{2}$

$m_2 = \dfrac{1-8}{5-2} = -\dfrac{7}{3}$

$y - 1 = -\dfrac{7}{3}(x-5)$

$y = -\dfrac{7}{3}x + \dfrac{38}{3}$

$m_3 = \dfrac{1-5}{5-(-2)} = -\dfrac{4}{7}$

$y - 1 = -\dfrac{4}{7}(x-5)$

$y = -\dfrac{4}{7}x + \dfrac{27}{7}$

$\begin{cases} y \le \dfrac{3}{4}x + \dfrac{13}{2} \\ y \le -\dfrac{7}{3}x + \dfrac{38}{3} \\ y \ge -\dfrac{4}{7}x + \dfrac{27}{7} \end{cases}$

81. $m_1 = \dfrac{4-3}{2-1} = 1$

$m_2 = \dfrac{-1-4}{3-2} = -5$

$m_1 \ne m_2$

83. Set slopes equal:

$\dfrac{-3.1-1}{2-a} = \dfrac{2.4-0}{3.8-(-1)}$

$\dfrac{-4.1}{2-a} = \dfrac{1}{2}$

$-8.2 = 2 - a$

$a = 10.2$

85. Solve $mx + b = m'x + b'$

$(m - m')x = b' - b$

$x = \dfrac{b'-b}{m-m'},$

which is defined if and only if $m \ne m'$.

87. Let x = Centigrade temperature

y = Fahrenheit temperature

$m = \dfrac{212-32}{100-0} = 1.8$

$y = 1.8x + 32$

$y = 1.8(30) + 32 = 86°F$

Answer (b) is correct.

89. Let x = number of T-shirts

profit = revenue − cost

$65,000 = 12.50x - (8x + 25,000)$

$90,000 = 4.50x$

$x = 20,000$

So 20,000 T-shirts must be produced and sold.

Answer (d) is correct.

91. $q = 800 - 4(150)$

$= 200$ bikes

revenue = $150(200) = \$30,000$

Answer (d) is correct.

93. Let x = variable costs

For 2008: profit = revenue − cost

$400,000 = 100(50,000) - (50,000x + 600,000)$

$50,000x = 4,000,000$

$x = \$80$ per unit

For 2009:

Let y = 2009 price

profit = revenue − cost

$$400{,}000 = 50{,}000y -$$
$$[80(50{,}000) + 600{,}000 + 200{,}000]$$
$$5{,}200{,}000 = 50{,}000y$$
$$y = \$104$$
Answer (d) is correct.

95.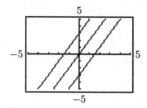

From left to right the lines are $y = 2x + 3$, $y = 2x$, and $y = 2x - 3$.
The lines are distinguished by their y-intercepts, which appear as b in the form $y = mx + b$.

97.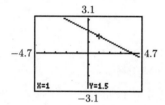

Since the slope equals $-\dfrac{1}{2}$, moving 2 units to the right requires moving $2 \cdot \left(-\dfrac{1}{2}\right) = -1$ unit up, or 1 unit down.

99.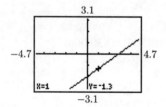

Since the slope equals 0.7, moving 2 units to the right requires moving $2 \cdot .7 = 1.4$ units up.

(2, 6)	(2, 7)	1
(3, 11)	(3, 10)	1
(4, 12)	(4, 13)	1

$1^2 + 1^2 + 1^2 + 1^2 = 4$

3. $E_1^2 = [1.1(1) + 3 - 3]^2 = 1.21$
$E_2^2 = [1.1(2) + 3 - 6]^2 = .64$
$E_3^2 = [1.1(3) + 3 - 8]^2 = 2.89$
$E_4^2 = [1.1(4) + 3 - 6]^2 = 1.96$
$E = 1.21 + .64 + 2.89 + 1.96 = 6.70$

Exercises 1.5

1.

Data Point	Point on Line	Vertical Distance
(1, 3)	(1, 4)	1

SSM: Finite Math **Chapter 1:** Linear Equations and Straight Lines

5.

x	y	xy	x^2
1	7	7	1
2	6	12	4
3	4	12	9
4	3	12	16
$\sum x = 10$	$\sum y = 20$	$\sum xy = 43$	$\sum x^2 = 30$

$$m = \frac{4 \cdot 43 - 10 \cdot 20}{4 \cdot 30 - 10^2} = -1.4$$

$$b = \frac{20 - (-1.4)(10)}{4} = 8.5$$

7. $\sum x = 6, \sum y = 18, \sum xy = 45, \sum x^2 = 14$

$$m = \frac{3 \cdot 45 - 6 \cdot 18}{3 \cdot 14 - 6^2} = 4.5$$

$$b = \frac{18 - (4.5)(6)}{3} = -3$$

$$y = 4.5x - 3$$

9. $\sum x = 10, \sum y = 26, \sum xy = 55,$
$\sum x^2 = 30$

$$m = \frac{4 \cdot 55 - 10 \cdot 26}{4 \cdot 30 - 10^2} = -2$$

$$b = \frac{26 - (-2)(10)}{4} = 11.5$$

$$y = -2x + 11.5$$

11. **a.** $\sum x = 12, \sum y = 7, \sum xy = 41,$
$\sum x^2 = 74$

$$m = \frac{2 \cdot 41 - 12 \cdot 7}{2 \cdot 74 - 12^2} = -.5$$

$$b = \frac{7 - (-.5)(12)}{2} = 6.5$$

$$y = -.5x + 6.5$$

b. $m = \frac{4-3}{5-7} = -\frac{1}{2} = -.5$

$y - 3 = -.5(x - 7)$

$y = -.5x + 6.5$

c. The least–squares error for the line in (b) is $E = 0$.

13. **a.** $\sum x = 7, \sum y = 20, \sum xy = 90,$
$\sum x^2 = 37$

$$m = \frac{2 \cdot 90 - 7 \cdot 20}{2 \cdot 37 - 7^2} = 1.6$$

$$b = \frac{20 - (1.6)(7)}{2} = 4.4$$

$$y = 1.6x + 4.4$$

b. $E = [1.6(4) + 4.4 - 5]^2 = 33.64$

15. **a.** Let x represent city and y represent highway, then $\sum x = 179, \sum y = 167, \sum xy = 7537,$
$\sum x^2 = 8067$

$$m = \frac{4 \cdot 7537 - 179 \cdot 167}{4 \cdot 8067 - 179^2} = 1.12335$$

$$b = \frac{167 - (1.12335)(179)}{4} = -8.51982$$

$$y = 1.12335x - 8.51982$$

b. $y = 1.12335(47) - 8.51982$

$y = 44.28$ mpg

c. $47 = 1.12335x - 8.51982$

$x = 49.42$ mpg

17. a.

   $y = .338x + 21.6$

   b. $.338(1100) + 21.6 = 393.4$
   About 393 deaths per million males

19. a. Let x be the number of years after 1985, then $y = 0.423x + 19.2$

   b. $.423(23) + 19.2 = 28.929$
   About 28.9%

   c. $32 = .423x + 19.2$
   $x \approx 30.26$
   The year 2015

21. a. $y = .1475x + 73.78$

   b. $.1475(30) + 73.78 = 78.205$
   About 78.2 years

   c. $.1475(50) + 73.78 = 81.155$
   About 81.2 years

   d. $.1475(90) + 73.78 = 87.055$
   About 87.1 years (This is an example of a fit that is not capable of extrapolating beyond the given data)

23. a. Let x be the number of years after 2000, then $y = .028x + .846$

   b. $.028(9) + .846 \approx 1.098$
   About $1.10

   c. $1.35 = .028x + .846$
   $x = 18$
   The year 2018

Chapter 1 Fundamental Concept Check

1. To determine the x-coordinate (y-coordinate), draw a straight line through the point perpendicular through the x-axis (y-axis) and read the number on the axis.

2. The graph is the collection of points in the plane whose coordinates satisfy the equation.

3. $ax + by = c$, where both a and b are not 0.

4. $y = mx + b$ or $x = a$.

5. The y-intercept is the point at which the graph of the line crosses the y-axis. To find the y-intercept, set $x = 0$ and solve for y. Then the y-intercept is the point (0, solution for y).

6. The x-intercept is the point at which the graph of the line crosses the x-axis. To find the x-intercept, set $y = 0$ and solve for x. Then the x-intercept is the point (solution for x, 0).

7. See the tinted box on page 5.

8. If $a < b$ then
   $a + c < b + c, a - c < b - c, ac < bc$ (when c is positive), and $ac > bc$ (when c is negative).

9. General forms: $cx + dy \leq e$ or $cx + dy \geq e$ where c and d are not both 0. Standard forms:
   $y \leq mx + b, y \geq mx + b, x \leq a$, and $x \geq a$.

10. Put the inequality into standard form, draw the related linear equation, and cross out the side that does not satisfy the inequality.

11. The collection of points that satisfy every inequality in the system.

12. First put the two linear equations into standard form. If both equations have the form $y =$ something, equate the two expressions for y, solve for x, substitute the value for x into one of the equations, and solve for y. Otherwise, substitute the value of x into the equation containing y and solve for y.

13. The slope of the line $y = mx + b$ is the number m. It is a measure of the steepness of the line.

14. Plot the given point, move one unit to the right the $|m|$ units in the y-direction (up if m is positive and down if m is negative), plot the second point, and draw a line through the two points.

15. $y - y_1 = m(x - x_1)$, where (x_1, y_1) is a point on the line and m is the slope of the line.

16. First calculate the slope $m = \dfrac{y_2 - y_1}{x_2 - x_1}$. Then, use m, either of the two points, and the point-slope formula to write the equation for the line.

17. One slope is the negative reciprocal of the other.

18. They are the same.

19. The straight line that gives the best fit to a collection of points in the sense that the sum of the squares of the vertical distances from the points to the line is as small as possible.

Chapter 1 Review Exercises

1. $x = 0$

2.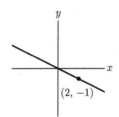

3. $\begin{cases} x - 5y = 6 \\ 3x = 6 \end{cases}$
$\begin{cases} x = 5y + 6 \\ x = 2 \end{cases}$
$5y + 6 = 2$
$y = -\dfrac{4}{5}$
$\left(2, -\dfrac{4}{5}\right)$

4. $3x - 4y = 8$
$y = \dfrac{3}{4}x - 2$
$m = \dfrac{3}{4}$

5. $m = \dfrac{0 - 5}{10 - 0} = -\dfrac{1}{2}, b = 5$
$y = -\dfrac{1}{2}x + 5$

6. $x - 3y \geq 12$
$y \leq \dfrac{1}{3}x - 4$

7. $3(1) + 4(2) \geq 11$
$3 + 8 \geq 11$
$11 \geq 11$
Yes

8. $\begin{cases} 2x - y = 1 \\ x + 2y = 13 \end{cases}$
$\begin{cases} y = 2x - 1 \\ y = -\dfrac{1}{2}x + \dfrac{13}{2} \end{cases}$
$2x - 1 = -\dfrac{1}{2}x + \dfrac{13}{2}$
$\dfrac{5}{2}x = \dfrac{15}{2}$
$x = 3$
$y = 2(3) - 1 = 5$
$(3, 5)$

9. $2x - 10y = 7$

$$y = \frac{1}{5}x - \frac{7}{10}$$

$$m = \frac{1}{5}$$

$$y - 16 = \frac{1}{5}(x - 15)$$

$$y = \frac{1}{5}x + 13$$

10. $y = 3(1) + 7 = 10$

11. $(5, 0)$

12.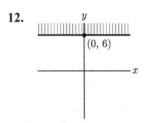

13. $\begin{cases} 3x - 2y = 1 \\ 2x + y = 24 \end{cases}$

$$\begin{cases} y = \frac{3}{2}x - \frac{1}{2} \\ y = -2x + 24 \end{cases}$$

$$\frac{3}{2}x - \frac{1}{2} = -2x + 24$$

$$\frac{7}{2}x = \frac{49}{2}$$

$$x = 7$$

$y = -2(7) + 24 = 10$
$(7, 10)$

14. $\begin{cases} 2y + 7x \geq 28 \\ 2y - x \geq 0 \\ y \leq 8 \end{cases}$

$\begin{cases} y \geq -\frac{7}{2}x + 14 \\ y \geq \frac{1}{2}x \\ y \leq 8 \end{cases}$

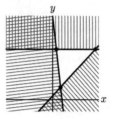

15. $y - 9 = \frac{1}{2}(x - 4)$

$$y = \frac{1}{2}x + 7$$

$b = 7$
$(0, 7)$

16. The rate is $35 per hour plus a flat fee of $20.

17. $m_1 = \dfrac{0 - 2}{2 - 1} = -2$

$m_2 = \dfrac{1 - 0}{3 - 2} = 1$

$m_1 \neq m_2$
No

18. $m = \dfrac{-2 - 0}{0 - 3} = \dfrac{2}{3}, b = -2$

$$y = \frac{2}{3}x - 2$$

19. $x + 7y = 30$
$-2y + 7y = 30$
$5y = 30$
$y = 6$
Answer (d) is correct.

20. $y \leq \dfrac{2}{3}x + \dfrac{3}{2}$

SSM: Finite Math   **Chapter 1:** Linear Equations and Straight Lines

21. $m = \dfrac{8.6-(-1)}{6-2} = 2.4$

    $y+1 \geq 2.4(x-2)$

    $y \geq 2.4x - 5.8$

22. $\begin{cases} 1.2x + 2.4y = .6 \\ 4.8y - 1.6x = 2.4 \end{cases}$

    $\begin{cases} y = -.5x + .25 \\ y = \dfrac{1}{3}x + .5 \end{cases}$

    $-.5x + .25 = \dfrac{1}{3}x + .5$

    $-\dfrac{5}{6}x = .25$

    $x = -.3$

    $y = \dfrac{1}{3}(-.3) + .5 = .4$

23. $\begin{cases} y = -x + 1 \\ y = 2x + 3 \end{cases}$

    $-x + 1 = 2x + 3$

    $-3x = 2$

    $x = -\dfrac{2}{3}$

    $y = -\left(-\dfrac{2}{3}\right) + 1 = \dfrac{5}{3}$

    $\left(-\dfrac{2}{3}, \dfrac{5}{3}\right)$

    $m = \dfrac{\frac{5}{3}-1}{-\frac{2}{3}-1} = -\dfrac{2}{5}$

    $y - 1 = -\dfrac{2}{5}(x-1)$

    $y = -\dfrac{2}{5}x + \dfrac{7}{5}$

24. $2x + 3(x-2) \geq 0$

    $5x \geq 6$

    $x \geq \dfrac{6}{5}$

25. $x + \dfrac{1}{2}y = 4$

    $y = -2x + 8$

    $m = -2$

    y-intercept: $(0, 8)$

    $0 = -2x + 8$

    $x = 4$

    x-intercept: $(4, 0)$

    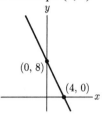

26. $\begin{cases} 5x + 2y = 0 \\ x + y = 1 \end{cases}$

    $\begin{cases} y = -\dfrac{5}{2}x \\ y = -x + 1 \end{cases}$

    $-\dfrac{5}{2}x = -x + 1$

    $-\dfrac{3}{2}x = 1$

    $x = -\dfrac{2}{3}$

    $y = -\left(-\dfrac{2}{3}\right) + 1 = \dfrac{5}{3}$

    Substitute $x = -\dfrac{2}{3}$ and $y = \dfrac{5}{3}$ in

    $2x - 3y = 1$

    $2\left(-\dfrac{2}{3}\right) - 3\left(\dfrac{5}{3}\right) = 1$

    $-\dfrac{19}{3} = 1$

    No

27. $\begin{cases} 2x - 3y = 1 \\ 3x + 2y = 4 \end{cases}$

$\begin{cases} y = \dfrac{2}{3}x - \dfrac{1}{3} \\ y = -\dfrac{3}{2}x + 2 \end{cases}$

$m_1 = -\dfrac{1}{m_2}$

28. a. $x + y \geq 1$
    $y \geq -x + 1$
    (C)

   b. $x + y \leq 1$
    $y \leq -x + 1$
    (A)

   c. $x - y \leq 1$
    $y \geq x - 1$
    (B)

   d. $y - x \leq -1$
    $y \leq x - 1$
    (D)

29. a. $4x + y = 17$
    $y = -4x + 17$
    L_3

   b. $y = x + 2$
    L_1

   c. $2x + 3y = 11$
    $y = -\dfrac{2}{3}x + \dfrac{11}{3}$
    L_2

30. $m_1 = \dfrac{\tfrac{3}{2} - 5}{4 - 0} = -\dfrac{7}{8}, b_1 = 5$

$y = -\dfrac{7}{8}x + 5$

$m_2 = -\dfrac{1}{m_1} = \dfrac{8}{7}$

$y - \dfrac{3}{2} = \dfrac{8}{7}(x - 4)$

$y = \dfrac{8}{7}x - \dfrac{43}{14}$

$\begin{cases} y \leq -\dfrac{7}{8}x + 5 \\ y \geq \dfrac{8}{7}x - \dfrac{43}{14} \\ x \geq 0, y \geq 0 \end{cases}$

$0 = \dfrac{8}{7}x - \dfrac{43}{14}$

$x = \dfrac{43}{16}$

$\left(\dfrac{43}{16}, 0\right)$

31. Supply curve is $p = .005q + .5$
    Demand curve is $p = -.01q + 5$
    $\begin{cases} p = .005q + .5 \\ p = -.01q + 5 \end{cases}$
    $.005q + .5 = -.01q + 5$
    $.015q = 4.5$
    $q = 300$ units
    $p = .005(300) + .5 = \$2$

32. $\begin{cases} x \geq 0 \\ y \geq 0 \end{cases}$
(0, 0)
$\begin{cases} y \geq 0 \\ 5x + y \leq 50 \end{cases}$
$\begin{cases} y \geq 0 \\ y \leq -5x + 50 \end{cases}$
$0 = -5x + 50$
$x = 10$
(10, 0)
$\begin{cases} 5x + y \leq 50 \\ 2x + 3y \leq 33 \end{cases}$
$\begin{cases} y \leq -5x + 50 \\ y \leq -\dfrac{2}{3}x + 11 \end{cases}$
$-5x + 50 = -\dfrac{2}{3}x + 11$
$-\dfrac{13}{3}x = -39$
$x = 9$
$y = -5(9) + 50 = 5$
(9, 5)
$\begin{cases} 2x + 3y \leq 33 \\ x - 2y \geq -8 \end{cases}$
$\begin{cases} y \leq -\dfrac{2}{3}x + 11 \\ y \leq \dfrac{1}{2}x + 4 \end{cases}$
$-\dfrac{2}{3}x + 11 = \dfrac{1}{2}x + 4$
$-\dfrac{7}{6}x = -7$
$x = 6$
$x = \dfrac{1}{2}(6) + 4 = 7$
(6, 7)
$\begin{cases} x - 2y \geq -8 \\ x \geq 0 \end{cases}$
$\begin{cases} x \geq 2y - 8 \\ x \geq 0 \end{cases}$
$2y - 8 = 0$
$y = 4$
(0, 4)

33. a. In 2000, 8.9% of college freshmen intended to obtain a medical degree.

b. $2011 - 2000 = 11$
$y = 0.1(11) + 8.9$
$y = 10$
10% of college freshmen in 2011 intended to obtain a medical degree. It is close to the actual value.

c. $9.3 = .1x + 8.9$
$x = 4$
$2000 + 4 = 2004$
In 2004, the percent of college freshmen that intended to obtain a medical degree was 9.3.

34. a. $m = 10$
$y - 4000 = 10(x - 1000)$
$y = 10x - 6000$

b. $0 = 10x - 6000$
$x = 600$
x-intercept: (600, 0)
y-intercept: (0, −6000)

c.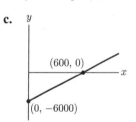

35. a. A: $y = .1x + 50$
B: $y = .2x + 40$

b. A: $.1(80) + 50 = 58$
B: $.2(80) + 40 = 56$
Company B

c. A: $.1(160) + 50 = 66$
B: $.2(160) + 40 = 72$
Company A

d. $.1x + 50 = .2x + 40$
$-.1x = -10$
$x = 100$ miles

36. a. $m = \dfrac{1.44 - .93}{11 - 0} = .046$

    $y - .93 = .046(x - 0)$

    $y = .046x + .93$

    b. $1.27 = .046x + .93$

    $x \approx 7.39$

    The year $2000 + 7 = 2007$

37. $x \le 3y + 2$

    $y \ge \dfrac{1}{3}x - \dfrac{2}{3}$

    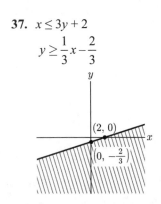

38. $.03x + 200 = .05x + 100$

    $-.02x = -100$

    $x = \$5000$

39. $m_1 = \dfrac{5 - 0}{0 - (-4)} = \dfrac{5}{4}, b_1 = 5$

    $y = \dfrac{5}{4}x + 5$

    $m_2 = \dfrac{0 - 2}{5 - 0} = -\dfrac{2}{5}, b_2 = 2$

    $y = -\dfrac{2}{5}x + 2$

    $m_3 = \dfrac{0 - (-3)}{5 - 0} = \dfrac{3}{5}, b_3 = -3$

    $y = \dfrac{3}{5}x - 3$

    $m_4 = \dfrac{-5 - 0}{0 - (-2)} = -\dfrac{5}{2}, b_4 = -5$

    $y = -\dfrac{5}{2}x - 5$

$\begin{cases} y \le \dfrac{5}{4}x + 5 \\ y \le -\dfrac{2}{5}x + 2 \\ y \ge \dfrac{3}{5}x - 3 \\ y \ge -\dfrac{5}{2} - 5 \end{cases}$

40. $m_1 = \dfrac{2 - 0}{0 - 3} = -\dfrac{2}{3}, b_1 = 2$

    $y = -\dfrac{2}{3}x + 2$

    The other lines are $x = -2$, $x = 4$, and $y = -3$.

    $\begin{cases} y \le -\dfrac{2}{3}x + 2 \\ x \ge -2 \\ x \le 4 \\ y \ge -3 \end{cases}$

41. $(0, 483{,}600)$; in 2018: $(10, 647{,}500)$

    $m = \dfrac{647{,}500 - 483{,}600}{10 - 0} = 16390$

    $y - 483{,}600 = 16390(x - 0)$

    $y = 16{,}390x + 483{,}600$

    For the year 2014, $x=6$:

    $y = 16390(6) + 483{,}600 = 581{,}940$.

42. Slope of line is –282.77. Equation of line is: $y = -282.77x + 105{,}384$. In 2014, $x = 18$ so $y = 100{,}294$.

43. Let $x = 0$ correspond to year 2000. Then $y = 20.4$. When $x = 10$, $y = 17.0$. The rate of change (slope) $= (17.0 - 20.4)/(10 - 0) = -.34$. The equation of the line that predicts the percentage of market is $y = -.34x + 20.4$. When $x = 8$, $y = 17.7\%$.

44. a. $y = .936x + 10.8$

    b. $.936(77.5) + 10.8 = 83.34$
       About 83.3 years

    c. $84.5 = .936x + 10.8$

       $x \approx 78.74$
       About 78.7 years

SSM: *Finite Math*   **Chapter 1:** *Linear Equations and Straight Lines*

45. a. $y = .2075x + 2.43$

    b. $.2075(14) + 2.43 = 5.34$
        About 5.3%

    c. $5.75 = .2075x + 2.43$
        $x = 16$
        16 years after 1999 or 2015

46. a.
```
LinReg
y=ax+b
a=.1517702501
b=-3.063197325
```
        $y = .152x - 3.063$

    b. $.152(160) - 3.063 = 21.257$
        About 21 deaths per 100,000

    c. $22 = .152x - 3.063$
        $x \approx 164.888$
        About 165 grams

47. Up; the value of *b* is the y-intercept

48. Counter - Clockwise

49. When the line passes through the origin.

50. A line with undefined sloe is a vertical line and a line with zero slope is a horizontal line.

51. a. No; A line that is parallel to the x axis and is not the *x* axis will not have an x intercept.

    b. No; A line that is parallel to the y axis and is not the *y* axis will not have a y intercept

52. Answers will vary.

Chapter 2

Exercises 2.1

1. $\begin{cases} \dfrac{1}{2}x - 3y = 2 \\ 5x + 4y = 1 \end{cases}$

   $\xrightarrow{2R_1} \begin{cases} x - 6y = 4 \\ 5x + 4y = 1 \end{cases}$

3. $\quad$ 5(first) $\quad 5x + 10y = 15$
   $+$ (second) $\underline{-5x + 4y = 1}$
   $\qquad\qquad\qquad 14y = 16$

   $\begin{cases} x + 2y = 3 \\ -5x + 4y = 1 \end{cases}$

   $\xrightarrow{R_2 + 5R_1} \begin{cases} x + 2y = 3 \\ 14y = 16 \end{cases}$

5. $\quad$ -4(first) $\quad -4x + 8y - 4z = 0$
   $+$ (third) $\quad \underline{4x + y + 3z = 5}$
   $\qquad\qquad\qquad 9y - z = 5$

   $\begin{cases} x - 2y + z = 0 \\ y - 2z = 4 \\ 4x + y + 3z = 5 \end{cases}$

   $\xrightarrow{R_3 + (-4)R_1} \begin{cases} x - 2y + z = 0 \\ y - 2z = 4 \\ 9y - z = 5 \end{cases}$

7. $\begin{bmatrix} 1 & -\dfrac{1}{2} & 3 \\ 0 & 1 & 4 \end{bmatrix} \xrightarrow{R_1 + \frac{1}{2}R_2} \begin{bmatrix} 1 & 0 & 5 \\ 0 & 1 & 4 \end{bmatrix}$

9. $\begin{bmatrix} -3 & 4 & -2 \\ 1 & -7 & 8 \end{bmatrix}$

11. $\begin{bmatrix} 1 & 13 & -2 & 0 \\ 2 & 0 & -1 & 3 \\ 0 & 1 & 0 & 5 \end{bmatrix}$

13. $\begin{cases} -2y = 3 \\ x + 7y = -4 \end{cases}$

15. $\begin{cases} 3x + 2y = -3 \\ y - 6z = 4 \\ -5x - y + 7z = 0 \end{cases}$

17. Multiply the second row of the matrix by $\dfrac{1}{3}$.

19. Change the first row of the matrix by adding to it 3 times the second row.

21. Interchange rows 2 and 3.

23. $\begin{bmatrix} 1 & 2 & 0 \\ 0 & 10 & 5 \end{bmatrix}$

25. $\begin{bmatrix} 1 & 2 & 3 \\ 3 & -2 & 0 \end{bmatrix}$

27. $\begin{bmatrix} 1 & 3 & -5 \\ 0 & 1 & 7 \end{bmatrix}$

29. Use $R_2 + 2R_1$ to change the -2 to a 0.

31. Use $R_1 + (-2) R_2$ to change the 2 to a 0.

33. Interchange rows 1 and 2 or rows 1 and 3 to make the first entry in row 1 nonzero.

35. Use $R_1 + (-3) R_3$ to change the 3 to a 0.

37. $\begin{bmatrix} 1 & 1 & -1 & 6 \\ -3 & 7 & 5 & 0 \\ 2 & -4 & 3 & -1 \end{bmatrix}$

   $\xrightarrow[R_3 + (-2)R_1]{R_2 + 3R_1} \begin{bmatrix} 1 & 1 & -1 & 6 \\ 0 & 10 & \boxed{2} & \boxed{18} \\ 0 & -6 & \boxed{5} & \boxed{-13} \end{bmatrix}$

39. $\begin{bmatrix} 1 & 9 & | & 8 \\ 2 & 8 & | & 6 \end{bmatrix}$

$\xrightarrow{R_2+(-2)R_1} \begin{bmatrix} 1 & 9 & | & 8 \\ 0 & -10 & | & -10 \end{bmatrix}$

$\xrightarrow{-\frac{1}{10}R_2} \begin{bmatrix} 1 & 9 & | & 8 \\ 0 & 1 & | & 1 \end{bmatrix}$

$\xrightarrow{R_1+(-9)R_2} \begin{bmatrix} 1 & 0 & | & -1 \\ 0 & 1 & | & 1 \end{bmatrix}$

$x = -1, y = 1$

41. $\begin{bmatrix} 1 & -3 & 4 & | & 1 \\ 4 & -10 & 10 & | & 4 \\ -3 & 9 & -5 & | & -6 \end{bmatrix}$

$\xrightarrow{R_2+(-4)R_1} \begin{bmatrix} 1 & -3 & 4 & | & 1 \\ 0 & 2 & -6 & | & 0 \\ -3 & 9 & -5 & | & -6 \end{bmatrix}$

$\xrightarrow{R_3+3R_1} \begin{bmatrix} 1 & -3 & 4 & | & 1 \\ 0 & 2 & -6 & | & 0 \\ 0 & 0 & 7 & | & -3 \end{bmatrix}$

$\xrightarrow{\frac{1}{2}R_2} \begin{bmatrix} 1 & -3 & 4 & | & 1 \\ 0 & 1 & -3 & | & 0 \\ 0 & 0 & 7 & | & -3 \end{bmatrix}$

$\xrightarrow{R_1+3R_2} \begin{bmatrix} 1 & 0 & -5 & | & 1 \\ 0 & 1 & -3 & | & 0 \\ 0 & 0 & 7 & | & -3 \end{bmatrix}$

$\xrightarrow{\frac{1}{7}R_3} \begin{bmatrix} 1 & 0 & -5 & | & 1 \\ 0 & 1 & -3 & | & 0 \\ 0 & 0 & 1 & | & -\frac{3}{7} \end{bmatrix}$

$\xrightarrow{R_1+5R_3} \begin{bmatrix} 1 & 0 & 0 & | & -\frac{8}{7} \\ 0 & 1 & -3 & | & 0 \\ 0 & 0 & 1 & | & -\frac{3}{7} \end{bmatrix}$

$\xrightarrow{R_2+3R_3} \begin{bmatrix} 1 & 0 & 0 & | & -\frac{8}{7} \\ 0 & 1 & 0 & | & -\frac{9}{7} \\ 0 & 0 & 1 & | & -\frac{3}{7} \end{bmatrix}$

$x = -\frac{8}{7}, y = -\frac{9}{7}, z = -\frac{3}{7}$

43. $\begin{bmatrix} 2 & -2 & | & -4 \\ 3 & 4 & | & 1 \end{bmatrix} \xrightarrow{\frac{1}{2}R_1} \begin{bmatrix} 1 & -1 & | & -2 \\ 3 & 4 & | & 1 \end{bmatrix}$

$\xrightarrow{R_2+(-3)R_1} \begin{bmatrix} 1 & -1 & | & -2 \\ 0 & 7 & | & 7 \end{bmatrix}$

$\xrightarrow{\frac{1}{7}R_2} \begin{bmatrix} 1 & -1 & | & -2 \\ 0 & 1 & | & 1 \end{bmatrix}$

$\xrightarrow{R_1+1R_2} \begin{bmatrix} 1 & 0 & | & -1 \\ 0 & 1 & | & 1 \end{bmatrix}$

$x = -1, y = 1$

45. $\begin{bmatrix} 4 & -4 & 4 & | & -8 \\ 1 & -2 & -2 & | & -1 \\ 2 & 1 & 3 & | & 1 \end{bmatrix}$

$\xrightarrow{\frac{1}{4}R_1} \begin{bmatrix} 1 & -1 & 1 & | & -2 \\ 1 & -2 & -2 & | & -1 \\ 2 & 1 & 3 & | & 1 \end{bmatrix}$

$\xrightarrow{R_2+(-1)R_1} \begin{bmatrix} 1 & -1 & 1 & | & -2 \\ 0 & -1 & -3 & | & 1 \\ 2 & 1 & 3 & | & 1 \end{bmatrix}$

$\xrightarrow{R_3+(-2)R_1} \begin{bmatrix} 1 & -1 & 1 & | & -2 \\ 0 & -1 & -3 & | & 1 \\ 0 & 3 & 1 & | & 5 \end{bmatrix}$

$\xrightarrow{(-1)R_2} \begin{bmatrix} 1 & -1 & 1 & | & -2 \\ 0 & 1 & 3 & | & -1 \\ 0 & 3 & 1 & | & 5 \end{bmatrix}$

$\xrightarrow{R_1+R_2} \begin{bmatrix} 1 & 0 & 4 & | & -3 \\ 0 & 1 & 3 & | & -1 \\ 0 & 3 & 1 & | & 5 \end{bmatrix}$

$\xrightarrow{R_3+(-3)R_2} \begin{bmatrix} 1 & 0 & 4 & | & -3 \\ 0 & 1 & 3 & | & -1 \\ 0 & 0 & -8 & | & 8 \end{bmatrix}$

$\xrightarrow{(-\frac{1}{8})R_3} \begin{bmatrix} 1 & 0 & 4 & | & -3 \\ 0 & 1 & 3 & | & -1 \\ 0 & 0 & 1 & | & -1 \end{bmatrix}$

$\xrightarrow{R_1+(-4)R_3} \begin{bmatrix} 1 & 0 & 0 & | & 1 \\ 0 & 1 & 3 & | & -1 \\ 0 & 0 & 1 & | & -1 \end{bmatrix}$

$\xrightarrow{R_2+(-3)R_3} \begin{bmatrix} 1 & 0 & 0 & | & 1 \\ 0 & 1 & 0 & | & 2 \\ 0 & 0 & 1 & | & -1 \end{bmatrix}$

$x = 1, y = 2, z = -1$

47. $\begin{bmatrix} .2 & .3 & | & 4 \\ .6 & 1.1 & | & 15 \end{bmatrix}$

$\xrightarrow{5R_1} \begin{bmatrix} 1 & 1.5 & | & 20 \\ .6 & 1.1 & | & 15 \end{bmatrix}$

$\xrightarrow{R_2+(-.6)R_1} \begin{bmatrix} 1 & 1.5 & | & 20 \\ 0 & .2 & | & 3 \end{bmatrix}$

$\xrightarrow{5R_2} \begin{bmatrix} 1 & 1.5 & | & 20 \\ 0 & 1 & | & 15 \end{bmatrix}$

$\xrightarrow{R_1+(-1.5)R_2} \begin{bmatrix} 1 & 0 & | & -2.5 \\ 0 & 1 & | & 15 \end{bmatrix}$

$x = -2.5, y = 15$

49. $\begin{bmatrix} 1 & 1 & 4 & | & 3 \\ 4 & 1 & -2 & | & -6 \\ -3 & 0 & 2 & | & 1 \end{bmatrix}$

$\xrightarrow{R_2+(-4)R_1} \begin{bmatrix} 1 & 1 & 4 & | & 3 \\ 0 & -3 & -18 & | & -18 \\ -3 & 0 & 2 & | & 1 \end{bmatrix}$

$\xrightarrow{R_3+3R_1} \begin{bmatrix} 1 & 1 & 4 & | & 3 \\ 0 & -3 & -18 & | & -18 \\ 0 & 3 & 14 & | & 10 \end{bmatrix}$

$\xrightarrow{\left(-\frac{1}{3}\right)R_2} \begin{bmatrix} 1 & 1 & 4 & | & 3 \\ 0 & 1 & 6 & | & 6 \\ 0 & 3 & 14 & | & 10 \end{bmatrix}$

$\xrightarrow{R_1+(-1)R_2} \begin{bmatrix} 1 & 0 & -2 & | & -3 \\ 0 & 1 & 6 & | & 6 \\ 0 & 3 & 14 & | & 10 \end{bmatrix}$

$\xrightarrow{R_3+(-3)R_2} \begin{bmatrix} 1 & 0 & -2 & | & -3 \\ 0 & 1 & 6 & | & 6 \\ 0 & 0 & -4 & | & -8 \end{bmatrix}$

$\xrightarrow{\left(-\frac{1}{4}\right)R_3} \begin{bmatrix} 1 & 0 & -2 & | & -3 \\ 0 & 1 & 6 & | & 6 \\ 0 & 0 & 1 & | & 2 \end{bmatrix}$

$\xrightarrow{R_1+2R_3} \begin{bmatrix} 1 & 0 & 0 & | & 1 \\ 0 & 1 & 6 & | & 6 \\ 0 & 0 & 1 & | & 2 \end{bmatrix}$

$\xrightarrow{R_2+(-6)R_3} \begin{bmatrix} 1 & 0 & 0 & | & 1 \\ 0 & 1 & 0 & | & -6 \\ 0 & 0 & 1 & | & 2 \end{bmatrix}$

$x = 1, y = -6, z = 2$

51. $\begin{bmatrix} -1 & 1 & 0 & | & -1 \\ 1 & 0 & 1 & | & 4 \\ 6 & -3 & 2 & | & 10 \end{bmatrix}$

$\xrightarrow{(-1)R_1} \begin{bmatrix} 1 & -1 & 0 & | & 1 \\ 1 & 0 & 1 & | & 4 \\ 6 & -3 & 2 & | & 10 \end{bmatrix}$

$\xrightarrow{R_2+(-1)R_1} \begin{bmatrix} 1 & -1 & 0 & | & 1 \\ 0 & 1 & 1 & | & 3 \\ 6 & -3 & 2 & | & 10 \end{bmatrix}$

$\xrightarrow{R_3+(-6)R_1} \begin{bmatrix} 1 & -1 & 0 & | & 1 \\ 0 & 1 & 1 & | & 3 \\ 0 & 3 & 2 & | & 4 \end{bmatrix}$

$\xrightarrow{R_1+1R_2} \begin{bmatrix} 1 & 0 & 1 & | & 4 \\ 0 & 1 & 1 & | & 3 \\ 0 & 3 & 2 & | & 4 \end{bmatrix}$

$\xrightarrow{R_3+(-3)R_2} \begin{bmatrix} 1 & 0 & 1 & | & 4 \\ 0 & 1 & 1 & | & 3 \\ 0 & 0 & -1 & | & -5 \end{bmatrix}$

$\xrightarrow{(-1)R_3} \begin{bmatrix} 1 & 0 & 1 & | & 4 \\ 0 & 1 & 1 & | & 3 \\ 0 & 0 & 1 & | & 5 \end{bmatrix}$

$\xrightarrow{R_1+(-1)R_3} \begin{bmatrix} 1 & 0 & 0 & | & -1 \\ 0 & 1 & 1 & | & 3 \\ 0 & 0 & 1 & | & 5 \end{bmatrix}$

$\xrightarrow{R_2+(-1)R_3} \begin{bmatrix} 1 & 0 & 0 & | & -1 \\ 0 & 1 & 0 & | & -2 \\ 0 & 0 & 1 & | & 5 \end{bmatrix}$

$x = -1, y = -2, z = 5$

53. Let x = grams of cheddar cheese
y = grams of potato
$$\begin{cases} x+y=180 \\ .25x+.02y=10.5 \end{cases}$$
$$\begin{bmatrix} 1 & 1 & | & 180 \\ .25 & .02 & | & 10.5 \end{bmatrix}$$
$$\xrightarrow{R_2+(-.25)R_1} \begin{bmatrix} 1 & 1 & | & 180 \\ 0 & -.23 & | & -34.5 \end{bmatrix}$$
$$\xrightarrow{-\frac{1}{.23}R_2} \begin{bmatrix} 1 & 1 & | & 180 \\ 0 & 1 & | & 150 \end{bmatrix}$$
$$\xrightarrow{R_1+(-1)R_2} \begin{bmatrix} 1 & 0 & | & 30 \\ 0 & 1 & | & 150 \end{bmatrix}$$
30 grams of cheddar cheese
Answer (b) is correct.

55. Let x = cost of golf balls
and y = cost of golf glove.
Then $x + y = 20$.
Using **Statement I**:
$y = 3x$
$x + 3x = 20$
$4x = 20$
$x = 5$.
Using **Statement II**:
$y = 15$
$x + 15 = 20$
$x = 5$.
The box of balls costs $5. Either statement is sufficient, so the answer is (d).

57. Let x = number of short sleeve shirts
y = number of long sleeve shirts
$x + y = 350$
$10x + 14y = 4300$
$$\begin{bmatrix} 1 & 1 & | & 350 \\ 10 & 14 & | & 4300 \end{bmatrix}$$
$$\xrightarrow{R_2+(-10)R_1} \begin{bmatrix} 1 & 1 & | & 350 \\ 0 & 4 & | & 800 \end{bmatrix}$$
$$\xrightarrow{\frac{1}{4}R_2} \begin{bmatrix} 1 & 1 & | & 350 \\ 0 & 1 & | & 200 \end{bmatrix}$$
$$\xrightarrow{R_1+(-1)R_2} \begin{bmatrix} 1 & 0 & | & 150 \\ 0 & 1 & | & 200 \end{bmatrix}$$
150 short sleeve, 200 long sleeve

59. Let x = adults, y = children
$$\begin{cases} x+y=350 \\ 10.50x+7.50y=3411 \end{cases}$$
$$\begin{bmatrix} 1 & 1 & | & 350 \\ 10.50 & 7.50 & | & 3411 \end{bmatrix}$$
$$\xrightarrow{R_2+(-10.5)R_1} \begin{bmatrix} 1 & 1 & | & 350 \\ 0 & -3 & | & -264 \end{bmatrix}$$
$$\xrightarrow{-\frac{1}{3}R_2} \begin{bmatrix} 1 & 1 & | & 350 \\ 0 & 1 & | & 88 \end{bmatrix}$$
$$\xrightarrow{R_1+(-1)R_2} \begin{bmatrix} 1 & 0 & | & 262 \\ 0 & 1 & | & 88 \end{bmatrix}$$
262 adults, 88 children

61. $$\begin{cases} x+y+z=9.5 \\ -x+y=0.2 \\ x-.5y=1.75 \end{cases}$$
$$\begin{bmatrix} 1 & 1 & 1 & | & 9.5 \\ -1 & 1 & 0 & | & 0.2 \\ 1 & -.5 & 0 & | & 1.75 \end{bmatrix}$$
$$\xrightarrow[R_3+(-1)R_1]{R_2+R_1} \begin{bmatrix} 1 & 1 & 1 & | & 9.5 \\ 0 & 2 & 1 & | & 9.7 \\ 0 & -1.5 & -1 & | & -7.75 \end{bmatrix}$$
$$\xrightarrow{\frac{1}{2}R_2} \begin{bmatrix} 1 & 1 & 1 & | & 9.5 \\ 0 & 1 & .5 & | & 4.85 \\ 0 & -1.5 & -1 & | & -7.75 \end{bmatrix}$$
$$\xrightarrow[R_3+1.5R_2]{R_1+(-1)R_2} \begin{bmatrix} 1 & 0 & .5 & | & 4.65 \\ 0 & 1 & .5 & | & 4.85 \\ 0 & 0 & -.25 & | & -.475 \end{bmatrix}$$
$$\xrightarrow{-4R_3} \begin{bmatrix} 1 & 0 & .5 & | & 4.65 \\ 0 & 1 & .5 & | & 4.85 \\ 0 & 0 & 1 & | & 1.9 \end{bmatrix}$$
$$\xrightarrow[R_2+(-0.5)R_3]{R_1+(-0.5)R_3} \begin{bmatrix} 1 & 0 & 0 & | & 3.7 \\ 0 & 1 & 0 & | & 3.9 \\ 0 & 0 & 1 & | & 1.9 \end{bmatrix}$$
United States is 3.7 million square miles, Canada is 3.9 million square miles and the other countries are 1.9 million square miles.

63. $\begin{cases} x+y+z=16 \\ .5x+.4y+.6z=8.10 \\ 2x-y=0 \end{cases}$

$\begin{bmatrix} 1 & 1 & 1 & | & 16 \\ .5 & .4 & .6 & | & 8.10 \\ 2 & -1 & 0 & | & 0 \end{bmatrix}$

$\xrightarrow[R_3+(-2)R_1]{R_2+(-0.5)R_1} \begin{bmatrix} 1 & 1 & 1 & | & 16 \\ 0 & -.1 & .1 & | & .1 \\ 0 & -3 & -2 & | & -32 \end{bmatrix}$

$\xrightarrow{-10R_2} \begin{bmatrix} 1 & 1 & 1 & | & 16 \\ 0 & 1 & -1 & | & -1 \\ 0 & -3 & -2 & | & -32 \end{bmatrix}$

$\xrightarrow[R_3+3R_2]{R_1+(-1)R_2} \begin{bmatrix} 1 & 0 & 2 & | & 17 \\ 0 & 1 & -1 & | & -1 \\ 0 & 0 & -5 & | & -35 \end{bmatrix}$

$\xrightarrow{-\frac{1}{5}R_3} \begin{bmatrix} 1 & 0 & 2 & | & 17 \\ 0 & 1 & -1 & | & -1 \\ 0 & 0 & 1 & | & 7 \end{bmatrix}$

$\xrightarrow[R_2+R_3]{R_1+(-2)R_3} \begin{bmatrix} 1 & 0 & 0 & | & 3 \\ 0 & 1 & 0 & | & 6 \\ 0 & 0 & 1 & | & 7 \end{bmatrix}$

3 ounces of the Brazilian, 6 ounces of the Columbian, and 7 ounces of the Peruvian.

65. $\begin{cases} x+y+z=100{,}000 \\ .08x+.07y+.1z=8000 \\ x+y-3z=0 \end{cases}$

$\begin{bmatrix} 1 & 1 & 1 & | & 100{,}000 \\ .08 & .07 & .1 & | & 8000 \\ 1 & 1 & -3 & | & 0 \end{bmatrix}$

$\xrightarrow{R_2+(-.08)R_1} \begin{bmatrix} 1 & 1 & 1 & | & 100{,}000 \\ 0 & -.01 & .02 & | & 0 \\ 1 & 1 & -3 & | & 0 \end{bmatrix}$

$\xrightarrow{R_3+(-1)R_1} \begin{bmatrix} 1 & 1 & 1 & | & 100{,}000 \\ 0 & -.01 & .02 & | & 0 \\ 0 & 0 & -4 & | & -100{,}000 \end{bmatrix}$

$\xrightarrow{(-100)R_2} \begin{bmatrix} 1 & 1 & 1 & | & 100{,}000 \\ 0 & 1 & -2 & | & 0 \\ 0 & 0 & -4 & | & -100{,}000 \end{bmatrix}$

$\xrightarrow{R_1+(-1)R_2} \begin{bmatrix} 1 & 0 & 3 & | & 100{,}000 \\ 0 & 1 & -2 & | & 0 \\ 0 & 0 & -4 & | & -100{,}000 \end{bmatrix}$

$\xrightarrow{\left(-\frac{1}{4}\right)R_3} \begin{bmatrix} 1 & 0 & 3 & | & 100{,}000 \\ 0 & 1 & -2 & | & 0 \\ 0 & 0 & 1 & | & 25{,}000 \end{bmatrix}$

$\xrightarrow{R_1+(-3)R_3} \begin{bmatrix} 1 & 0 & 0 & | & 25{,}000 \\ 0 & 1 & -2 & | & 0 \\ 0 & 0 & 1 & | & 25{,}000 \end{bmatrix}$

$\xrightarrow{R_2+2R_3} \begin{bmatrix} 1 & 0 & 0 & | & 25{,}000 \\ 0 & 1 & 0 & | & 50{,}000 \\ 0 & 0 & 1 & | & 25{,}000 \end{bmatrix}$

$x = \$25{,}000$, $y = \$50{,}000$, $z = \$25{,}000$

67. Let x = pounds of first type
y = pounds of second type
z = pounds of third type.

$.4x + .4z = 90$
$.6x + .3y + .3z = 100$
$.7y + .3z = 120$

$\begin{bmatrix} .4 & 0 & .4 & | & 90 \\ .6 & .3 & .3 & | & 100 \\ 0 & .7 & .3 & | & 120 \end{bmatrix}$

$\xrightarrow{\frac{1}{.4}R_1} \begin{bmatrix} 1 & 0 & 1 & | & 225 \\ .6 & .3 & .3 & | & 100 \\ 0 & .7 & .3 & | & 120 \end{bmatrix}$

$\xrightarrow{R_2+(-.6)R_1} \begin{bmatrix} 1 & 0 & 1 & | & 225 \\ 0 & .3 & -.3 & | & -35 \\ 0 & .7 & .3 & | & 120 \end{bmatrix}$

$\xrightarrow{\frac{1}{.3}R_2} \begin{bmatrix} 1 & 0 & 1 & | & 225 \\ 0 & 1 & -1 & | & -\frac{350}{3} \\ 0 & .7 & .3 & | & 120 \end{bmatrix}$

$\xrightarrow{R_3+(-.7)R_2} \begin{bmatrix} 1 & 0 & 1 & | & 225 \\ 0 & 1 & -1 & | & -\frac{350}{3} \\ 0 & 0 & 1 & | & \frac{605}{3} \end{bmatrix}$

$\xrightarrow{R_1+(-1)R_3} \begin{bmatrix} 1 & 0 & 0 & | & \frac{70}{3} \\ 0 & 1 & -1 & | & \frac{-350}{3} \\ 0 & 0 & 1 & | & \frac{605}{3} \end{bmatrix}$

$$\xrightarrow{R_2+R_3} \begin{bmatrix} 1 & 0 & 0 & | & \frac{70}{3} \\ 0 & 1 & 0 & | & 85 \\ 0 & 0 & 1 & | & \frac{605}{3} \end{bmatrix}$$

$\frac{70}{3}$ pounds of the first type, 85 pounds of the second type, and $\frac{605}{3}$ pounds of the third type

69. $\begin{bmatrix} 1 & 0 & | & -5 \\ 0 & 1 & | & 4 \end{bmatrix}$

71. $\begin{bmatrix} 1 & 0 & 0 & | & \frac{175}{54} \\ 0 & 1 & 0 & | & \frac{16}{9} \\ 0 & 0 & 1 & | & \frac{26}{27} \end{bmatrix}$

73.

	A	B
1	-2.5	4
2	15	15
3		
4	CELL	CONTENT
5	B1	=.2x+.3y
6	B2	=.6x+1.1y

75.

	A	B
1	1	3
2	-6	-6
3	2	1
4		
5	Cell	Content
6	B1	=x+y+4*z
7	B2	=4*x+y-2*z
8	B3	=-3*x+2*z

Exercises 2.2

1. $\begin{bmatrix} 2 & -4 & 6 \\ 3 & 7 & 1 \end{bmatrix} \xrightarrow[R_2+(-3)R_1]{\frac{1}{2}R_1} \begin{bmatrix} 1 & -2 & 3 \\ 0 & 13 & -8 \end{bmatrix}$

3. $\begin{bmatrix} 7 & 1 & 4 & 5 \\ -1 & 1 & 2 & 6 \\ 4 & 0 & 2 & 3 \end{bmatrix}$

$\xrightarrow[\substack{\frac{1}{2}R_2 \\ R_1+(-4)R_2 \\ R_3+(-2)R_2}]{} \begin{bmatrix} 9 & -1 & 0 & -7 \\ -\frac{1}{2} & \frac{1}{2} & 1 & 3 \\ 5 & -1 & 0 & -3 \end{bmatrix}$

5. $\begin{bmatrix} 2 & 3 \\ 6 & 0 \\ 1 & 5 \end{bmatrix} \xrightarrow[\substack{\frac{1}{2}R_1 \\ R_2+(-6)R_1 \\ R_3+(-1)R_1}]{} \begin{bmatrix} 1 & \frac{3}{2} \\ 0 & -9 \\ 0 & \frac{7}{2} \end{bmatrix}$

7. $\begin{bmatrix} 4 & 3 & 0 \\ \frac{2}{3} & 0 & -2 \\ 1 & 3 & 6 \end{bmatrix} \xrightarrow[R_2+2R_3]{\frac{1}{6}R_3} \begin{bmatrix} 4 & 3 & 0 \\ 1 & 1 & 0 \\ \frac{1}{6} & \frac{1}{2} & 1 \end{bmatrix}$

9. $\begin{cases} x+y+4z=6 \\ 2x+y+z=10 \end{cases}$

$z =$ any value, $y = 2-7z$, $x = 4+3z$

11. $\begin{cases} -5x+15y-10z=5 \\ x-3y+2z=0 \end{cases}$

no solution

13. $\begin{cases} 2x-y+5z=12 \\ -x-4y+2z=3 \\ 8x+5y+11z=30 \end{cases}$

$z =$ any value, $y = z-2$, $x = 5-2z$

15. $\begin{cases} x+2y+3z-w=4 \\ 2x+3y+w=-3 \\ 4x+7y+6z-w=5 \end{cases}$

$z =$ any value, $w =$ any value,
$y = 11-6z+3w$, $x = 9z-5w-18$

SSM: Finite Math   Chapter 2: Matrices

17. $\begin{bmatrix} 2 & -4 & | & 6 \\ -1 & 2 & | & -3 \end{bmatrix}$

$\begin{bmatrix} 1 & -2 & | & 3 \\ 0 & 0 & | & 0 \end{bmatrix}$

$\begin{cases} x - 2y = 3 \\ 0 = 0 \end{cases}$

y = any value, $x = 2y + 3$

19. $\begin{bmatrix} -1 & 3 & | & 11 \\ 3 & -9 & | & -30 \end{bmatrix}$

$\begin{bmatrix} 1 & -3 & | & 0 \\ 0 & 0 & | & 1 \end{bmatrix}$

$\begin{cases} x - 3y = 0 \\ 0 = 1 \end{cases}$

No solution

21. $\begin{bmatrix} 1 & 2 & | & 5 \\ 3 & -1 & | & 1 \\ -1 & 3 & | & 5 \end{bmatrix}$

$\begin{bmatrix} 1 & 2 & | & 5 \\ 0 & -7 & | & -14 \\ 0 & 5 & | & 10 \end{bmatrix}$

$\begin{bmatrix} 1 & 0 & | & 1 \\ 0 & 1 & | & 2 \\ 0 & 0 & | & 0 \end{bmatrix}$

$x = 1, y = 2$

23. $\begin{bmatrix} 4 & 5 & | & 3 \\ 3 & 6 & | & 1 \\ 2 & -3 & | & 7 \end{bmatrix}$

$\begin{bmatrix} 1 & \frac{5}{4} & | & \frac{3}{4} \\ 0 & \frac{9}{4} & | & -\frac{5}{4} \\ 0 & -\frac{11}{2} & | & \frac{11}{2} \end{bmatrix}$

$\begin{bmatrix} 1 & 0 & | & 0 \\ 0 & 1 & | & 0 \\ 0 & 0 & | & 1 \end{bmatrix}$

no solution

25. $\begin{bmatrix} 1 & -1 & 3 & | & 3 \\ -2 & 3 & -11 & | & -4 \\ 1 & -2 & 8 & | & 6 \end{bmatrix}$

$\begin{bmatrix} 1 & -1 & 3 & | & 3 \\ 0 & 1 & -5 & | & 2 \\ 0 & -1 & 5 & | & 3 \end{bmatrix}$

$\begin{bmatrix} 1 & 0 & -2 & | & 5 \\ 0 & 1 & -5 & | & 2 \\ 0 & 0 & 0 & | & 5 \end{bmatrix}$

$\begin{cases} x - 2z = 5 \\ y - 5z = 2 \\ 0 = 5 \end{cases}$

No solution

27. $\begin{bmatrix} 1 & 1 & 1 & | & -1 \\ 2 & 3 & 2 & | & 3 \\ 2 & 1 & 2 & | & -7 \end{bmatrix}$

$\begin{bmatrix} 1 & 1 & 1 & | & -1 \\ 0 & 1 & 0 & | & 5 \\ 0 & -1 & 0 & | & -5 \end{bmatrix}$

$\begin{bmatrix} 1 & 0 & 1 & | & -6 \\ 0 & 1 & 0 & | & 5 \\ 0 & 0 & 0 & | & 0 \end{bmatrix}$

$\begin{cases} x + z = -6 \\ y = 5 \\ 0 = 0 \end{cases}$

z = any value, $x = -z - 6, y = 5$

29. $\begin{bmatrix} 6 & -2 & 2 & | & 4 \\ 3 & -1 & 2 & | & 2 \\ -12 & 4 & -8 & | & 8 \end{bmatrix}$

$\begin{bmatrix} 1 & -\frac{1}{3} & 0 & | & 0 \\ 0 & 0 & 1 & | & 0 \\ 0 & 0 & 0 & | & 1 \end{bmatrix}$

$\begin{cases} x - \frac{1}{3}y = 0 \\ z = 0 \\ 0 = 1 \end{cases}$

No solution

31. $\begin{bmatrix} 1 & 2 & 8 & | & 1 \\ 3 & -1 & 4 & | & 10 \\ -1 & 5 & 10 & | & -8 \\ 1 & 1 & 1 & | & 3 \end{bmatrix}$

$\begin{bmatrix} 1 & 0 & 0 & | & 0 \\ 0 & 1 & 0 & | & 0 \\ 0 & 0 & 1 & | & 0 \\ 0 & 0 & 0 & | & 1 \end{bmatrix}$

$\begin{cases} x = 0 \\ y = 0 \\ z = 0 \\ 0 = 1 \end{cases}$

no solution

33. $\begin{bmatrix} 1 & 1 & -2 & 2 & | & 5 \\ 2 & 1 & -4 & 1 & | & 5 \\ 3 & 4 & -6 & 9 & | & 20 \\ 4 & 4 & -8 & 8 & | & 20 \end{bmatrix}$

$\begin{bmatrix} 1 & 1 & -2 & 2 & | & 5 \\ 0 & -1 & 0 & -3 & | & -5 \\ 0 & 1 & 0 & 3 & | & 5 \\ 0 & 0 & 0 & 0 & | & 0 \end{bmatrix}$

$\begin{bmatrix} 1 & 0 & -2 & -1 & | & 0 \\ 0 & 1 & 0 & 3 & | & 5 \\ 0 & 0 & 0 & 0 & | & 0 \\ 0 & 0 & 0 & 0 & | & 0 \end{bmatrix}$

$\begin{cases} x - 2z - w = 0 \\ y + 3w = 5 \\ 0 = 0 \\ 0 = 0 \end{cases}$

z = any value, w = any value, $x = 2z + w$, $y = -3w + 5$

35. $\begin{bmatrix} 1 & -1 & 1 & 1 & | & 1 \\ 0 & 1 & 3 & 2 & | & -7 \\ 0 & 1 & -1 & -3 & | & 1 \\ 1 & 0 & 4 & 3 & | & 0 \end{bmatrix}$

$\begin{bmatrix} 1 & 0 & 0 & -2 & | & 0 \\ 0 & 1 & 0 & -1.75 & | & 0 \\ 0 & 0 & 1 & 1.25 & | & 0 \\ 0 & 0 & 0 & 0 & | & 1 \end{bmatrix}$

$\begin{cases} x - 2w = 0 \\ y - \dfrac{7}{4}w = 0 \\ z + \dfrac{5}{4}w = 0 \\ 0 = 1 \end{cases}$

no solution

37. $\begin{bmatrix} 1 & 2 & 1 & | & 5 \\ 0 & 1 & 3 & | & 9 \end{bmatrix}$

$\begin{bmatrix} 1 & 2 & 1 & | & 5 \\ 0 & 1 & 3 & | & 9 \end{bmatrix}$

$\begin{bmatrix} 1 & 0 & -5 & | & -13 \\ 0 & 1 & 3 & | & 9 \end{bmatrix}$

$\begin{cases} x - 5z = -13 \\ y + 3z = 9 \end{cases}$

z = any value, $x = 5z - 13$, $y = -3z + 9$
Possible answers: $z = 0$, $x = -13$, $y = 9$;
$z = 1$, $x = -8$, $y = 6$; $z = 2$, $x = -3$, $y = 3$

39. $\begin{bmatrix} 1 & 7 & -3 & | & 8 \\ 0 & 0 & 1 & | & 5 \end{bmatrix}$

$\begin{bmatrix} 1 & 7 & -3 & | & 8 \\ 0 & 0 & 1 & | & 5 \end{bmatrix}$

$\begin{bmatrix} 1 & 7 & 0 & | & 23 \\ 0 & 0 & 1 & | & 5 \end{bmatrix}$

$\begin{cases} x + 7y = 23 \\ z = 5 \end{cases}$

y = any value, $x = -7y + 23$, $z = 5$
Possible answers: $y = 0$, $x = 23$, $z = 5$;
$y = 1$, $x = 16$, $z = 5$; $y = 2$, $x = 9$, $z = 5$

41. $\begin{bmatrix} 2 & 4 & 6 & | & 1000 \\ 3 & 7 & 10 & | & 1600 \\ 5 & 9 & 14 & | & 2400 \end{bmatrix}$

$\begin{bmatrix} 1 & 0 & 1 & | & 300 \\ 0 & 1 & 1 & | & 100 \\ 0 & 0 & 0 & | & 0 \end{bmatrix}$

$\begin{cases} x + z = 300 \\ y + z = 100 \\ 0 = 0 \end{cases}$

Food 3: z = any value between 0 and 100, Food 2: $y = 100 - z$, Food 1: $x = 300 - z$

43. $\begin{bmatrix} 3 & 10 & 15 & | & 72 \\ 4 & 12 & 8 & | & 68 \\ 5 & 14 & 1 & | & 60 \end{bmatrix}$

$\begin{bmatrix} 1 & 0 & -25 & | & 0 \\ 0 & 1 & 9 & | & 0 \\ 0 & 0 & 0 & | & 1 \end{bmatrix}$

$\begin{cases} x - 25z = 0 \\ y + 9z = 0 \\ 0 = 1 \end{cases}$

There would be no solution

45. $\begin{bmatrix} 1 & 3 & 6 & | & 380 \\ 3 & 6 & 3 & | & 450 \end{bmatrix}$

$\begin{bmatrix} 1 & 0 & -9 & | & -310 \\ 0 & 1 & 5 & | & 230 \end{bmatrix}$

$\begin{cases} x - 9z = -310 \\ y + 5z = 230 \end{cases}$

Possible answers: 50 ottomans, 30 sofas, 40 chairs; 5 ottomans, 55 sofas, 35 chairs; 95 ottomans, 5 sofas, 45 chairs

47. $\begin{cases} g + b + f = 8 \times 12 \\ g + b - 15f = 0 \\ 3g + 3b + 5f = 300 \end{cases}$

$\begin{bmatrix} 1 & 1 & 1 & | & 96 \\ 1 & 1 & -15 & | & 0 \\ 3 & 3 & 5 & | & 300 \end{bmatrix}$

$\begin{bmatrix} 1 & 1 & 1 & | & 96 \\ 0 & 0 & -16 & | & -96 \\ 0 & 0 & 2 & | & 12 \end{bmatrix}$

$\begin{bmatrix} 1 & 1 & 0 & | & 90 \\ 0 & 0 & 1 & | & 6 \\ 0 & 0 & 0 & | & 0 \end{bmatrix}$

$\begin{cases} g + b = 90 \\ f = 6 \\ 0 = 0 \end{cases}$

6 floral squares, the other 90 any mix of solid green and blue

49. $\begin{bmatrix} 2 & -3 & | & 4 \\ -6 & 9 & | & k \end{bmatrix}$

$\begin{bmatrix} 1 & -\frac{3}{2} & | & 2 \\ 0 & 0 & | & 12+k \end{bmatrix}$

No solution if $0 \neq 12 + k$, which happens when $k \neq -12$.
Infinitely many if $0 = 12 + k$, which happens when $k = -12$.

51. (b); There is no point that satisfies all three equations at the same time.

53.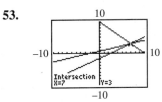

One solution when $x = 7$ and $y = 3$

55.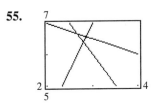

No Solution

57. There has been a pivot about the bottom right entry.

Exercises 2.3

1. 2 by 3

3. 1 by 3, row matrix

5. 2 by 2, square matrix

7. $a_{12} = -4;\ a_{21} = 0$

9. $i = 1;\ j = 3$

11. $\begin{bmatrix} 4+5 & -2+5 \\ 3+4 & 0+(-1) \end{bmatrix} = \begin{bmatrix} 9 & 3 \\ 7 & -1 \end{bmatrix}$

13. $\begin{bmatrix} 1.3+.7 & 5-1 & 2.3+.2 \\ -6+.5 & 0+1 & .7+.5 \end{bmatrix} = \begin{bmatrix} 2 & 4 & 2.5 \\ -5.5 & 1 & 1.2 \end{bmatrix}$

15. $\begin{bmatrix} 2-1 & 8-5 \\ \frac{4}{3}-\frac{1}{3} & 4-2 \\ 1-(-3) & -2-0 \end{bmatrix} = \begin{bmatrix} 1 & 3 \\ 1 & 2 \\ 4 & -2 \end{bmatrix}$

17. $\begin{bmatrix} -5-2 \\ \frac{1}{2}-\frac{1}{3} \end{bmatrix} = \begin{bmatrix} -7 \\ \frac{1}{6} \end{bmatrix}$

19. $[5 \cdot 1 + 3 \cdot 2] = [11]$

21. $\left[6 \cdot \frac{1}{2} + 1(-3) + 5 \cdot 2\right] = [10]$

23. $\begin{bmatrix} \frac{2}{3} \cdot 6 & \frac{2}{3} \cdot 0 & \frac{2}{3} \cdot -1 \\ \frac{2}{3} \cdot -9 & \frac{2}{3} \cdot \frac{3}{4} & \frac{2}{3} \cdot \frac{1}{2} \end{bmatrix} = \begin{bmatrix} 4 & 0 & -\frac{2}{3} \\ -6 & \frac{1}{2} & \frac{1}{3} \end{bmatrix}$

25. $\begin{bmatrix} 2 \cdot .5 & 2 \cdot -1 \\ 2 \cdot 4 & 2 \cdot 0 \end{bmatrix} = \begin{bmatrix} 1 & -2 \\ 8 & 0 \end{bmatrix}$

$\begin{bmatrix} 3 \cdot \frac{2}{3} & 3 \cdot 7 \\ 3 \cdot 5 & 3 \cdot 1 \end{bmatrix} = \begin{bmatrix} 2 & 21 \\ 15 & 3 \end{bmatrix}$

$\begin{bmatrix} 1+2 & -2+21 \\ 8+15 & 0+3 \end{bmatrix} = \begin{bmatrix} 3 & 19 \\ 23 & 3 \end{bmatrix}$

27. Yes, columns of A = rows of B; 4, therefore the product will be size 3×5

29. No, columns of $A \neq$ rows of B

31. Yes, columns of A = rows of B; 3, therefore the product will be size 3×1

33. $\begin{bmatrix} 3 \cdot 1 + 1 \cdot 3 & 3 \cdot 4 + 1 \cdot 5 \\ 0 \cdot 1 + 2 \cdot 3 & 0 \cdot 4 + 2 \cdot 5 \end{bmatrix} = \begin{bmatrix} 6 & 17 \\ 6 & 10 \end{bmatrix}$

35. $\begin{bmatrix} 4 \cdot 5 + 1 \cdot 1 + 0 \cdot 2 \\ -2 \cdot 5 + 0 \cdot 1 + 3 \cdot 2 \\ 1 \cdot 5 + 5 \cdot 1 + (-1)2 \end{bmatrix} = \begin{bmatrix} 21 \\ -4 \\ 8 \end{bmatrix}$

37. Multiplication by identity matrix: $\begin{bmatrix} 5 & 6 \\ 7 & 8 \end{bmatrix}$

39. $\begin{bmatrix} (.6)(.6)+(.3)(.4) & (.6)(.3)+(.3)(.7) \\ (.4)(.6)+(.7)(.4) & (.4)(.3)+(.7)(.7) \end{bmatrix}$
$= \begin{bmatrix} .48 & .39 \\ .52 & .61 \end{bmatrix}$

41. $\begin{bmatrix} 2 \cdot 4+(-1)3+4 \cdot 5 & 2 \cdot 8+(-1)(-1)+4 \cdot 0 & 2 \cdot 0+(-1)2+4 \cdot 1 \\ 0 \cdot 4+1 \cdot 3+0 \cdot 5 & 0 \cdot 8+1(-1)+0 \cdot 0 & 0 \cdot 0+1 \cdot 2+0 \cdot 1 \\ \frac{1}{2} \cdot 4+3 \cdot 3+(-2)5 & \frac{1}{2} \cdot 8+3(-1)+(-2)0 & \frac{1}{2} \cdot 0+3 \cdot 2+(-2) \cdot 1 \end{bmatrix} = \begin{bmatrix} 25 & 17 & 2 \\ 3 & -1 & 2 \\ 1 & 1 & 4 \end{bmatrix}$

43. $\begin{cases} 2x+3y=6 \\ 4x+5y=7 \end{cases}$

45. $\begin{cases} x+2y+3z=10 \\ 4x+5y+6z=11 \\ 7x+8y+9z=12 \end{cases}$

47. $\begin{bmatrix} 3 & 2 \\ 7 & -1 \end{bmatrix} \begin{bmatrix} x \\ y \end{bmatrix} = \begin{bmatrix} -1 \\ 2 \end{bmatrix}$

49. $\begin{bmatrix} 1 & -2 & 3 \\ 0 & 1 & 1 \\ 0 & 0 & 1 \end{bmatrix} \begin{bmatrix} x \\ y \\ z \end{bmatrix} = \begin{bmatrix} 5 \\ 6 \\ 2 \end{bmatrix}$

51. $\left(\begin{bmatrix} 1 & 2 \\ 0 & 3 \end{bmatrix} + \begin{bmatrix} 3 & -2 \\ 4 & 5 \end{bmatrix}\right)\begin{bmatrix} 1 & 6 \\ 2 & 0 \end{bmatrix} = \begin{bmatrix} 4 & 0 \\ 4 & 8 \end{bmatrix}\begin{bmatrix} 1 & 6 \\ 2 & 0 \end{bmatrix} = \begin{bmatrix} 4 & 24 \\ 20 & 24 \end{bmatrix}$

$\begin{bmatrix} 1 & 2 \\ 0 & 3 \end{bmatrix}\begin{bmatrix} 1 & 6 \\ 2 & 0 \end{bmatrix} + \begin{bmatrix} 3 & -2 \\ 4 & 5 \end{bmatrix}\begin{bmatrix} 1 & 6 \\ 2 & 0 \end{bmatrix} = \begin{bmatrix} 5 & 6 \\ 6 & 0 \end{bmatrix} + \begin{bmatrix} -1 & 18 \\ 14 & 24 \end{bmatrix} = \begin{bmatrix} 4 & 24 \\ 20 & 24 \end{bmatrix}$

53. $\begin{bmatrix} 3\cdot 1 + (-1)2 & 3\cdot 2 + (-1)6 \\ -1\cdot 1 + \frac{1}{2}\cdot 2 & -1\cdot 2 + \frac{1}{2}\cdot 6 \end{bmatrix} = \begin{bmatrix} 1 & 0 \\ 0 & 1 \end{bmatrix}$

$\begin{bmatrix} 1\cdot 3 + 2(-1) & 1(-1)+(2)\frac{1}{2} \\ 2\cdot 3 + 6(-1) & 2(-1)+(6)\frac{1}{2} \end{bmatrix} = \begin{bmatrix} 1 & 0 \\ 0 & 1 \end{bmatrix}$

55. a. $\begin{bmatrix} 6 & 8 & 2 \\ 2 & 5 & 3 \end{bmatrix}\begin{bmatrix} 20 \\ 15 \\ 50 \end{bmatrix} = \begin{bmatrix} 340 \\ 265 \end{bmatrix}$

b. Mike's clothes are worth $340; Don's clothes are worth $265.

c. $\begin{bmatrix} 1.25 \cdot 20 \\ 1.25 \cdot 15 \\ 1.25 \cdot 50 \end{bmatrix} = \begin{bmatrix} 25 \\ 18.75 \\ 62.50 \end{bmatrix}$

d. The matrix represents the costs of the three items of clothing after a 25% increase.

57. a. $\begin{bmatrix} 210 & 175 & 135 \end{bmatrix}\begin{bmatrix} 3 & 3 & 5.8 \\ 2.5 & 3.5 & 6 \\ 9 & 8 & 9.5 \end{bmatrix}$

$= \begin{bmatrix} 2282.50 & 2322.50 & 3550.50 \end{bmatrix}$

The total value of the store's plain items was $2282.50, of the milk chocolate-covered items was $2322.50, and of the dark chocolate covered items was $3550.50.

b. $\begin{bmatrix} 3 & 3 & 5.8 \\ 2.5 & 3.5 & 6 \\ 9 & 8 & 9.5 \end{bmatrix}\begin{bmatrix} 105 \\ 390 \\ 285 \end{bmatrix} = \begin{bmatrix} 3138.00 \\ 3337.50 \\ 6772.50 \end{bmatrix}$

The store's weekly sales of peanuts was $3138.00, of raisins was $3337.50, and of espresso beans was $6772.50.

c. $\begin{bmatrix} .9 \cdot 105 \\ .9 \cdot 390 \\ .9 \cdot 285 \end{bmatrix} = \begin{bmatrix} 94.50 \\ 351 \\ 256.50 \end{bmatrix}$

The store's weekly sales if there were a 10% decrease in sales.

59. a. $\begin{bmatrix} .25 & .35 & .30 & .10 & 0 \\ .10 & .20 & .40 & .20 & .10 \\ .05 & .10 & .20 & .40 & .25 \end{bmatrix}\begin{bmatrix} 4 \\ 3 \\ 2 \\ 1 \\ 0 \end{bmatrix} = \begin{bmatrix} 2.75 \\ 2.00 \\ 1.30 \end{bmatrix}$

I: 2.75; II: 2, III: 1.3

b. $\begin{bmatrix} 240 & 120 & 40 \end{bmatrix}\begin{bmatrix} .25 & .35 & .30 & .10 & 0 \\ .10 & .20 & .40 & .20 & .10 \\ .05 & .10 & .20 & .40 & .25 \end{bmatrix}$

$= \begin{bmatrix} 74 & 112 & 128 & 64 & 22 \end{bmatrix}$

A: 74, B: 112, C: 128, D: 64, F: 22

61. $[6000 \ 8000 \ 4000]\begin{bmatrix} .65 & .35 \\ .55 & .45 \\ .45 & .55 \end{bmatrix} = [10,100 \ 7900]$

10,100 voting Democratic, 7900 voting Republican

63. $\begin{bmatrix} 50 & 20 & 10 \\ 30 & 30 & 15 \\ 20 & 20 & 5 \end{bmatrix} \begin{bmatrix} 10 \\ 15 \\ 20 \end{bmatrix} = \begin{bmatrix} 1000 \\ 1050 \\ 600 \end{bmatrix}$

Carpenters: $1000, bricklayers: $1050, plumbers: $600

65. a. $BN = [162 \ 150 \ 143]$, number of units of each nutrient consumed at breakfast

  b. $LN = [186 \ 200 \ 239]$, number of units of each nutrient consumed at lunch

  c. $DN = [288 \ 300 \ 344]$, number of units of each nutrient consumed at dinner

  d. $B + L + D = [5 \ 8]$, total number of ounces of each food that Mikey eats during a day

  e. $(B+L+D)N = [636 \ 650 \ 726]$, number of units of each nutrient consumed per day

67. a. $AP = \begin{bmatrix} 720 \\ 646 \end{bmatrix}$, the average amount taken in daily by the pool and the weight room

  b. $720

69. a.
$$T = \begin{bmatrix} 30 & 45 \\ 30 & 50 \\ 15 & 10 \end{bmatrix} \begin{matrix} \text{Preparation} \\ \text{Baking} \\ \text{Finishing} \end{matrix}$$
with columns Boston Cream Pie, Carrot Cake

  b. $S = \begin{bmatrix} 20 \\ 8 \end{bmatrix} \begin{matrix} \text{Boston Cream Pie} \\ \text{Carrot Cake} \end{matrix}$

  $TS = \begin{bmatrix} 960 \\ 1000 \\ 380 \end{bmatrix} \begin{matrix} \text{Preparation} \\ \text{Baking} \\ \text{Finishing} \end{matrix}$

  c. Total baking time is 1000 minutes or $16\frac{2}{3}$ hours. Total finishing time is 380 minutes or $6\frac{1}{3}$ hours.

SSM: Finite Math   **Chapter 2: Matrices**

71. **a.** $T = \begin{bmatrix} \text{Cutting} & \text{Sewing} & \text{Finishing} \\ 2 & 3 & 2 \\ 1.5 & 2 & 1 \end{bmatrix} \begin{matrix} \text{Huge One} \\ \text{Regular Joe} \end{matrix}$

   b. $S = \begin{bmatrix} 32 \\ 24 \end{bmatrix} \begin{matrix} \text{Huge One} \\ \text{Regular Joe} \end{matrix}$

   c. $A = \begin{bmatrix} \text{Huge One} & \text{Regular Joe} \\ 27 & 56 \end{bmatrix}$

$AT = \begin{bmatrix} \text{Cutting} & \text{Sewing} & \text{Finishing} \\ 138 & 193 & 110 \end{bmatrix}$

$AS = [2208]$

   d. 193 hours are needed for sewing.

   e. The total revenue would be $2208.

73. answers will vary.

75. $(A+B) - A = \begin{bmatrix} 3 & -2 & 1 \\ -5 & 6 & 7 \end{bmatrix}$

77. 4×4

79. The matrix product:
$\begin{bmatrix} 8322 & 58{,}940 & 71{,}415 \end{bmatrix} \begin{bmatrix} 15.1 \\ 14.3 \\ 17.3 \end{bmatrix}$ gives the total number of pupils in the three states.

81. $A + B = \begin{bmatrix} 6.4 & -2 & -2.7 \\ 20.5 & 22.5 & -2.4 \\ -14 & 17.6 & 16 \end{bmatrix}$

83. $BA = \begin{bmatrix} -171.3 & 40.8 & -31.8 \\ 454.6 & -22.5 & 22.7 \\ -2.6 & 122.3 & 53.56 \end{bmatrix}$

85. $3A = \begin{bmatrix} 1.2 & 21 & -9 \\ 57 & 1.5 & 4.8 \\ -27 & 33 & 6 \end{bmatrix}$

87. Answer may vary. One possibility is with the message ERR:DIM MISMATCH

Exercises 2.4

1. $\begin{bmatrix} 1 & -2 \\ -\frac{1}{2} & 2 \end{bmatrix} \begin{bmatrix} 4 \\ 1 \end{bmatrix} = \begin{bmatrix} 2 \\ 0 \end{bmatrix}$

   $x = 2, y = 0$

3. $D = 7 \cdot 1 - 3 \cdot 2 = 1$

   $\begin{bmatrix} \frac{1}{1} & -\frac{2}{1} \\ -\frac{3}{1} & \frac{7}{1} \end{bmatrix} = \begin{bmatrix} 1 & -2 \\ -3 & 7 \end{bmatrix}$

5. $D = 6 \cdot 2 - 5 \cdot 2 = 2$

   $\begin{bmatrix} \frac{2}{2} & -\frac{2}{2} \\ -\frac{5}{2} & \frac{6}{2} \end{bmatrix} = \begin{bmatrix} 1 & -1 \\ -\frac{5}{2} & 3 \end{bmatrix}$

7. $D = 0.7 \cdot 0.8 - 0.3 \cdot 0.2 = 0.5$

   $\begin{bmatrix} \frac{.8}{.5} & -\frac{.2}{.5} \\ -\frac{.3}{.5} & \frac{.7}{.5} \end{bmatrix} = \begin{bmatrix} 1.6 & -.4 \\ -.6 & 1.4 \end{bmatrix}$

9. For a 1×1 matrix $[a]$ $(a \neq 0)$, $[a]^{-1} = \begin{bmatrix} \frac{1}{a} \end{bmatrix}$.

   $\begin{bmatrix} \frac{1}{3} \end{bmatrix}$

11. $\begin{bmatrix} 1 & 2 \\ 2 & 6 \end{bmatrix}^{-1} \begin{bmatrix} 3 \\ 5 \end{bmatrix} = \begin{bmatrix} 3 & -1 \\ -1 & \frac{1}{2} \end{bmatrix} \begin{bmatrix} 3 \\ 5 \end{bmatrix} = \begin{bmatrix} 4 \\ -\frac{1}{2} \end{bmatrix}$

  $x = 4, \ y = -\dfrac{1}{2}$

13. $\begin{bmatrix} \frac{1}{2} & 2 \\ 3 & 16 \end{bmatrix}^{-1} \begin{bmatrix} 4 \\ 0 \end{bmatrix} = \begin{bmatrix} 8 & -1 \\ -\frac{3}{2} & \frac{1}{4} \end{bmatrix} \begin{bmatrix} 4 \\ 0 \end{bmatrix} = \begin{bmatrix} 32 \\ -6 \end{bmatrix}$

  $x = 32, \ y = -6$

15. **a.** $\begin{bmatrix} .8 & .3 \\ .2 & .7 \end{bmatrix} \begin{bmatrix} x \\ y \end{bmatrix} = \begin{bmatrix} m \\ s \end{bmatrix}$

  b. $\begin{bmatrix} x \\ y \end{bmatrix} = \begin{bmatrix} .8 & .3 \\ .2 & .7 \end{bmatrix}^{-1} \begin{bmatrix} m \\ s \end{bmatrix}$
  $= \begin{bmatrix} 1.4 & -.6 \\ -.4 & 1.6 \end{bmatrix} \begin{bmatrix} m \\ s \end{bmatrix}$

  c. $\begin{bmatrix} 1.4 & -.6 \\ -.4 & 1.6 \end{bmatrix} \begin{bmatrix} 100{,}000 \\ 50{,}000 \end{bmatrix} = \begin{bmatrix} 110{,}000 \\ 40{,}000 \end{bmatrix}$
  110,000 married; 40,000 single

  d. $\begin{bmatrix} 1.4 & -.6 \\ -.4 & 1.6 \end{bmatrix} \begin{bmatrix} 110{,}000 \\ 40{,}000 \end{bmatrix} = \begin{bmatrix} 130{,}000 \\ 20{,}000 \end{bmatrix}$
  130,000 married; 20,000 single

17. **a.** $\begin{bmatrix} .7 & .1 \\ .3 & .9 \end{bmatrix} \begin{bmatrix} x \\ y \end{bmatrix} = \begin{bmatrix} u \\ v \end{bmatrix}$

  b. $\begin{bmatrix} x \\ y \end{bmatrix} = \begin{bmatrix} .7 & .1 \\ .3 & .9 \end{bmatrix}^{-1} \begin{bmatrix} u \\ v \end{bmatrix} = \begin{bmatrix} \frac{3}{2} & -\frac{1}{6} \\ -\frac{1}{2} & \frac{7}{6} \end{bmatrix} \begin{bmatrix} u \\ v \end{bmatrix}$

  c. $\begin{bmatrix} \frac{3}{2} & -\frac{1}{6} \\ -\frac{1}{2} & \frac{7}{6} \end{bmatrix} \begin{bmatrix} 6000 \\ 3000 \end{bmatrix} = \begin{bmatrix} 8500 \\ 500 \end{bmatrix}$

  $\begin{bmatrix} .7 & .1 \\ .3 & .9 \end{bmatrix} \begin{bmatrix} 6000 \\ 3000 \end{bmatrix} = \begin{bmatrix} 4500 \\ 4500 \end{bmatrix}$
  8500; 4500

19. $\begin{bmatrix} 5 & -2 & -2 \\ -1 & 1 & 0 \\ -1 & 0 & 1 \end{bmatrix} \begin{bmatrix} 1 \\ -1 \\ -1 \end{bmatrix} = \begin{bmatrix} 9 \\ -2 \\ -2 \end{bmatrix}$
  $x = 9, \ y = -2, \ z = -2$

21. $\begin{bmatrix} 1 & 2 & 2 \\ 1 & 3 & 2 \\ 1 & 2 & 3 \end{bmatrix} \begin{bmatrix} 3 \\ 4 \\ 5 \end{bmatrix} = \begin{bmatrix} 21 \\ 25 \\ 26 \end{bmatrix}$
  $x = 21, \ y = 25, \ z = 26$

23. $\begin{bmatrix} 1 & 0 & -2 & 0 \\ 0 & 1 & 0 & -5 \\ -4 & 0 & 9 & 0 \\ 0 & 2 & 1 & -9 \end{bmatrix} \begin{bmatrix} 1 \\ 0 \\ 0 \\ -1 \end{bmatrix} = \begin{bmatrix} 1 \\ 5 \\ -4 \\ 9 \end{bmatrix}$
  $x = 1, \ y = 5, \ z = -4, \ w = 9$

25. $\begin{bmatrix} 9 & 0 & 2 & 0 \\ -20 & -9 & -5 & 5 \\ 4 & 0 & 1 & 0 \\ -4 & -2 & -1 & 1 \end{bmatrix} \begin{bmatrix} 0 \\ 1 \\ 2 \\ 0 \end{bmatrix} = \begin{bmatrix} 4 \\ -19 \\ 2 \\ -4 \end{bmatrix}$
  $x = 4, \ y = -19, \ z = 2, \ w = -4$

27. Suppose $\begin{bmatrix} 6 & 3 \\ 2 & 1 \end{bmatrix}^{-1} = \begin{bmatrix} s & t \\ u & v \end{bmatrix}$.

  Then $\begin{bmatrix} 6s+3u & 6t+3v \\ 2s+u & 2t+v \end{bmatrix} = \begin{bmatrix} 1 & 0 \\ 0 & 1 \end{bmatrix}$.

  Then $\dfrac{6s+3u}{3} = 2s+u = \dfrac{1}{3}$, which contradicts $2s + u = 0$.

29. **a.** $\begin{cases} x + 2y = a \\ .9x = b \end{cases}$

  $\begin{bmatrix} 1 & 2 \\ .9 & 0 \end{bmatrix} \begin{bmatrix} x \\ y \end{bmatrix} = \begin{bmatrix} a \\ b \end{bmatrix}$

  b. $\begin{bmatrix} 1 & 2 \\ .9 & 0 \end{bmatrix} \begin{bmatrix} 450{,}000 \\ 360{,}000 \end{bmatrix} = \begin{bmatrix} 1{,}170{,}000 \\ 405{,}000 \end{bmatrix}$
  After 1 year:
  1,170,000 in group I,
  405,000 in group II

  $\begin{bmatrix} 1 & 2 \\ .9 & 0 \end{bmatrix} \begin{bmatrix} 1{,}170{,}000 \\ 405{,}000 \end{bmatrix} = \begin{bmatrix} 1{,}980{,}000 \\ 1{,}053{,}000 \end{bmatrix}$
  After 2 years:
  1,980,000 in group I,
  1,053,000 in group II

  c. $\begin{bmatrix} 1 & 2 \\ .9 & 0 \end{bmatrix}^{-1} \begin{bmatrix} 810{,}000 \\ 630{,}000 \end{bmatrix} = \begin{bmatrix} 700{,}000 \\ 55{,}000 \end{bmatrix}$
  700,000 in group I, 55,000 in group II

31. If AB = 0 (zero matrix) and A has an inverse the B is a matrix of all zeros:
Proof : Assume AB = 0 and A has an inverse.
$A^{-1}(AB) = A^{-1}(0) \rightarrow$
$(A^{-1}A)B = A^{-1}(0) \rightarrow$
$(I_n)B = 0 \rightarrow$
$B = 0$

33. One example:
$AX = B$
Let $A = \begin{bmatrix} 1 & 1 \\ 1 & 1 \end{bmatrix}$ $X = \begin{bmatrix} x \\ y \end{bmatrix}$ $B = \begin{bmatrix} 2 \\ 3 \end{bmatrix}$

The system $\begin{cases} x+y=2 \\ x+y=3 \end{cases}$ has no solution.

35.
$\begin{bmatrix} -\frac{10}{73} & \frac{75}{292} \\ \frac{25}{73} & -\frac{5}{292} \end{bmatrix}$

37.
$\begin{bmatrix} \frac{1020}{8887} & \frac{2910}{8887} & -\frac{500}{8887} \\ \frac{3050}{8887} & \frac{860}{8887} & \frac{1990}{8887} \\ \frac{125}{8887} & \frac{618}{8887} & \frac{810}{8887} \end{bmatrix}$

39.
$x = -\frac{4}{5}, y = \frac{28}{5}, z = 5$

41.
$x = 0, y = 2, z = 0, w = 2$

43. Answer may vary. One possibility is with the message ERR:INVALID DIM

Exercises 2.5

1. $\begin{bmatrix} 7 & 3 & | & 1 & 0 \\ 5 & 2 & | & 0 & 1 \end{bmatrix}$

$\begin{bmatrix} 1 & \frac{3}{7} & | & \frac{1}{7} & 0 \\ 0 & -\frac{1}{7} & | & -\frac{5}{7} & 1 \end{bmatrix}$

$\begin{bmatrix} 1 & 0 & | & -2 & 3 \\ 0 & 1 & | & 5 & -7 \end{bmatrix}$

$\begin{bmatrix} -2 & 3 \\ 5 & -7 \end{bmatrix}$

3. $\begin{bmatrix} 2 & 3 & | & 1 & 0 \\ -4 & -7 & | & 0 & 1 \end{bmatrix}$

$\begin{bmatrix} 1 & \frac{3}{2} & | & \frac{1}{2} & 0 \\ 0 & -1 & | & 2 & 1 \end{bmatrix}$

$\begin{bmatrix} 1 & 0 & | & \frac{7}{2} & \frac{3}{2} \\ 0 & 1 & | & -2 & -1 \end{bmatrix}$

$\begin{bmatrix} \frac{7}{2} & \frac{3}{2} \\ -2 & -1 \end{bmatrix}$

5. $\begin{bmatrix} 2 & -4 & | & 1 & 0 \\ -1 & 2 & | & 0 & 1 \end{bmatrix}$

$\begin{bmatrix} 1 & -2 & | & \frac{1}{2} & 0 \\ 0 & 0 & | & \frac{1}{2} & 1 \end{bmatrix}$

No inverse

7. $\begin{bmatrix} \underline{1} & 2 & -2 & | & 1 & 0 & 0 \\ 1 & 1 & 1 & | & 0 & 1 & 0 \\ 0 & 0 & 1 & | & 0 & 0 & 1 \end{bmatrix}$

$\begin{bmatrix} 1 & 2 & -2 & | & 1 & 0 & 0 \\ 0 & \underline{-1} & 3 & | & -1 & 1 & 0 \\ 0 & 0 & 1 & | & 0 & 0 & 1 \end{bmatrix}$

$\begin{bmatrix} 1 & 0 & 4 & | & -1 & 2 & 0 \\ 0 & 1 & -3 & | & 1 & -1 & 0 \\ 0 & 0 & \underline{1} & | & 0 & 0 & 1 \end{bmatrix}$

$\begin{bmatrix} 1 & 0 & 0 & | & -1 & 2 & -4 \\ 0 & 1 & 0 & | & 1 & -1 & 3 \\ 0 & 0 & 1 & | & 0 & 0 & 1 \end{bmatrix}$

$\begin{bmatrix} -1 & 2 & -4 \\ 1 & -1 & 3 \\ 0 & 0 & 1 \end{bmatrix}$

9. $\begin{bmatrix} \underline{-2} & 5 & 2 & | & 1 & 0 & 0 \\ 1 & -3 & -1 & | & 0 & 1 & 0 \\ -1 & 2 & 1 & | & 0 & 0 & 1 \end{bmatrix}$

$\begin{bmatrix} 1 & -\frac{5}{2} & -1 & | & -\frac{1}{2} & 0 & 0 \\ 0 & \underline{-\frac{1}{2}} & 0 & | & \frac{1}{2} & 1 & 0 \\ 0 & -\frac{1}{2} & 0 & | & -\frac{1}{2} & 0 & 1 \end{bmatrix}$

$\begin{bmatrix} 1 & 0 & -1 & | & -3 & -5 & 0 \\ 0 & 1 & 0 & | & -1 & -2 & 0 \\ 0 & 0 & 0 & | & -1 & -1 & 1 \end{bmatrix}$

No inverse

11. $\begin{bmatrix} \underline{1} & 6 & 0 & 0 & | & 1 & 0 & 0 & 0 \\ 1 & 5 & 0 & 0 & | & 0 & 1 & 0 & 0 \\ 0 & 0 & 4 & 2 & | & 0 & 0 & 1 & 0 \\ 0 & 0 & 50 & 2 & | & 0 & 0 & 0 & 1 \end{bmatrix}$

$\begin{bmatrix} 1 & 6 & 0 & 0 & | & 1 & 0 & 0 & 0 \\ 0 & \underline{-1} & 0 & 0 & | & -1 & 1 & 0 & 0 \\ 0 & 0 & 4 & 2 & | & 0 & 0 & 1 & 0 \\ 0 & 0 & 50 & 2 & | & 0 & 0 & 0 & 1 \end{bmatrix}$

$\begin{bmatrix} 1 & 0 & 0 & 0 & | & -5 & 6 & 0 & 0 \\ 0 & 1 & 0 & 0 & | & 1 & -1 & 0 & 0 \\ 0 & 0 & \underline{4} & 2 & | & 0 & 0 & 1 & 0 \\ 0 & 0 & 50 & 2 & | & 0 & 0 & 0 & 1 \end{bmatrix}$

$\begin{bmatrix} 1 & 0 & 0 & 0 & | & -5 & 6 & 0 & 0 \\ 0 & 1 & 0 & 0 & | & 1 & -1 & 0 & 0 \\ 0 & 0 & 1 & \frac{1}{2} & | & 0 & 0 & \frac{1}{4} & 0 \\ 0 & 0 & 0 & \underline{-23} & | & 0 & 0 & -\frac{25}{2} & 1 \end{bmatrix}$

$\begin{bmatrix} 1 & 0 & 0 & 0 & | & -5 & 6 & 0 & 0 \\ 0 & 1 & 0 & 0 & | & 1 & -1 & 0 & 0 \\ 0 & 0 & 1 & 0 & | & 0 & 0 & -\frac{1}{46} & \frac{1}{46} \\ 0 & 0 & 0 & 1 & | & 0 & 0 & \frac{25}{46} & -\frac{1}{23} \end{bmatrix}$

$\begin{bmatrix} -5 & 6 & 0 & 0 \\ 1 & -1 & 0 & 0 \\ 0 & 0 & -\frac{1}{46} & \frac{1}{46} \\ 0 & 0 & \frac{25}{46} & -\frac{1}{23} \end{bmatrix}$

13. Find the inverse of $\begin{bmatrix} 1 & 1 & 2 \\ 3 & 2 & 2 \\ 1 & 1 & 3 \end{bmatrix}$.

$\begin{bmatrix} 1 & 1 & 2 & | & 1 & 0 & 0 \\ 3 & 2 & 2 & | & 0 & 1 & 0 \\ 1 & 1 & 3 & | & 0 & 0 & 1 \end{bmatrix}$

$\begin{bmatrix} 1 & 1 & 2 & | & 1 & 0 & 0 \\ 0 & -1 & -4 & | & -3 & 1 & 0 \\ 0 & 0 & 1 & | & -1 & 0 & 1 \end{bmatrix}$

$\begin{bmatrix} 1 & 0 & -2 & | & -2 & 1 & 0 \\ 0 & 1 & 4 & | & 3 & -1 & 0 \\ 0 & 0 & 1 & | & -1 & 0 & 1 \end{bmatrix}$

$\begin{bmatrix} 1 & 0 & 0 & | & -4 & 1 & 2 \\ 0 & 1 & 0 & | & 7 & -1 & -4 \\ 0 & 0 & 1 & | & -1 & 0 & 1 \end{bmatrix}$

$\begin{bmatrix} 1 & 1 & 2 \\ 3 & 2 & 2 \\ 1 & 1 & 3 \end{bmatrix}^{-1} \begin{bmatrix} 3 \\ 4 \\ 5 \end{bmatrix} = \begin{bmatrix} -4 & 1 & 2 \\ 7 & -1 & -4 \\ -1 & 0 & 1 \end{bmatrix} \begin{bmatrix} 3 \\ 4 \\ 5 \end{bmatrix} = \begin{bmatrix} 2 \\ -3 \\ 2 \end{bmatrix}$

$x = 2, y = -3, z = 2$

15. Find the inverse of $\begin{bmatrix} 1 & 0 & -2 & -2 \\ 0 & 1 & 0 & -5 \\ -4 & 0 & 9 & 9 \\ 0 & 2 & 1 & -8 \end{bmatrix}$.

$\left[\begin{array}{cccc|cccc} 1 & 0 & -2 & -2 & 1 & 0 & 0 & 0 \\ 0 & 1 & 0 & -5 & 0 & 1 & 0 & 0 \\ -4 & 0 & 9 & 9 & 0 & 0 & 1 & 0 \\ 0 & 2 & 1 & -8 & 0 & 0 & 0 & 1 \end{array}\right]$

$\left[\begin{array}{cccc|cccc} 1 & 0 & -2 & -2 & 1 & 0 & 0 & 0 \\ 0 & 1 & 0 & -5 & 0 & 1 & 0 & 0 \\ 0 & 0 & 1 & 1 & 4 & 0 & 1 & 0 \\ 0 & 2 & 1 & -8 & 0 & 0 & 0 & 1 \end{array}\right]$

$\left[\begin{array}{cccc|cccc} 1 & 0 & -2 & -2 & 1 & 0 & 0 & 0 \\ 0 & 1 & 0 & -5 & 0 & 1 & 0 & 0 \\ 0 & 0 & 1 & 1 & 4 & 0 & 1 & 0 \\ 0 & 0 & 1 & 2 & 0 & -2 & 0 & 1 \end{array}\right]$

$\left[\begin{array}{cccc|cccc} 1 & 0 & 0 & 0 & 9 & 0 & 2 & 0 \\ 0 & 1 & 0 & -5 & 0 & 1 & 0 & 0 \\ 0 & 0 & 1 & 1 & 4 & 0 & 1 & 0 \\ 0 & 0 & 0 & 1 & -4 & -2 & -1 & 1 \end{array}\right]$

$\left[\begin{array}{cccc|cccc} 1 & 0 & 0 & 0 & 9 & 0 & 2 & 0 \\ 0 & 1 & 0 & 0 & -20 & -9 & -5 & 5 \\ 0 & 0 & 1 & 0 & 8 & 2 & 2 & -1 \\ 0 & 0 & 0 & 1 & -4 & -2 & -1 & 1 \end{array}\right]$

$\begin{bmatrix} 1 & 0 & -2 & -2 \\ 0 & 1 & 0 & -5 \\ -4 & 0 & 9 & 9 \\ 0 & 2 & 1 & -8 \end{bmatrix}^{-1} \begin{bmatrix} 0 \\ 1 \\ 2 \\ 3 \end{bmatrix}$

$= \begin{bmatrix} 9 & 0 & 2 & 0 \\ -20 & -9 & -5 & 5 \\ 8 & 2 & 2 & -1 \\ -4 & -2 & -1 & 1 \end{bmatrix} \begin{bmatrix} 0 \\ 1 \\ 2 \\ 3 \end{bmatrix}$

$= \begin{bmatrix} 4 \\ -4 \\ 3 \\ -1 \end{bmatrix}$

$x = 4, y = -4, z = 3, w = -1$

17. If $A^{-1} = \begin{bmatrix} 2 & 7 \\ 1 & 3 \end{bmatrix}$,

Fact: $A^{-1}A = I = A \cdot A^{-1}$

Let $A = \begin{bmatrix} a & b \\ c & d \end{bmatrix}$. Then $\begin{bmatrix} a & b \\ c & d \end{bmatrix} \cdot \begin{bmatrix} 2 & 7 \\ 1 & 3 \end{bmatrix} = \begin{bmatrix} 1 & 0 \\ 0 & 1 \end{bmatrix}$

$2a + b = 1$ and $7a + 3b = 0$
Also: $2c + d = 0$ and $7c + 3d = 1$
Solving for a, b, c, d, we have $a = -3$, $b = 7$, $c = 1$ and $d = -2$

$A = \begin{bmatrix} -3 & 7 \\ 1 & -2 \end{bmatrix}$

19. $A \cdot \begin{bmatrix} 2 & 5 \\ 1 & 3 \end{bmatrix} = \begin{bmatrix} -1 & 0 \\ 4 & 2 \end{bmatrix}$

$A = A \cdot \begin{bmatrix} 2 & 5 \\ 1 & 3 \end{bmatrix} \begin{bmatrix} 2 & 5 \\ 1 & 3 \end{bmatrix}^{-1}$

$= \begin{bmatrix} -1 & 0 \\ 4 & 2 \end{bmatrix} \begin{bmatrix} 2 & 5 \\ 1 & 3 \end{bmatrix}^{-1}$

$= \begin{bmatrix} -1 & 0 \\ 4 & 2 \end{bmatrix} \begin{bmatrix} 3 & -5 \\ -1 & 2 \end{bmatrix}$

$= \begin{bmatrix} -3 & 5 \\ 10 & -16 \end{bmatrix}$

21. $\begin{cases} x + y + z = 100 \\ x - 2y = 0 \\ x + y - z = 26 \end{cases}$

$\left[\begin{array}{ccc|ccc} 1 & 1 & 1 & 1 & 0 & 0 \\ 1 & -2 & 0 & 0 & 1 & 0 \\ 1 & 1 & -1 & 0 & 0 & 1 \end{array}\right]$

$\left[\begin{array}{ccc|ccc} 1 & 1 & 1 & 1 & 0 & 0 \\ 0 & -3 & -1 & -1 & 1 & 0 \\ 0 & 0 & -2 & -1 & 0 & 1 \end{array}\right]$

$\left[\begin{array}{ccc|ccc} 1 & 1 & 1 & 1 & 0 & 0 \\ 0 & 1 & \frac{1}{3} & \frac{1}{3} & -\frac{1}{3} & 0 \\ 0 & 0 & -2 & -1 & 0 & 1 \end{array}\right]$

$$\begin{bmatrix} 1 & 0 & \frac{2}{3} & \frac{2}{3} & \frac{1}{3} & 0 \\ 0 & 1 & \frac{1}{3} & \frac{1}{3} & -\frac{1}{3} & 0 \\ 0 & 0 & -2 & -1 & 0 & 1 \end{bmatrix}$$

$$\begin{bmatrix} 1 & 0 & \frac{2}{3} & \frac{2}{3} & \frac{1}{3} & 0 \\ 0 & 1 & \frac{1}{3} & \frac{1}{3} & -\frac{1}{3} & 0 \\ 0 & 0 & 1 & \frac{1}{2} & 0 & -\frac{1}{2} \end{bmatrix}$$

$$\begin{bmatrix} 1 & 0 & 0 & \frac{1}{3} & \frac{1}{3} & \frac{1}{3} \\ 0 & 1 & 0 & \frac{1}{6} & -\frac{1}{3} & \frac{1}{6} \\ 0 & 0 & 1 & \frac{1}{2} & 0 & -\frac{1}{2} \end{bmatrix}$$

$$\begin{bmatrix} \frac{1}{3} & \frac{1}{3} & \frac{1}{3} \\ \frac{1}{6} & -\frac{1}{3} & \frac{1}{6} \\ \frac{1}{2} & 0 & -\frac{1}{2} \end{bmatrix} \begin{bmatrix} 100 \\ 0 \\ 26 \end{bmatrix} = \begin{bmatrix} 42 \\ 21 \\ 37 \end{bmatrix}$$

$x = 42, y = 21, z = 37$

Exercises 2.6

1. $A_{21} = .2$, 20 cents of energy are required to produce $1 worth of manufactured goods.

3. $A_{31} = .1, A_{32} = .2, A_{33} = .15$

   $A_{32} > A_{33} > A_{31}$

   The energy sector uses the greatest amount of services in order to produce $1 worth of output.

5. $D_{\text{new}} = \begin{bmatrix} 4 \\ 1.5 \\ 9 \end{bmatrix}$

   $(I-A)^{-1} D_{\text{new}} = \begin{bmatrix} 1.01 & .20 & .50 \\ .02 & 1.05 & .23 \\ .01 & .09 & 1.08 \end{bmatrix} \begin{bmatrix} 4 \\ 1.5 \\ 9 \end{bmatrix}$

   $= \begin{bmatrix} 8.84 \\ 3.725 \\ 9.895 \end{bmatrix}$

   Coal: $8.84 billion, steel: $3.725 billion, electricity: $9.895 billion

SSM: Finite Math   Chapter 2: Matrices

7. $D_{new} = \begin{bmatrix} 3 \\ 1 \\ 4 \end{bmatrix}$

$(I-A)^{-1}D_{new} = \begin{bmatrix} 1.04 & .02 & .10 \\ .21 & 1.01 & .02 \\ .11 & .02 & 1.02 \end{bmatrix} \begin{bmatrix} 3 \\ 1 \\ 4 \end{bmatrix} = \begin{bmatrix} 3.54 \\ 1.72 \\ 4.43 \end{bmatrix}$

Computers: $354 million,
semiconductors: $172 million

9. $D_{new} = \begin{bmatrix} 4 \\ 2 \\ 5 \end{bmatrix}$

$(I-A)D_{new} = \begin{bmatrix} 1 & -.15 & -.43 \\ -.02 & .97 & -.20 \\ -.01 & -.08 & .95 \end{bmatrix} \begin{bmatrix} 4 \\ 2 \\ 5 \end{bmatrix} = \begin{bmatrix} 1.55 \\ 0.86 \\ 4.55 \end{bmatrix}$

$1.55 billion worth of coal; $.86 billion worth of steel; $4.55 billion worth of electricity

11. $A = \begin{bmatrix} .3 & 0 & .1 \\ .2 & .3 & .2 \\ .1 & .2 & .05 \end{bmatrix}, D = \begin{bmatrix} 1 \\ 4 \\ 2 \end{bmatrix}$

$(I-A)^{-1}D = \begin{bmatrix} 1.47 & .05 & .16 \\ .49 & 1.54 & .38 \\ .26 & .33 & 1.15 \end{bmatrix} \begin{bmatrix} 1 \\ 4 \\ 2 \end{bmatrix} = \begin{bmatrix} 1.98 \\ 7.39 \\ 3.87 \end{bmatrix}$

Wood: $1.98, steel: $7.39, coal $3.87

13. **a.** $\quad\quad\quad T \quad E$
    $A = \begin{bmatrix} .25 & .30 \\ .20 & .15 \end{bmatrix} \begin{matrix} T \\ E \end{matrix}$

b. $(I-A)^{-1} = \left(\begin{bmatrix} 1 & 0 \\ 0 & 1 \end{bmatrix} - \begin{bmatrix} .25 & .30 \\ .20 & .15 \end{bmatrix} \right)^{-1} \approx \begin{bmatrix} 1.47 & .52 \\ .35 & 1.30 \end{bmatrix}$

c. $(I-A)^{-1}D \approx \begin{bmatrix} 1.47 & .52 \\ .35 & 1.30 \end{bmatrix} \begin{bmatrix} 5 \\ 3 \end{bmatrix} = \begin{bmatrix} 8.91 \\ 5.65 \end{bmatrix}$

Transportation should produce $8.91 billion worth of output and Energy should produce $5.65 billion.

15. $\quad\quad\quad P \quad I$
$A = \begin{bmatrix} .02 & .01 \\ .10 & .05 \end{bmatrix} \begin{matrix} P \\ I \end{matrix}$

$D = \begin{bmatrix} 930 \\ 465 \end{bmatrix}$

$(I-A)^{-1}D = \begin{bmatrix} 955 \\ 590 \end{bmatrix}$

Plastics: $955,000,
industrial equipment: $590,000

17. $(I-A)^{-1} \begin{bmatrix} 100 \\ 80 \\ 200 \end{bmatrix} \approx \begin{bmatrix} 398 \\ 313 \\ 452 \end{bmatrix}$

manufacturing: $398 million,
transportation: $313 million,
agriculture: $452 million

19. $(I-A)^{-1}\begin{bmatrix}3\\1\\3\end{bmatrix} = (I-A)^{-1}\left(\begin{bmatrix}2\\1\\3\end{bmatrix}+\begin{bmatrix}1\\0\\0\end{bmatrix}\right)$ When the

matrix $(I-A)^{-1}$ is multiplied by the column

matrix $\begin{bmatrix}1\\0\\0\end{bmatrix}$, the entries give the first column of

the matrix $(I-A)^{-1}$, which represents the additional amounts that must be produced by the three industries.

$= (I-A)^{-1}D_{new} - (I-A)^{-1}D_{old}$

$= (I-A)^{-1}\left(\begin{bmatrix}3\\1\\3\end{bmatrix}-\begin{bmatrix}2\\1\\3\end{bmatrix}\right)$

$= (I-A)^{-1}\begin{bmatrix}1\\0\\0\end{bmatrix}$

$= \begin{bmatrix}1.01 & .20 & .50\\.02 & 1.05 & .23\\.01 & .09 & 1.08\end{bmatrix}\begin{bmatrix}1\\0\\0\end{bmatrix} = \begin{bmatrix}1.01\\.02\\.01\end{bmatrix}$

21.
```
round((identity(
3)-[A])^-1[D],2)
     [[10.25]
      [13.82]
      [8.65 ]]
```

$\begin{bmatrix}10.25\\13.82\\8.65\end{bmatrix}$

Chapter 2 Fundamental Concept Check

1. Values of $x, y, z, \ldots$ that satisfy each equation in the system

2. Rectangular array of numbers

3. (a) interchange any two equations (or rows) (b) multiply an equation (or row) by a nonzero number (c) change an equation (or row) by adding to it a multiple of another equation (or row)

4. System of equations: $x = c_1; y = c_2; \cdots$

   matrix: all entries on the main diagonal are 1; all entries off the main diagonal are zero.

5. Use elementary row operations to make the entry have value 1 and make the other entries in the column have value 0

6. (a) create a matrix corresponding to the system of linear equations (b) attempt to put the matrix into diagonal form as described in the box following Example 1 of Section 2.2 (c) if the matrix cannot be put into diagonal form follow the first step in the box following Example 3 of Section 2.2 (d) write the system of linear equations corresponding to the matrix and read off the solution(s)

7. Row matrix: a matrix consisting of a single row (that is, a $1 \times n$ matrix)
   column matrix: a matrix consisting of a single column (that is, a $m \times 1$ matrix)
   square matrix: a matrix having the same number of columns as rows (that is, a $n \times n$ matrix)
   identity matrix: a square matrix having 1s on the main diagonal and 0s elsewhere

8. The entry in the ith row and the jth column

9. For two matrices of the same size, the sum (difference) is the matrix obtained by adding (subtracting) the corresponding entries of the two matrices

10. For two matrices A and B, where the number of columns of A is the same as the number of rows of B, the matrix AB is the matrix having the same number of rows as A and the same number of columns as B whose ij^{th} entry is obtained by adding the products of the corresponding entries of the ith row of A with the jth column of B.

11. The scalar product of the number c and the matrix A is the matrix obtained by multiplying each element of A by c.

SSM: Finite Math **Chapter 2: Matrices**

12. The inverse of the square matrix A is the matrix whose product with A is an identity matrix

13. The inverse of $\begin{bmatrix} a & b \\ c & d \end{bmatrix}$ is $\dfrac{1}{D}\begin{bmatrix} d & -b \\ -c & a \end{bmatrix}$, where $D = ad - bc$ and $D \neq 0$

14. Write the matrix form ($AX = B$) of the system of linear equations. If the matrix A has an inverse, then the solution of the system of linear equations is given by the entries of the matrix $A^{-1}B$.

15. Adjoin an identity matrix to the right of the invertible matrix A and then apply the Gauss-Jordan elimination method to the entire matrix until its left side is an identity matrix. The new right side of the matrix will be the inverse of A.

16. A square matrix whose ij^{th} entry is the amount of input from the ith industry required to produce one unit of the jth industry; a column matrix whose i^{th} element is the amount of units demanded from the ith industry

17. If A is an input-output matrix and D is a final-demand matrix, then the ith entry of the matrix $(I-A)^{-1}D$ gives the amount of input required from the ith industry to meet the final demand

Chapter 2 Review Exercises

1. $\begin{bmatrix} 3 & -6 & 1 \\ 2 & 4 & 6 \end{bmatrix} \xrightarrow{\frac{1}{3}R_1} \begin{bmatrix} 1 & -2 & \frac{1}{3} \\ 2 & 4 & 6 \end{bmatrix}$

   $\xrightarrow{R_2+(-2)R_1} \begin{bmatrix} 1 & -2 & \frac{1}{3} \\ 0 & 8 & \frac{16}{3} \end{bmatrix}$

2. $\begin{bmatrix} -5 & -3 & 1 \\ 4 & 2 & 0 \\ 0 & 6 & 7 \end{bmatrix} \xrightarrow{\frac{1}{2}R_2} \begin{bmatrix} -5 & -3 & 1 \\ 2 & 1 & 0 \\ 0 & 6 & 7 \end{bmatrix}$

   $\xrightarrow{R_3+(-6)R_2} \begin{bmatrix} 1 & 0 & 1 \\ 2 & 1 & 0 \\ -12 & 0 & 7 \end{bmatrix}$

3. $\begin{bmatrix} \frac{1}{2} & -1 & | & -3 \\ 4 & -5 & | & -9 \end{bmatrix}$

   $\begin{bmatrix} 1 & -2 & | & -6 \\ 0 & 3 & | & 15 \end{bmatrix}$

   $\begin{bmatrix} 1 & 0 & | & 4 \\ 0 & 1 & | & 5 \end{bmatrix}$

   $x = 4, y = 5$

4. $\begin{bmatrix} 3 & 0 & 9 & | & 42 \\ 2 & 1 & 6 & | & 30 \\ -1 & 3 & -2 & | & -20 \end{bmatrix}$

   $\begin{bmatrix} 1 & 0 & 3 & | & 14 \\ 0 & 1 & 0 & | & 2 \\ 0 & 3 & 1 & | & -6 \end{bmatrix}$

   $\begin{bmatrix} 1 & 0 & 3 & | & 14 \\ 0 & 1 & 0 & | & 2 \\ 0 & 0 & 1 & | & -12 \end{bmatrix}$

   $\begin{bmatrix} 1 & 0 & 0 & | & 50 \\ 0 & 1 & 0 & | & 2 \\ 0 & 0 & 1 & | & -12 \end{bmatrix}$

   $x = 50, y = 2, z = -12$

5. $\begin{bmatrix} 3 & -6 & 6 & | & -5 \\ -2 & 3 & -5 & | & \frac{7}{3} \\ 1 & 1 & 10 & | & 3 \end{bmatrix}$

   $\begin{bmatrix} 1 & -2 & 2 & | & -\frac{5}{3} \\ 0 & -1 & -1 & | & -1 \\ 0 & 3 & 8 & | & \frac{14}{3} \end{bmatrix}$

   $\begin{bmatrix} 1 & 0 & 4 & | & \frac{1}{3} \\ 0 & 1 & 1 & | & 1 \\ 0 & 0 & 5 & | & \frac{5}{3} \end{bmatrix}$

   $\begin{bmatrix} 1 & 0 & 0 & | & -1 \\ 0 & 1 & 0 & | & \frac{2}{3} \\ 0 & 0 & 1 & | & \frac{1}{3} \end{bmatrix}$

   $x = -1, y = \dfrac{2}{3}, z = \dfrac{1}{3}$

6. $\begin{bmatrix} 3 & 6 & -9 & | & 1 \\ 2 & 4 & -6 & | & 1 \\ 3 & 4 & 5 & | & 0 \end{bmatrix}$

$\begin{bmatrix} 1 & 2 & -3 & | & \frac{1}{3} \\ 0 & 0 & 0 & | & \frac{1}{3} \\ 0 & -2 & 14 & | & -1 \end{bmatrix}$

$\begin{bmatrix} 1 & 2 & -3 & | & \frac{1}{3} \\ 0 & -2 & 14 & | & -1 \\ 0 & 0 & 0 & | & \frac{1}{3} \end{bmatrix}$

$\begin{bmatrix} 1 & 0 & 11 & | & -\frac{2}{3} \\ 0 & 1 & -7 & | & \frac{1}{2} \\ 0 & 0 & 0 & | & \frac{1}{3} \end{bmatrix}$

No solution

7. $\begin{bmatrix} 1 & 2 & -5 & 3 & | & 16 \\ -5 & -7 & 13 & -9 & | & -50 \\ -1 & 1 & -7 & 2 & | & 9 \\ 3 & 4 & -7 & 6 & | & 33 \end{bmatrix}$

$\begin{bmatrix} 1 & 2 & -5 & 3 & | & 16 \\ 0 & 3 & -12 & 6 & | & 30 \\ 0 & 3 & -12 & 5 & | & 25 \\ 0 & -2 & 8 & -3 & | & -15 \end{bmatrix}$

$\begin{bmatrix} 1 & 0 & 3 & -1 & | & -4 \\ 0 & 1 & -4 & 2 & | & 10 \\ 0 & 0 & 0 & -1 & | & -5 \\ 0 & 0 & 0 & 1 & | & 5 \end{bmatrix}$

$\begin{bmatrix} 1 & 0 & 3 & 0 & | & 1 \\ 0 & 1 & -4 & 0 & | & 0 \\ 0 & 0 & 0 & 1 & | & 5 \\ 0 & 0 & 0 & 0 & | & 0 \end{bmatrix}$

$\begin{cases} x+3z=1 \\ y-4z=0 \\ w=5 \\ 0=0 \end{cases}$

z = any value, $x = 1 - 3z$, $y = 4z$, $w = 5$

8. $\begin{bmatrix} 5 & -10 & | & 5 \\ 3 & -8 & | & -3 \\ -3 & 7 & | & 0 \end{bmatrix}$

$\begin{bmatrix} 1 & -2 & | & 1 \\ 0 & -2 & | & -6 \\ 0 & 1 & | & 3 \end{bmatrix}$

$\begin{bmatrix} 1 & 0 & | & 7 \\ 0 & 1 & | & 3 \\ 0 & 0 & | & 0 \end{bmatrix}$

$x = 7, y = 3$

9. $\begin{bmatrix} 2+3 \\ -1+4 \\ 0+7 \end{bmatrix} = \begin{bmatrix} 5 \\ 3 \\ 7 \end{bmatrix}$

10. $\begin{bmatrix} 1\cdot3+3\cdot1+(-2)0 & 1\cdot5+3\cdot0+(-2)(-6) \\ 4\cdot3+0\cdot1+(-1)0 & 4\cdot5+0\cdot0+(-1)(-6) \end{bmatrix}$

$= \begin{bmatrix} 6 & 17 \\ 12 & 26 \end{bmatrix}$

11. $\begin{bmatrix} \frac{3}{4}\cdot8 & \frac{3}{4}\cdot(-6) \\ \frac{3}{4}\cdot\frac{2}{3} & \frac{3}{4}\cdot0 \end{bmatrix} = \begin{bmatrix} 6 & \frac{-9}{2} \\ \frac{1}{2} & 0 \end{bmatrix}$

12. $\begin{bmatrix} 1.4-.8 & -3-7 \\ 8.2-1.6 & 0-(-2) \\ 4-0 & 5.5-(-5.5) \end{bmatrix} = \begin{bmatrix} .6 & -10 \\ 6.6 & 2 \\ 4 & 11 \end{bmatrix}$

13. $AB = \begin{bmatrix} 23 & -10+5k \\ -9 & 15+k \end{bmatrix}$

$BA = \begin{bmatrix} 23 & 15 \\ 6-3k & 15+k \end{bmatrix}$

Therefore,
$-10+5k = 15$
$5k = 25$
$k = 5$
$6-3k = -9$
$-3k = -15$
$k = 5$

Since any value of k will work in the $a_{2,2}$ position $k = 5$ is the only solution.

14. $AB = \begin{bmatrix} 3k+5 & -10+k \\ 9 & 17 \end{bmatrix}$

$BA = \begin{bmatrix} 17 & -10+k \\ 9 & 3k+5 \end{bmatrix}$

Therefore,
$3k + 5 = 17$
$3k = 12$
$k = 4$
$3k + 5 = 17$
$3k = 12$
$k = 4$

Since any value of k will work in the $a_{1,2}$ position, $k = 4$ is the only solution.

15. $\begin{bmatrix} 3 & 2 \\ 5 & 4 \end{bmatrix}^{-1} = \begin{bmatrix} \frac{4}{2} & -\frac{2}{2} \\ -\frac{5}{2} & \frac{3}{2} \end{bmatrix} = \begin{bmatrix} 2 & -1 \\ -\frac{5}{2} & \frac{3}{2} \end{bmatrix}$

$\begin{bmatrix} 2 & -1 \\ -\frac{5}{2} & \frac{3}{2} \end{bmatrix}\begin{bmatrix} 0 \\ 2 \end{bmatrix} = \begin{bmatrix} -2 \\ 3 \end{bmatrix}$

$x = -2, y = 3$

16. a. $\begin{bmatrix} 4 & -2 & 3 \\ 8 & -3 & 5 \\ 7 & -2 & 4 \end{bmatrix}\begin{bmatrix} 1 \\ 0 \\ 3 \end{bmatrix} = \begin{bmatrix} 13 \\ 23 \\ 19 \end{bmatrix}$

$x = 13, y = 23, z = 19$

b. $\begin{bmatrix} -2 & 2 & -1 \\ 3 & -5 & 4 \\ 5 & -6 & 4 \end{bmatrix}\begin{bmatrix} 0 \\ -1 \\ 2 \end{bmatrix} = \begin{bmatrix} -4 \\ 13 \\ 14 \end{bmatrix}$

$x = -4, y = 13, z = 14$

17. $\begin{bmatrix} 2 & 6 & | & 1 & 0 \\ 1 & 2 & | & 0 & 1 \end{bmatrix}$

$\begin{bmatrix} 1 & 3 & | & \frac{1}{2} & 0 \\ 0 & -1 & | & -\frac{1}{2} & 1 \end{bmatrix}$

$\begin{bmatrix} 1 & 0 & | & -1 & 3 \\ 0 & 1 & | & \frac{1}{2} & -1 \end{bmatrix}$

$\begin{bmatrix} -1 & 3 \\ \frac{1}{2} & -1 \end{bmatrix}$

18. $\begin{bmatrix} 1 & 1 & 1 & | & 1 & 0 & 0 \\ 3 & 4 & 3 & | & 0 & 1 & 0 \\ 1 & 1 & 2 & | & 0 & 0 & 1 \end{bmatrix}$

$\begin{bmatrix} 1 & 1 & 1 & | & 1 & 0 & 0 \\ 0 & 1 & 0 & | & -3 & 1 & 0 \\ 0 & 0 & 1 & | & -1 & 0 & 1 \end{bmatrix}$

$\begin{bmatrix} 1 & 0 & 1 & | & 4 & -1 & 0 \\ 0 & 1 & 0 & | & -3 & 1 & 0 \\ 0 & 0 & 1 & | & -1 & 0 & 1 \end{bmatrix}$

$\begin{bmatrix} 1 & 0 & 0 & | & 5 & -1 & -1 \\ 0 & 1 & 0 & | & -3 & 1 & 0 \\ 0 & 0 & 1 & | & -1 & 0 & 1 \end{bmatrix}$

$\begin{bmatrix} 5 & -1 & -1 \\ -3 & 1 & 0 \\ -1 & 0 & 1 \end{bmatrix}$

19. $\begin{cases} c + w + s = 1000 \\ 206c + 85w + 97s = 151{,}500 \\ c - w - s = 0 \end{cases}$

$\begin{bmatrix} 1 & 1 & 1 \\ 206 & 85 & 97 \\ 1 & -1 & -1 \end{bmatrix}^{-1}\begin{bmatrix} 1000 \\ 151{,}500 \\ 0 \end{bmatrix} = \begin{bmatrix} 500 \\ 0 \\ 500 \end{bmatrix}$

Corn: 500 acres, wheat: 0 acres, soybeans: 500 acres

20. a. $AC = \begin{bmatrix} 15 & 20 & 8 \\ 10 & 17 & 12 \end{bmatrix}\begin{bmatrix} 165 \\ 65 \\ 210 \end{bmatrix} = \begin{bmatrix} 5455 \\ 5275 \end{bmatrix}$

The total cost to make the equipment in the first store was $5455 and in the second store $5275.

b. $AS = \begin{bmatrix} 15 & 20 & 8 \\ 10 & 17 & 12 \end{bmatrix}\begin{bmatrix} 200 \\ 80 \\ 250 \end{bmatrix} = \begin{bmatrix} 6600 \\ 6360 \end{bmatrix}$

The total revenue of the equipment in the first store was $6600 and in the second store $6360.

c. $S - C = \begin{bmatrix} 200 \\ 80 \\ 250 \end{bmatrix} - \begin{bmatrix} 165 \\ 65 \\ 210 \end{bmatrix} = \begin{bmatrix} 35 \\ 15 \\ 40 \end{bmatrix}$

The entries represent the profit per unit of each item.

d. $A(S - C) = \begin{bmatrix} 15 & 20 & 8 \\ 10 & 17 & 12 \end{bmatrix} \begin{bmatrix} 35 \\ 15 \\ 40 \end{bmatrix} = \begin{bmatrix} 1145 \\ 1085 \end{bmatrix}$

The total profit for the first store was $1145 and for the second store $1085.

21. a. $BA = \begin{bmatrix} 5000 & 8000 & 10{,}000 \end{bmatrix} \begin{bmatrix} .50 & .43 & .07 \\ .45 & .26 & .29 \\ .40 & .40 & .20 \end{bmatrix} = \begin{bmatrix} 10{,}100 & 8230 & 4670 \end{bmatrix}$

Total amount invested in bonds, stocks, and the conservative fixed income fund, respectively.

b. $BC = \begin{bmatrix} 5000 & 8000 & 10{,}000 \end{bmatrix} \begin{bmatrix} .0032 & .1119 \\ .0233 & .0976 \\ .0320 & .0467 \end{bmatrix} = \begin{bmatrix} 522.40 & 1807.30 \end{bmatrix}$

total return on the investments for one year and five years, respectively.

c. $2B = \begin{bmatrix} 2(5000) & 2(8000) & 2(10{,}000) \end{bmatrix} = \begin{bmatrix} 10{,}000 & 16{,}000 & 20{,}000 \end{bmatrix}$ The result of doubling the amounts invested

d. $8230 is the total amount invested in stocks.

e. $522.40 is the total return after one year.

22. a. $AB = \begin{bmatrix} 11 & 7 & 12 \\ 9 & 5 & 16 \\ 13 & 8 & 9 \\ 13 & 7 & 10 \end{bmatrix} \begin{bmatrix} 8 \\ 6 \\ 9 \end{bmatrix} = \begin{bmatrix} 238 \\ 246 \\ 233 \\ 236 \end{bmatrix} \begin{matrix} Sara \\ Quinn \\ Tamia \\ Zack \end{matrix}$

total amount earned by each person for the week.

b. Quinn earned the most and Tamia earned the least.

c. $AB = \begin{bmatrix} 11 & 7 & 12 \\ 9 & 5 & 16 \\ 13 & 8 & 9 \\ 13 & 7 & 10 \end{bmatrix} \begin{bmatrix} 9 \\ 6 \\ 8 \end{bmatrix} = \begin{bmatrix} 237 \\ 239 \\ 237 \\ 239 \end{bmatrix} \begin{matrix} Sara \\ Quinn \\ Tamia \\ Zack \end{matrix}$

Quinn and Zack both earn $239.

d. Sara worked 11 hours in concessions, 7 hours at the front desk, and 12 hours cleaning for a total of 30 hours.

23. Let x = number of apples
y = number of bananas
z = number of oranges
$$x+y+z=18$$
$$.85x+.20y+.76z=9$$
$$-x+y-z=0$$

$$\begin{bmatrix} 1 & 1 & 1 & | & 18 \\ .85 & .20 & .76 & | & 9 \\ -1 & 1 & -1 & | & 0 \end{bmatrix}$$

$\xrightarrow{R_3+R_1}$ $\begin{bmatrix} 1 & 1 & 1 & | & 18 \\ .85 & .20 & .76 & | & 9 \\ 0 & 2 & 0 & | & 18 \end{bmatrix}$

$\xrightarrow{R_2+(-.85)R_1}$ $\begin{bmatrix} 1 & 1 & 1 & | & 18 \\ 0 & -.65 & -.09 & | & -6.3 \\ 0 & 2 & 0 & | & 18 \end{bmatrix}$

$\xrightarrow{R_2 \leftrightarrow R_3}$ $\begin{bmatrix} 1 & 1 & 1 & | & 18 \\ 0 & 2 & 0 & | & 18 \\ 0 & -.65 & -.09 & | & -6.3 \end{bmatrix}$

$\xrightarrow{\frac{1}{2}R_2}$ $\begin{bmatrix} 1 & 1 & 1 & | & 18 \\ 0 & 1 & 0 & | & 9 \\ 0 & -.65 & -.09 & | & -6.3 \end{bmatrix}$

$\xrightarrow{R_1+(-1)R_2}$ $\begin{bmatrix} 1 & 0 & 1 & | & 9 \\ 0 & 1 & 0 & | & 9 \\ 0 & -.65 & -.09 & | & -6.3 \end{bmatrix}$

$\xrightarrow{R_3+(.65)R_2}$ $\begin{bmatrix} 1 & 0 & 1 & | & 9 \\ 0 & 1 & 0 & | & 9 \\ 0 & 0 & -.09 & | & -.45 \end{bmatrix}$

$\xrightarrow{-\frac{1}{0.09}R_3}$ $\begin{bmatrix} 1 & 0 & 1 & | & 9 \\ 0 & 1 & 0 & | & 9 \\ 0 & 0 & 1 & | & 5 \end{bmatrix}$

$\xrightarrow{R_1+(-1)R_3}$ $\begin{bmatrix} 1 & 0 & 0 & | & 4 \\ 0 & 1 & 0 & | & 9 \\ 0 & 0 & 1 & | & 5 \end{bmatrix}$

4 apples, 9 bananas, 5 oranges

24. x = A's current stockpile,
y = B's current stockpile,
a = A's next-year's stockpile,
b = B's next-year's stockpile
$$.8x+.2y=a$$
$$.1x+.9y=b$$

a. Next year: $\begin{bmatrix} .8 & .2 \\ .1 & .9 \end{bmatrix}\begin{bmatrix} 10,000 \\ 7000 \end{bmatrix} = \begin{bmatrix} 9400 \\ 7300 \end{bmatrix}$

Two years: $\begin{bmatrix} .8 & .2 \\ .1 & .9 \end{bmatrix}^2 \begin{bmatrix} 10,000 \\ 7000 \end{bmatrix} = \begin{bmatrix} 8980 \\ 7510 \end{bmatrix}$

A: 9400, 8980; B: 7300, 7510

b. Previous year: $\begin{bmatrix} .8 & .2 \\ .1 & .9 \end{bmatrix}^{-1} \begin{bmatrix} 10,000 \\ 7000 \end{bmatrix} \approx \begin{bmatrix} 10,857 \\ 6571 \end{bmatrix}$

Two years ago:

$\left(\begin{bmatrix} .8 & .2 \\ .1 & .9 \end{bmatrix}^{-1}\right)^2 \begin{bmatrix} 10,000 \\ 7000 \end{bmatrix} \approx \begin{bmatrix} 12,082 \\ 5959 \end{bmatrix}$

A: 10,857, 12,082; B: 6571, 5959

c. $a - b = (.8x + .2y) - (.1x + .9y)$
$= .7x - .7y = .7(x - y)$
Thus, the "missile gap" of the following year is 70% of the previous, so the "missile gap" decreases 30% each year.
$a + b = .9x + 1.1y < x + y$
when $.1y < .1x$ or $y < x$.
$a + b = .9x + 1.1y > x + y$ when $y > x$.

25. $\left(\begin{bmatrix} 1 & 0 \\ 0 & 1 \end{bmatrix} - \begin{bmatrix} .4 & .2 \\ .1 & .3 \end{bmatrix}\right)^{-1} \begin{bmatrix} 8 \\ 12 \end{bmatrix} = \begin{bmatrix} 20 \\ 20 \end{bmatrix}$

Industry I: 20; industry II: 20

26. Let x = number of nickels
    y = number of dimes
    z = number of quarters
    $$\begin{cases} x+y+z=30 \\ .05x+.10y+.25z=3.30 \\ y=5z \end{cases}$$

    $$\begin{bmatrix} \underline{1} & 1 & 1 & | & 30 \\ .05 & .10 & .25 & | & 3.30 \\ 0 & 1 & -5 & | & 0 \end{bmatrix}$$

    $$\begin{bmatrix} 1 & 1 & 1 & | & 30 \\ 0 & \underline{.05} & .20 & | & 1.8 \\ 0 & 1 & -5 & | & 0 \end{bmatrix}$$

    $$\begin{bmatrix} 1 & 1 & 1 & | & 30 \\ 0 & \underline{1} & 4 & | & 36 \\ 0 & 1 & -5 & | & 0 \end{bmatrix}$$

    $$\begin{bmatrix} 1 & 0 & -3 & | & -6 \\ 0 & 1 & 4 & | & 36 \\ 0 & 0 & \underline{-9} & | & -36 \end{bmatrix}$$

    $$\begin{bmatrix} 1 & 0 & -3 & | & -6 \\ 0 & 1 & 4 & | & 36 \\ 0 & 0 & \underline{1} & | & 4 \end{bmatrix}$$

    $$\begin{bmatrix} 1 & 0 & 0 & | & 6 \\ 0 & 1 & 0 & | & 20 \\ 0 & 0 & 1 & | & 4 \end{bmatrix}$$

    Joe has 4 quarters.
    Answer (a) is correct.

27. **a.** True; a system of equations has no solution, exactly one solution, or infinitely many solutions.

    b. False; a system of equations could have two or more equations that are multiples of each other.

    c. True; At least one variable must be dependent on another when there are less equations than variables.

28. **a.** True; The matrix and itself will have the same dimensions and therefore can be added.

    b. False; In order to be multiplied by itself, the number of rows must be the same as the number of columns. In other words, only a square matrix can be multiplied by itself.

29. One possible system with infinitely many solutions is: $\begin{cases} 2x+3y=4 \\ 4x+6y=8 \end{cases}$.

30. One possible system with no solution is:
    $\begin{cases} 2x+3y=4 \\ 2x+3y=5 \end{cases}$.

31. Answers may vary, but two possible matrices are: $A = \begin{bmatrix} 1 & 1 \\ 1 & 1 \end{bmatrix}$ and $B = \begin{bmatrix} -1 & 1 \\ 1 & -1 \end{bmatrix}$.

32. If the matrix has no inverse, there will be a row of zeros in the resulting Gauss-Jordan matrix.

33. The column values should add to less than one in an input-output matrix because they reflect the amount of input required from each industry and all the industries are dependent on each other.

Chapter 3

Exercises 3.1

1. (8, 7)

$$\begin{cases} 6(8)+3(7) \le 96 \\ 8+7 \le 18 \\ 2(8)+6(7) \le 72 \\ 8 \ge 0,\ 7 \ge 0 \end{cases}$$

$$\begin{cases} 69 \le 96 & \text{true} \\ 15 \le 18 & \text{true} \\ 58 \le 72 & \text{true} \\ 8 \ge 0,\ 7 \ge 0 & \text{true} \end{cases}$$

Yes

3. (9, 10)

$$\begin{cases} 6(9)+3(10) \le 96 \\ 9+10 \le 18 \\ 2(9)+6(10) \le 72 \\ 9 \ge 0,\ 3 \ge 0 \end{cases}$$

$$\begin{cases} 84 \le 96 & \text{true} \\ 19 \le 18 & \text{false} \\ 78 \le 72 & \text{false} \\ 9 \ge 0,\ 3 \ge 0 & \text{true} \end{cases}$$

No

5. $\begin{cases} 6x \le 96 \\ x \le 18 \\ 2x \le 72 \\ x \ge 0 \end{cases} \Rightarrow \begin{cases} x \le 16 \\ x \le 18 \\ x \le 36 \\ x \ge 0 \end{cases}$

Therefore, 16 chairs could be produced

7. **a.**

	A	B	Available
Candy bars	2	1	500
Suckers	2	2	600
Profit (in cents)	40	30	

  b. Candy Bars: $2x + y \le 500$
  Suckers: $2x + 2y \le 600$

  c. $x \ge 0,\ y \ge 0$

  d. $40x + 30y$

e.

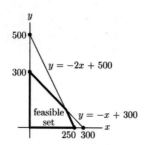

9. a.

	A	B	Truck capacity
Volume	4 cubic feet	3 cubic feet	300 cubic feet
Weight	100 pounds	200 pounds	10,000 pounds
Earnings	$13	$9	

b. Volume: $4x + 3y \leq 300$
Weight: $100x + 200y \leq 10,000$

c. $y \leq 2x$, $x \geq 0$, $y \geq 0$

d. $13x + 9y$

e. In standard form, the inequalities from (b) and (c) are:
$$\begin{cases} y \leq -\dfrac{4}{3}x + 100 \\ y \leq -\dfrac{1}{2}x + 50 \\ y \leq 2x \\ x \geq 0, \ y \geq 0 \end{cases}$$

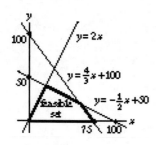

11. a.

	Essay questions	Short-answer Questions	Available
Time to answer	10 minutes	2 minutes	90 minutes
Quantity	10	50	
Required	3	10	
Worth	20 points	5 points	

b. $10x + 2y \leq 90$

c. $3 \leq x \leq 10$, $10 \leq y \leq 50$

d. $20x + 5y$

e. In standard form, the inequality from (b) is $y \leq -5x + 45$. Graph the system:
$$\begin{cases} y \leq -5x + 45 \\ 3 \leq x \leq 10 \\ 10 \leq y \leq 50 \end{cases}$$

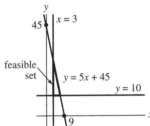

Note that the conditions $x \leq 10$ and $y \leq 50$ are superfluous because they are automatically assured if the other inequalities hold.

13. a.

	Alfalfa	Corn	Requirements
Protein	.13 pound	.065 pound	4550 pounds
TDN	.48 pound	.96 pound	26,880 pounds
Vitamin A	2.16 IUs	0 IUs	43,200 IUs
Cost/lb	$.08	$.13	

b. Protein: $.13x + .065y \geq 4550$
TDN: $.48x + .96y \geq 26{,}880$
Vitamin A: $2.16x \geq 43{,}200$
Other: $y \geq 0$
(The condition $x \geq 0$ is unnecessary because it is automatically assured if the inequality for Vitamin A holds.)

c. In standard form, the inequalities are:
$$\begin{cases} y \geq -2x + 70,000 \\ y \geq -.5x + 28,000 \\ x \geq 20,000 \\ y \geq 0 \end{cases}$$

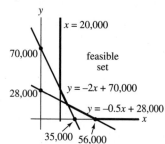

d. $.08x + .13y$

Exercises 3.2

1. | Vertex | $4x + 3y$ |
    | --- | --- |
    | (0, 0) | $4(0) + 3(0) = 0$ |
    | (0, 20) | $4(0) + 3(20) = 60$ |
    | (20, 0) | $4(20) + 3(0) = 80$ |

    The objective function is maximized at (20, 0).

3. Find the vertices.
    Lower left corner: (0, 0)
    y-intercept of $y = -\frac{1}{2}x + 4$: (0, 4)

    Intersection of $y = -\frac{1}{2}x + 4$ and $y = -x + 6$:

    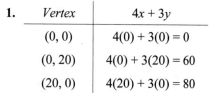

    (4, 2)
    x-intercept of $y = -x + 6$: (6, 0)

Vertex	$4x + 3y$
(0, 0)	$4(0) + 3(0) = 0$
(0, 4)	$4(0) + 3(4) = 12$
(4, 2)	$4(4) + 3(2) = 22$
(6, 0)	$4(6) + 3(0) = 24$

    The objective function is maximized at (6, 0).

5. | Vertex | $x + 2y$ |
    | --- | --- |
    | (0, 0) | $0 + 2(0) = 0$ |
    | (0, 5) | $0 + 2(5) = 10$ |
    | (3, 3) | $3 + 2(3) = 9$ |
    | (4, 0) | $4 + 2(0) = 4$ |

    The objective function is maximized at (0, 5).

7. | Vertex | $2x + y$ |
    | --- | --- |
    | (0, 0) | $2(0) + 0 = 0$ |
    | (0, 5) | $2(0) + 5 = 5$ |
    | (3, 3) | $2(3) + 3 = 9$ |
    | (4, 0) | $2(4) + 0 = 8$ |

    The objective function is maximized at (3, 3).

9. | Vertex | $8x + y$ |
    | --- | --- |
    | (0, 7) | $8(0) + 7 = 7$ |
    | (1, 2) | $8(1) + 2 = 10$ |
    | (2, 1) | $8(2) + 1 = 17$ |
    | (6, 0) | $8(6) + 0 = 48$ |

    The objective function is minimized at (0, 7).

11.

Vertex	$2x + 3y$
(0, 7)	$2(0) + 3(7) = 21$
(1, 2)	$2(1) + 3(2) = 8$
(2, 1)	$2(2) + 3(1) = 7$
(6, 0)	$2(6) + 3(0) = 12$

The objective function is minimized at (2, 1).

13. $\begin{cases} 15x \geq 90 \\ 810x \geq 1620 \\ \frac{1}{9}x \geq 1 \\ x \geq 0 \end{cases} \Rightarrow \begin{cases} x \geq 6 \\ x \geq 2 \\ x \geq 9 \\ x \geq 0 \end{cases}$

Therefore, 9 cups would be required

15. Find the vertices.

y-intercept of $y = -x + 300$ (0, 300)
Intersection of
$y = -x + 300$ and $y = -2x + 500$
$-x + 300 = -2x + 500$
$\quad x = 200$
$y = -200 + 300$
$y = 100$
(200, 100)
x-intercept of $y = -2x + 500$: (250, 0)

Vertex	Profit = $40x + 30y$
(0, 300)	$40(0) + 30(300) = 9000$
(200, 100)	$40(200) + 30(100) = 11000$
(250, 0)	$40(250) + 30(0) = 10000$

The profit is maximized at (200, 100).
Joe should make 200 Assortment A's and 100 Assortment B's for a profit of $110.

17. Find the vertices.
Lower left corner: (0, 0)
Intersection of $y = 2x$ and $y = -\frac{1}{2}x + 50$:

$2x = -\frac{1}{2}x + 50$

$\frac{5}{2}x = 50$

$x = 20$
$y = 2x = 2(20) = 40$
(20, 40)

Intersection of $y = -\frac{1}{2}x + 50$ and
$y = -\frac{4}{3}x + 100$:

$-\frac{1}{2}x + 50 = -\frac{4}{3}x + 100$

$\frac{5}{6}x = 50$

$x = 60$
$y = -\frac{1}{2}x + 50 = -\frac{1}{2}(60) + 50 = 20$
(60, 20)

x-intercept of $y = -\frac{4}{3}x + 100$: (75, 0)

Vertex	Earnings = $13x + 9y$
(0, 0)	$13(0) + 9(0) = 0$
(20, 40)	$13(20) + 9(40) = 620$
(60, 20)	$13(60) + 9(20) = 960$
(75, 0)	$13(75) + 9(0) = 975$

The earnings are maximized at (75, 0).
Ship 75 crates of cargo A and no crates of cargo B.

19.

Vertex	Score = $20x + 5y$
(3, 10)	$20(3) + 5(10) = 110$
(3, 30)	$20(3) + 5(30) = 210$
(7, 10)	$20(7) + 5(10) = 190$

The score is maximized at (3, 30).
Answer 3 essay questions and 30 short-answer questions.

21. Find the vertices.
x-intercept of $y = -.5x + 28{,}000$: $(56{,}000, 0)$

Intersection of $y = -.5x + 28{,}000$ and
$y = -2x + 70{,}000$:
$$-.5x + 28{,}000 = -2x + 70{,}000$$
$$1.5x = 42{,}000$$
$$x = 28{,}000$$
$y = -2x + 70{,}000 = -2(28{,}000) + 70{,}000 = 14{,}000$
$(28{,}000, 14{,}000)$

Intersection of $y = -2x + 70{,}000$ and
$x = 20{,}000$:
$y = -2x + 70{,}000 = -2(20{,}000) + 70{,}000 = 30{,}000$
$(20{,}000, 30{,}000)$

Vertex	cost = .08x + .13y
(56,000, 0)	.08(56,000) + .13(0) = 4480
(28,000, 14,000)	.08(28,000) + .13(14,000) = 4060
(20,000, 30,000)	.08(20,000) + .13(30,000) = 5500

The cost is minimized at $(28{,}000, 14{,}000)$. Buy 28,000 pounds of alfalfa and 14,000 pounds of corn at a cost of $4060.

23.

Vertex	Profit = 150x + 70y
(0, 0)	150(0) + 70(0) = 0
(0, 12)	150(0) + 70(12) = 840
(9, 9)	150(9) + 70(9) = 1980
(14, 4)	150(14) + 70(4) = 2380
(16, 0)	150(16) + 70(0) = 2400

The profit is maximized at $(16, 0)$.
Make 16 chairs and no sofas.

25. In standard form the inequalities are:
$$\begin{cases} y \geq -2x + 10 \\ y \geq -\dfrac{1}{2}x + 7 \\ x \geq 0,\ y \geq 0 \end{cases}$$

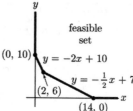

Find the vertices.
y-intercept of $y = -2x + 10$: $(0, 10)$

Intersection of $y = -2x + 10$ and $y = -\dfrac{1}{2}x + 7$:
$$-\dfrac{1}{2}x + 7 = -2x + 10$$
$$\dfrac{3}{2}x = 3$$
$$x = 2$$
$y = -2x + 10 = -2(2) + 10 = 6$
$(2, 6)$

x-intercept of $y = -\dfrac{1}{2}x + 7$: $(14, 0)$

Vertex	3x + 4y
(0, 10)	3(0) + 4(10) = 40
(2, 6)	3(2) + 4(6) = 30
(14, 0)	3(14) + 4(0) = 42

The minimum value is 30 and occurs at $(2, 6)$.

27. In standard form the inequalities are:

$$\begin{cases} y \leq -\dfrac{1}{2}x + 10 \\ y \geq -\dfrac{3}{2}x + 12 \\ x \leq 6 \\ x \geq 0, \; y \geq 0 \end{cases}$$

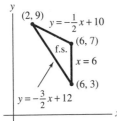

Find the vertices.

Intersection of $y = -\dfrac{1}{2}x + 10$ and $y = -\dfrac{3}{2}x + 12$:

$-\dfrac{1}{2}x + 10 = -\dfrac{3}{2}x + 12$

$x = 2$

$y = -\dfrac{1}{2}x + 10 = -\dfrac{1}{2}(2) + 10 = 9$

(2, 9)

Intersection of $y = -\dfrac{1}{2}x + 10$ and $x = 6$:

$x = 6$

$y = -\dfrac{1}{2}x + 10 = -\dfrac{1}{2}(6) + 10 = 7$

(6, 7)

Intersection of $y = -\dfrac{3}{2}x + 12$ and $x = 6$:

$x = 6$

$y = -\dfrac{3}{2}x + 12 = -\dfrac{3}{2}(6) + 12 = 3$

(6, 3)

Vertex	$2x + 5y$
(2, 9)	2(2) + 5(9) = 49
(6, 7)	2(6) + 5(7) = 47
(6, 3)	2(6) + 5(3) = 27

The maximum value is 49 and occurs at (2, 9).

29. In standard form the inequalities are:

$$\begin{cases} y \leq -\dfrac{1}{3}x + 40 \\ y \leq -\dfrac{7}{2}x + 78 \\ x \leq 20 \\ x \geq 0, \; y \geq 0 \end{cases}$$

Find the vertices.
Lower left corner: (0, 0)

y-intercept of $y = -\dfrac{1}{3}x + 40$: (0, 40)

Intersection of $y = -\dfrac{1}{3}x + 40$ and $y = -\dfrac{7}{2}x + 78$:

$-\dfrac{1}{3}x + 40 = -\dfrac{7}{2}x + 78$

$\dfrac{19}{6}x = 38$

$x = 12$

$y = -\dfrac{1}{3}x + 40 = -\dfrac{1}{3}(12) + 40 = 36$

(12, 36)

Intersection of $y = -\dfrac{7}{2}x + 78$ and $x = 20$:

$x = 20$; $y = -\dfrac{7}{2}x + 78 = -\dfrac{7}{2}(20) + 78 = 8$

(20, 8)

Intersection of $x = 20$ and $y = 0$: (20, 0)

Vertex	$100x + 150y$
(0, 0)	100(0) + 150(0) = 0
(0, 40)	100(0) + 150(40) = 6000
(12, 36)	100(12) + 150(36) = 6600
(20, 8)	100(20) + 150(8) = 3200
(20, 0)	100(20) + 150(0) = 2000

The maximum value is 6600 and occurs at (12, 36).

31. The inequalities are given in standard form.

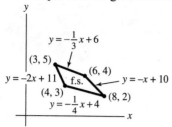

Find the vertices.

Intersection of $y = -2x + 11$ and $y = -\frac{1}{3}x + 6$:

$$-\frac{1}{3}x + 6 = -2x + 11$$

$$\frac{5}{3}x = 5$$

$$x = 3$$

$$y = -\frac{1}{3}x + 6 = -\frac{1}{3}(3) + 6 = 5$$

(3, 5)

Intersection of $y = -\frac{1}{3}x + 6$ and $y = -x + 10$:

$$-\frac{1}{3}x + 6 = -x + 10$$

$$\frac{2}{3}x = 4$$

$$x = 6$$

$$y = -x + 10 = -6 + 10 = 4$$

(6, 4)

Intersection of $y = -x + 10$ and $y = -\frac{1}{4}x + 4$:

$$-\frac{1}{4}x + 4 = -x + 10$$

$$\frac{3}{4}x = 6$$

$$x = 8$$

$$y = -x + 10 = -8 + 10 = 2$$

(8, 2)

Intersection of $y = -\frac{1}{4}x + 4$ and $y = -2x + 11$:

$$-\frac{1}{4}x + 4 = -2x + 11$$

$$\frac{7}{4}x = 7$$

$$x = 4$$

$$y = -2x + 11 = -2(4) + 11 = 3$$

(4, 3)

Vertex	$7x + 4y$
(3, 5)	$7(3) + 4(5) = 41$
(6, 4)	$7(6) + 4(4) = 58$
(8, 2)	$7(8) + 4(2) = 64$
(4, 3)	$7(4) + 4(3) = 40$

The minimum value is 40 and occurs at (4, 3).

33.

	Hockey games	Soccer games	Available
Assembly	2 labor-hours	3 labor-hours	42 labor-hours
Testing	2 labor-hours	1 labor-hour	26 labor-hours

Let x be the number of hockey games produced each day, and let y be the number of soccer games produced each day.

Assembly: $2x + 3y \leq 42$
Testing: $2x + y \leq 26$
In standard form the equations are:

$$\begin{cases} y \leq -\frac{2}{3}x + 14 \\ y \leq -2x + 26 \\ x \geq 0, y \geq 0 \end{cases}$$

SSM: Finite Math **Chapter 3:** Linear Programming, A Geometric Approach

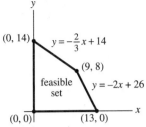

Find the vertices.
Lower left corner: (0, 0)

y-intercept of $y = -\frac{2}{3}x + 14$: (0, 14)

Intersection of $y = -\frac{2}{3}x + 14$ and $y = -2x + 26$:

$$-\frac{2}{3}x + 14 = -2x + 26$$

$$\frac{4}{3}x = 12$$

$$x = 9$$

$y = -2x + 26 = -2(9) + 26 = 8$
(9, 8)
x-intercept of $y = -2x + 26$: (13, 0)
The total daily output is simply the total number of games produced, or $x + y$.

Vertex	Output = $x + y$
(0, 0)	0 + 0 = 0
(0, 14)	0 + 14 = 14
(9, 8)	9 + 8 = 17
(13, 0)	13 + 0 = 13

The maximum output occurs at (9, 8). Produce 9 hockey games and 8 soccer games each day.

35. Since the farmer can spend $2400 for labor at $8 per hour, the available labor is $2400 \div 8 = 300$ hours.

	Oats	Corn	Available
Capital	$18	$36	$2100
Labor	2 hours	6 hours	300 hours
Land	1 acre	1 acre	100 acres
Revenue	$55	$125	

Let x be the number of acres of oats, and let y be the number of acres of corn.
Capital: $18x + 36y \leq 2100$
Labor: $2x + 6y \leq 300$
Land: $x + y \leq 100$

3-9
Copyright © 2014 Pearson Education, Inc.

In standard form the inequalities are:
$$\begin{cases} y \leq -\frac{1}{2}x + \frac{175}{3} \\ y \leq -\frac{1}{3}x + 50 \\ y \leq -x + 100 \\ x \geq 0, y \geq 0 \end{cases}$$

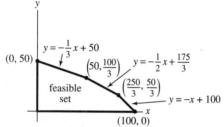

Find the vertices.
Lower left corner: $(0, 0)$

y-intercept of $y = -\frac{1}{3}x + 50$: $(0, 50)$

Intersection of $y = -\frac{1}{3}x + 50$ and $y = -\frac{1}{2}x + \frac{175}{3}$:

$$-\frac{1}{3}x + 50 = -\frac{1}{2}x + \frac{175}{3}$$

$$\frac{1}{6}x = \frac{25}{3}$$

$$x = 50$$

$$y = -\frac{1}{3}x + 50 = -\frac{1}{3}(50) + 50 = \frac{100}{3}$$

$\left(50, \frac{100}{3}\right)$

Intersection of $y = -\frac{1}{2}x + \frac{175}{3}$ and $y = -x + 100$:

$$-\frac{1}{2}x + \frac{175}{3} = -x + 100$$

$$\frac{1}{2}x = \frac{125}{3}$$

$$x = \frac{250}{3}$$

$$y = -x + 100 = -\frac{250}{3} + 100 = \frac{50}{3}$$

$\left(\frac{250}{3}, \frac{50}{3}\right)$

x-intercept of $y = -x + 100$: $(100, 0)$
Find the objective function for the profit.
Revenue: $55x + 125y$
Leftover capital: $2100 - 18x - 36y$

Leftover labor cash reserve: $2400 - 8(2x + 6y) = 2400 - 16x - 48y$
The profit is the sum of the above: $21x + 41y + 4500$

Vertex	Profit = $21x + 41y + 4500$
$(0, 0)$	$21(0) + 41(0) + 4500 = 4500$
$(0, 50)$	$21(0) + 41(50) + 4500 = 6550$
$\left(50, \frac{100}{3}\right)$	$21(50) + 41\left(\frac{100}{3}\right) + 4500 \approx 6916.67$
$\left(\frac{250}{3}, \frac{50}{3}\right)$	$21\left(\frac{250}{3}\right) + 41\left(\frac{50}{3}\right) + 4500 \approx 6933.33$
$(100, 0)$	$21(100) + 41(0) + 4500 = 6600$

The maximum profit is achieved at $\left(\frac{250}{3}, \frac{50}{3}\right)$, or $\left(83\frac{1}{3}, 16\frac{2}{3}\right)$. The farmer should plant $83\frac{1}{3}$ acres of oats and $16\frac{2}{3}$ acres of corn to make a profit of $6933.33.

37.

	Regular	Deluxe	Available
Capital	$32	$38	$2100
Labor	4 hours	6 hours	280 hours
Price	$46	$55	

Let x be the number of regular bags made each day, and let y be the number of deluxe

$\text{capital}: 32x + 38y \leq 2100$
$\text{Labor}: 4x + 6y \leq 280$
$x \geq 0, y \geq 0$

In standard form the inequalities are:
$$\begin{cases} y \leq -\frac{16}{19}x + \frac{1050}{19} \\ y \leq -\frac{2}{3}x + \frac{140}{3} \\ x \geq 0, y \geq 0 \end{cases}$$

Find the vertices:
y-intercept of $y = -\frac{2}{3}x + \frac{140}{3}$: $\left(0, \frac{140}{3}\right)$

Intersection of $y = -\frac{16}{19}x + \frac{1050}{19}$ and $y = -\frac{2}{3}x + \frac{140}{3}$:

$$-\frac{16}{19}x + \frac{1050}{19} = -\frac{2}{3}x + \frac{140}{3}$$

$$\frac{10}{57}x = \frac{490}{57}$$

$$x = 49$$

$$y = -\frac{16}{19}x + \frac{1050}{19} = -\frac{16}{19}(49) + \frac{1050}{19} = 14$$

(49, 14)

x-intercept of $y = -\frac{16}{19}x + \frac{1050}{19}$: $\left(\frac{525}{8}, 0\right)$

Vertex	Revenue = 46x + 55y
$\left(0, \frac{140}{3}\right)$	$46(0) + 55(46.67) \approx 2566.67$
(49, 14)	$46(49) + 55(14) = 3024$
$\left(\frac{525}{8}, 0\right)$	$46(65.625) + 55(0) = 3018.75$

The maximum profit is achieved at (49, 14). Make 49 regular bags and 14 deluxe.

39.

	Food A	Food B	Requirement
Protein	4 units	3 units	42 units
Carbohydrates	2 units	6 units	30 units
Fat	2 units	1 unit	18 units
Weight	3 pounds	2 pounds	

Let x be the number of tubes of food A, and let y be the number of tubes of food B.
Protein: $4x + 3y \geq 42$
Carbohydrates: $2x + 6y \geq 30$
Fat: $2x + y \geq 18$
In standard form the inequalities are:

$$\begin{cases} y \geq -\frac{4}{3}x + 14 \\ y \geq -\frac{1}{3}x + 5 \\ y \geq -2x + 18 \\ x \geq 0, \ y \geq 0 \end{cases}$$

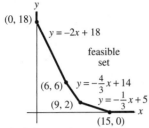

Find the vertices.
y-intercept of $y = -2x + 18$: $(0, 18)$

Intersection of $y = -2x + 18$ and $y = -\frac{4}{3}x + 14$:

$$-\frac{4}{3}x + 14 = -2x + 18$$
$$\frac{2}{3}x = 4$$
$$x = 6$$
$y = -2x + 18 = -2(6) + 18 = 6$
$(6, 6)$

Intersection of $y = -\frac{4}{3}x + 14$ and $y = -\frac{1}{3}x + 5$:

$$-\frac{1}{3}x + 5 = -\frac{4}{3}x + 14$$
$$x = 9$$
$$y = -\frac{1}{3}x + 5 = -\frac{1}{3}(9) + 5 = 2$$
$(9, 2)$

x-intercept of $y = -\frac{1}{3}x + 5$: $(15, 0)$

Vertex	Weight = $3x + 2y$
(0, 18)	$3(0) + 2(18) = 36$
(6, 6)	$3(6) + 2(6) = 30$
(9, 2)	$3(9) + 2(2) = 31$
(15, 0)	$3(15) + 2(0) = 45$

The minimum weight is achieved at (6, 6).
Send 6 tubes of food A and 6 tubes of food B.

41.

	Fruit Delight	Heavenly Punch	Available
Pineapple juice	10 ounces	10 ounces	9000 ounces
Orange juice	3 ounces	2 ounces	2400 ounces
Apricot juice	1 ounce	2 ounces	1400 ounces
Profit	$0.40	$0.60	

Let x be the number of cans of Fruit Delight, and let y be the number of cans of Heavenly Punch produced each week.
Pineapple: $10x + 10y \leq 9000$
Orange: $3x + 2y \leq 2400$
Apricot: $x + 2y \leq 1400$
In standard form the inequalities are:
$$\begin{cases} y \leq -x + 900 \\ y \leq -\frac{3}{2}x + 1200 \\ y \leq -\frac{1}{2}x + 700 \\ x \geq 0, \ y \geq 0 \end{cases}$$

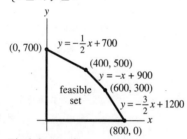

Find the vertices.
Lower left corner: $(0, 0)$

y-intercept of $y = -\frac{1}{2}x + 700$: $(0, 700)$

Intersection of $y = -\frac{1}{2}x + 700$ and $y = -x + 900$:

$-\frac{1}{2}x + 700 = -x + 900$

$\frac{1}{2}x = 200$

$x = 400$

$y = -x + 900 = -400 + 900 = 500$
$(400, 500)$

Intersection of $y = -x + 900$ and $y = -\dfrac{3}{2}x + 1200$:

$$-x + 900 = -\dfrac{3}{2}x + 1200$$

$$\dfrac{1}{2}x = 300$$

$$x = 600$$

$y = -x + 900 = -600 + 900 = 300$
(600, 300)

x-intercept of $y = -\dfrac{3}{2}x + 1200$: (800, 0)

Vertex	Profit = .4x + .6y
(0, 0)	.4(0) + .6(0) = 0
(0, 700)	.4(0) + .6(700) = 420
(400, 500)	.4(400) + .6(500) = 460
(600, 300)	.4(600) + .6(300) = 420
(800, 0)	.4(800) + .6(0) = 320

The maximum profit is achieved at (400, 500).
Make 400 cans of Fruit Delight and 500 cans of Heavenly Punch.

43. a. Since the conditions are less restrictive than they were in Exercise 35, the optimal solution found in Exercise 35 will still be in the feasible set.

b. The total cost to plant an acre of oats is $18 capital plus $16 for labor, or $34.
The total cost to plant an acre of corn is $36 capital plus $48 for labor, or $84.

	Oats	Corn	Available
Cap. And Lab.	$34	$84	$4500
Land	1 acre	1 acre	100 acres
Revenue	$55	$125	

Let x be the number of acres of oats, and let y be the number of acres of corn.
Capital and Labor: $34x + 84y \leq 4500$
Land: $x + y \leq 100$
In standard form the inequalities are:
$$\begin{cases} y \leq -\dfrac{17}{42}x + \dfrac{375}{7} \\ y \leq -x + 100 \\ x \geq 0, y \geq 0 \end{cases}$$

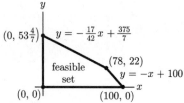

Find the vertices.
Lower left corner: $(0, 0)$

y-intercept of $y = -\dfrac{17}{42}x + \dfrac{375}{7}$: $\left(0, 53\dfrac{4}{7}\right)$

Intersection of $y = -\dfrac{17}{42}x + \dfrac{375}{7}$ and $y = -x + 100$:

$$-\dfrac{17}{42}x + \dfrac{375}{7} = -x + 100$$

$$\dfrac{25}{42}x = \dfrac{325}{7}$$

$$x = 78$$

$y = -x + 100 = -(78) + 100 = 22$

$(78, 22)$

x-intercept of $y = -x + 100$: $(100, 0)$

Note that the objective function for the profit is the same as in Exercise 35.
Find the objective function for the profit.
Revenue: $55x + 125y$
Leftover capital: $2100 - 18x - 36y$
Leftover labor cash reserve: $2400 - 8(2x + 6y) = 2400 - 16x - 48y$
The profit is the sum of the above: $21x + 41y + 4500$

Vertex	Profit = $21x + 41y + 4500$
$(0, 0)$	$21(0) + 41(0) + 4500 = 4500$
$(78, 22)$	$21(78) + 41(22) + 4500 = 7040$
$\left(0, \dfrac{375}{7}\right)$	$21(0) + 41\left(\dfrac{375}{7}\right) + 4500 \approx 6696.43$
$(100, 0)$	$21(100) + 41(0) + 4500 = 6600$

The maximum profit is achieved at $(78, 22)$. The farmer should plant 78 acres of oats and 22 acres of corn to make a profit of \$7040. Yes, it provides more profit.

45.

	Cupid	Patriotic	Available
Red	60	30	2400
White	40	35	1750
Blue	0	35	1470
Profit	8	6	

Let x be the number of Cupid assortments produced, and let y be the number of Patriotic assortments produced.

Red: $60x + 30y \leq 2400$

White: $40x + 35y \leq 1750$

Blue: $35y \leq 1470$

In standard form the inequalities are:
$$\begin{cases} y \leq -2x+80 \\ y \leq -\dfrac{8}{7}x+50 \\ y \leq 42 \\ x \geq 0,\ y \geq 0 \end{cases}$$

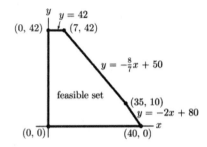

Find the vertices.
Intersection of $y = 0$ and $x = 0$ (0, 0)

Intersection of $y = 42$ and $x = 0$ (0, 42)

Intersection of $y = 42$ and $y = -\dfrac{8}{7}x + 50$

$42 = -\dfrac{8}{7}x + 50$

$\dfrac{8}{7}x = 8$

$x = 7$

(7, 42)

Intersection of $y = -2x + 80$ and $y = -\dfrac{8}{7}x + 50$

$-2x + 80 = -\dfrac{8}{7}x + 50$

$30 = \dfrac{6}{7}x$

$x = 35$

$y = -2(35) + 80$

$y = 10$

(35, 10)

Intersection of $y = 0$ and $y = -2x + 80$
$0 = -2x + 80$
$2x = 80$
$x = 40$
(40, 0)

Vertex	Profit = $8x + 6y$
(0, 0)	$8(0) + 6(0) = 0$
(0, 42)	$8(0) + 6(42) = 252$
(7, 42)	$8(7) + 6(42) = 308$
(35, 10)	$8(35) + 6(10) = 340$
(40, 0)	$8(40) + 6(0) = 320$

The maximum profit occurs at (35, 10).
Produce 35 Cupid assortments and 10 Patriotic assortments.

47.

	Pamper Me	Best Friends	Available
Shower Gel	1	2	400
Bubble Bath	2	2	550
Candles	2	0	400
Profit	15	12	

Let x be the number of Pamper Me baskets produced, and let y be the number of Best Friends baskets produced.
Shower Gel: $x + 2y \leq 400$
Bubble Bath: $2x + 2y \leq 550$
Candles: $2x \leq 400$
In standard form the inequalities are:
$$\begin{cases} y \leq -0.5x + 200 \\ y \leq -x + 275 \\ x \leq 200 \\ x \geq 0, \ y \geq 0 \end{cases}$$

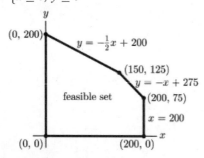

Find the vertices.
Intersection of $y = 0$ and $x = 0$ (0, 0)

Intersection of $y = -.5x + 200$ and $x = 0$
$y = -.5(0) + 200$
$y = 200$
(0, 200)
Intersection of $y = -.5x + 200$ and $y = -x + 275$
$-.5x + 200 = -x + 275$
$.5x = 75$
$x = 150$
$y = -150 + 275$
$y = 125$
(150, 125)
Intersection of $y = -x + 275$ and $x = 200$
$y = -200 + 275$
$y = 75$
(200, 75)
Intersection of $y = 0$ and $x = 200$
(200, 0)

Vertex	Profit = 15x + 12y
(0, 0)	15(0) + 12(0) = 0
(0, 200)	15(0) + 12(200) = 2400
(150, 125)	15(150) + 12(125) = 3750
(200, 75)	15(200) + 12(75) = 3900
(200, 0)	15(200) + 12(0) = 3000

The maximum profit occurs at (200, 75).
Produce 200 Pamper Me baskets and 75 Best Friends baskets.

49. Since $x \geq 0$ and $y \geq 0$, $4x + 3y \geq x + y$.
Therefore, the inequality $x + y \geq 6$ implies that $4x + 3y \geq 6$, which contradicts $4x + 3y \leq 4$.
The feasible set contains no points.

51.

	A	B
1	2	10
2	6	14
3		2
4		6
5		30
6		
7	Cell	Content
8	B1	=2*x+y
9	B2	=x+2*y
10	B3	=x
11	B4	=y
12	B5	=3*x+4*y

Exercises 3.3

1. a. $21x + 14y = c$

   $y = -\dfrac{3}{2}x + \dfrac{c}{14}$

   b. Up

   c. B

3. The objective function $ax + by$ has constant value on any line of slope $-\dfrac{a}{b}$. The slope of the line containing (8, 3) and (9, 0) is $\dfrac{0-3}{9-8} = -3$.

   Therefore, if $-\dfrac{a}{b} < 0$, we require

   $-\dfrac{a}{b} \leq -3$, or $\dfrac{a}{b} \geq 3$, where $a > 0$ and $b > 0$. One possibility is $a = 5$ and $b = 1$, giving the objective function $5x + y$.

   (We could also have chosen an objective function that is constant on lines of nonnegative or undefined slope of the form $ax + by$ with $a \geq 0$ and $b \leq 0$.)

5. The objective function $ax + by$ has constant value on any line of slope $-\dfrac{a}{b}$. Since the slope of the line containing (3, 8) and (8, 3) is $\dfrac{3-8}{8-3} = -1$, and the slope of the line containing (8, 3) and (9, 0) is $\dfrac{0-3}{9-8} = -3$, we require

   $-1 \geq -\dfrac{a}{b} \geq -3$, or $1 \leq \dfrac{a}{b} \leq 3$.

   One possibility is $a = 2$ and $b = 1$, giving the objective function $2x + y$.

7. The objective function $ax + by$ has constant value on any line of slope $-\dfrac{a}{b}$. The slope of the line containing (6, 1) and (9, 0) is $\dfrac{0-1}{9-6} = -\dfrac{1}{3}$.

   We require $0 \geq -\dfrac{a}{b} \geq -\dfrac{1}{3}$, or $0 \leq \dfrac{a}{b} \leq \dfrac{1}{3}$, where $a \geq 0$ and $b > 0$. One possibility is $a = 1$ and $b = 5$, giving the objective function $x + 5y$.

9. The objective function $ax + by$ has constant value on any line of slope $-\dfrac{a}{b}$. The slope of the line containing (1, 6) and (6, 1) is $\dfrac{1-6}{6-1} = -1$, and the slope of the line containing (6, 1) and (9, 0) is $\dfrac{0-1}{9-6} = -\dfrac{1}{3}$.

   We require $-\dfrac{1}{3} \geq -\dfrac{a}{b} \geq -1$, or

   $\dfrac{1}{3} \leq \dfrac{a}{b} \leq 1$, where $a > 0$ and $b > 0$. One possibility is $a = 2$ and $b = 3$, giving the objective function $2x + 3y$.

11. The objective function $3x + 2y$ has constant value on any line of slope $-\dfrac{3}{2}$. Since this is between -1 and -4, the objective function is maximized at C.

13. The objective function $10x + 2y$ has constant value on any line of slope -5. Since this is less (steeper) than -4, the objective function is maximized at D.

15. The objective function $2x + 10y$ has constant value on any line of slope $-\dfrac{1}{5}$.

    Since this is between 0 and $-\dfrac{1}{4}$, the objective function is minimized at D.

17. The objective function $2x + 3y$ has constant value on any line of slope $-\dfrac{2}{3}$.

Since this is between $-\dfrac{1}{4}$ and -1, the objective function is minimized at C.

19. The objective function $x + ky$ has constant value on any line of slope $-\dfrac{1}{k}$. The slope of the line containing $(0, 5)$ and $(3, 4)$ is $\dfrac{4-5}{3-0} = -\dfrac{1}{3}$, and the slope of the line containing $(3, 4)$ and $(4, 0)$ is $\dfrac{0-4}{4-3} = -4$.

We require $-4 \leq -\dfrac{1}{k} \leq -\dfrac{1}{3}$. This is equivalent to $4 \geq \dfrac{1}{k} \geq \dfrac{1}{3}$, or $\dfrac{1}{4} \leq k \leq 3$.

21.

	Brand A	Brand B	Requirement
Protein	3 units	1 unit	6 units
Carbohydrate	1 unit	1 unit	4 units
Fat	2 units	6 units	12 units
Cost	$.80	$.50	

Let x be the number of units of brand A, and let y be the number of units of brand B.
Protein: $3x + y \geq 6$
Carbohydrate: $x + y \geq 4$
Fat: $2x + 6y \geq 12$
In standard form the inequalities are:
$$\begin{cases} y \geq -3x + 6 \\ y \geq -x + 4 \\ y \geq -\dfrac{1}{3}x + 2 \\ x \geq 0, \, y \geq 0 \end{cases}$$

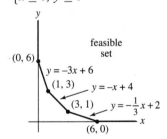

Vertex	Cost = .8x + .5y
(0, 6)	3
(1, 3)	2.3
(3, 1)	2.9
(6, 0)	4.8

The minimum cost of $2.30 is obtained at (1, 3).
Feed 1 can of brand A and 3 cans of brand B.

23. Let the variables represent amounts in thousands of dollars.

	Low-risk	Medium-risk	High-risk
Yield	.06	.07	.08
Variables	x	y	$9-x-y$

The required inequalities are:
$$\begin{cases} x \le y+1 \\ x+y \ge 5 \\ y+(9-x-y) \le 7 \\ x \ge 0, y \ge 0 \\ 9-x-y \ge 0 \end{cases} \text{ or } \begin{cases} y \ge x-1 \\ y \ge -x+5 \\ x \ge 2 \\ x \ge 0, y \ge 0 \\ y \le -x+9 \end{cases}$$

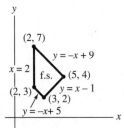

(The inequalities $x \ge 0$ and $y \ge 0$ do not appear in the graph, as they are assured by the other inequalities.)

The objective function for the expected yield is $.06x + .07y + .08(9 - x - y)$, or $.72 - .02x - .01y$.

Vertex	Yield = .72 − .02x − .01y
(2, 3)	.65
(2, 7)	.61
(5, 4)	.58
(3, 2)	.64

The maximum yield of $650 is achieved at (2, 3). Then $9 - x - y = 9 - 2 - 3 = 4$.
Invest $2000 in low-risk stocks, $3000 in medium-risk stocks, and $4000 in high-risk stocks.

25.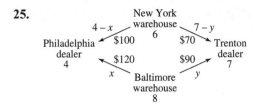

The required inequalities are:

$$\begin{cases} x \geq 0,\ y \geq 0 \\ 4 - x \geq 0 \\ 7 - y \geq 0 \\ x + y \leq 8 \\ (4-x) + (7-y) \leq 6 \end{cases} \quad \text{or} \quad \begin{cases} x \geq 0,\ y \geq 0 \\ x \leq 4 \\ y \leq 7 \\ y \leq -x + 8 \\ y \geq -x + 5 \end{cases}$$

(The inequality $y \geq 0$ does not appear in the graph, as it is assured by the other inequalities.)

The objective function for the cost is $120x + 90y + 100(4 - x) + 70(7 - y)$, or $890 + 20x + 20y$.

Vertex	Cost = 890 + 20x + 20y
(0, 5)	990
(0, 7)	1030
(1, 7)	1050
(4, 4)	1050
(4, 1)	990

The minimum cost of $990 is achieved at (0, 5) or (4, 1)—or anywhere along the segment connecting these points.

There are several ways to minimize costs, as summarized below.

Baltimore to Philadelphia	Baltimore to Trenton	New York to Philadelphia	New York to Trenton
0	5	4	2
1	4	3	3
2	3	2	4
3	2	1	5

27. Let the variables represent the number of thousands of gallons.

	Gasoline	Jet fuel	Diesel fuel
Profit	$.15	$.12	$.10
Variables	x	y	$100-x-y$

The required inequalities are:

$$\begin{cases} x \geq 5,\ y \geq 5 \\ 100-x-y \geq 5 \\ x+y \geq 20 \\ x+(100-x-y) \geq 50 \end{cases} \text{ or } \begin{cases} x \geq 5,\ y \geq 5 \\ y \geq -x+95 \\ y \geq -x+20 \\ y \leq 50 \end{cases}$$

(Note that the inequalities above have ignored the somewhat subtle issue of whether there will be enough gasoline for *both* the airline and the trucking firm at the same time. This is only a potential issue if both suppliers require gasoline—that is, if $y \leq 20$ and $100-x-y \leq 50$. In this case, we require that the gasoline requirements of each firm be less than the amount of gasoline actually produced—that is,
$(20-y) + (50-(100-x-y)) \leq x$. This inequality is equivalent to $-30 \leq 0$, so it is always satisfied.)

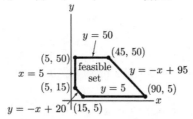

The objective function for the profit is $.15x + .12y + .1(100-x-y)$, or $10 + .05x + .02y$.

Vertex	Profit = $10 + .05x + .02y$
(5, 50)	11.25
(45, 50)	13.25
(90, 5)	14.6
(15, 5)	10.85
(5, 15)	10.55

The maximum profit of $14,600 is achieved at (90, 5). Then $100-x-y = 100-90-5 = 5$. Produce 90,000 gallons of gasoline, 5000 gallons of jet fuel, and 5000 gallons of diesel fuel.

29.

	High-capacity	Low-capacity	Available
Cost ($thousands)	50	30	1080
Drivers	1	1	30
Capacity	320 cases	200 cases	

Let x be the number of high-capacity trucks, and let y be the number of low-capacity trucks. The required inequalities are:

$$\begin{cases} 50x + 30y \leq 1080 \\ x + y \leq 30 \\ x \leq 15 \\ x \geq 0,\ y \geq 0 \end{cases} \text{ or } \begin{cases} y \leq -\dfrac{5}{3}x + 36 \\ y \leq -x + 30 \\ x \leq 15 \\ x \geq 0,\ y \geq 0 \end{cases}$$

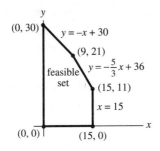

Vertex	Capacity = $320x + 200y$
(0, 0)	0
(0, 30)	6000
(9, 21)	7080
(15, 11)	7000
(15, 0)	4800

The maximum capacity of 7080 cases is achieved at (9, 21).
Buy 9 high-capacity trucks and 21 low-capacity trucks.

31. Let x = pounds of coffee shipped from San Jose to Salt Lake City, and
    y = pounds of coffee shipped from San Jose to Reno.

    The required inequalities are:
    $$\begin{cases} x+y \leq 500 \\ (400-x)+(350-y) \leq 700 \\ x \geq 0, y \geq 0 \\ 400-x \geq 0, 350-y \geq 0 \end{cases} \text{ or } \begin{cases} y \leq -x+500 \\ y \geq -x+50 \\ x \geq 0, y \geq 0 \\ x \leq 400, y \leq 350 \end{cases}$$

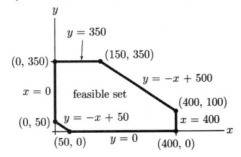

Objective function: [cost] = $2x + 2.5y + 2.5(400 - x) + 3.5(350 - y) = 2225 - .5x - y$.

Vertex	Cost = $2225 - .5x - y$
(0, 50)	2175
(0, 350)	1875
(150, 350)	1800
(400, 100)	1925
(400, 0)	2025
(50, 0)	2200

The minimum cost of $1800 is at (150, 350).
Ship 150 pounds of coffee from San Jose to Salt Lake City, 350 pounds from San Jose to Reno, and 250 pounds from Seattle to Salt Lake City.

33.

	Kit I	Kit II	Kit III	Available
Filters	1	2	1	54
Gravel (pounds)	2	2	3	100
Fish food	1	0	2	53
Profit	$7	$10	$13	

Let x be the number of Kit I's, and let y be the number of Kit II's. Then $x - y$ is the number of Kit III's. The required inequalities are:

$$\begin{cases} x + 2y + (x - y) \leq 54 \\ 2x + 2y + 3(x - y) \leq 100 \\ x + 2(x - y) \leq 53 \\ x - y \geq 0 \\ x \geq 0, y \geq 0 \end{cases} \text{ or } \begin{cases} y \leq -2x + 54 \\ y \geq 5x - 100 \\ y \geq \frac{3}{2}x - \frac{53}{2} \\ y \leq x \\ x \geq 0, y \geq 0 \end{cases}$$

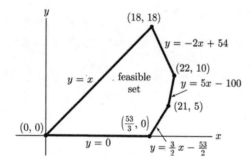

The objective function is [profit] $= 7x + 10y + 13(x - y)$
$$= 20x - 3y$$

Vertex	Profit = 20x – 3y
(0, 0)	0
(18, 18)	306
(22, 10)	410
(21, 5)	405
$\left(\frac{53}{3}, 0\right)$	$353\frac{1}{3}$

The maximum profit of $410 is acheived at (22, 10). Create 22 Kit I, 10 Kit II; and 12 Kit III.

35. a. $Ax + 70y = p$ or $y = -\frac{A}{70}x + \frac{p}{70}$ The slope of the line is $-\frac{A}{70}$.

b. $-2 \le -\frac{A}{70} \le -1$

$-140 \le -A \le -70$

$140 \ge A \ge 70$

c. $80x + Ay = p$

$y = -\frac{80}{A}x + \frac{p}{A}$

$-2 \le -\frac{80}{A} \le -1$

$-\frac{1}{2} \ge -\frac{A}{80} \ge -1$

$40 \le A \le 80$

37.

Cell	Name	Final Value	Reduced Cost	Objective Coefficient	Allowable Increase	Allowable Decrease
A1	x	.857142857	0	21	21	11.66666667
A2	y	3.428571429	0	14	17.5	7

The range of optimality for the cost of rice is [21 – 11.67, 21 + 21] or [9.33, 42].
The range of optimality for the cost of soybeans is [14 – 7, 14 + 17.5] or [7, 31.5].

Chapter 3 Fundamental Concept Check

1. Linear Programming is a method for determining a way to achieve the best outcome (such as maximum profit or lowest cost) subject to requirements given by a list of linear inequalities.

2. A linear function that specifies the outcome to be optimized.

3. The set of points in the plane whose coordinates satisfy the list of linear inequalities.

4. The maximum (or minimum) value of the objective function is achieved at one of the vertices of the feasible set.

5. See the tinted box on page 117.

Chapter 3 Review Exercises

1.

	Type A	Type B	Required or available
Passengers	50	300	1400
Flight attendants	3	4	42
Cost	$14,000	$90,000	

Let x be the number of type A planes, and let y be the number of type B planes. The required inequalities are:

$$\begin{cases} 50x + 300y \geq 1400 \\ 3x + 4y \leq 42 \\ x \geq y \\ x \geq 0, y \geq 0 \end{cases} \text{ or } \begin{cases} y \geq -\frac{1}{6}x + \frac{14}{3} \\ y \leq -\frac{3}{4}x + \frac{21}{2} \\ y \leq x \\ x \geq 0, y \geq 0 \end{cases}$$

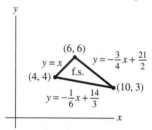

(The inequalities $x \geq 0$ and $y \geq 0$ are not shown in the graph because they are assured by the other inequalities.)

Vertex	Cost = 14,000x + 90,000y
(4, 4)	416,000
(6, 6)	624,000
(10, 3)	410,000

The minimum cost of $410,000 is achieved at (10, 3). Use 10 type A planes and 3 type B planes.

2.

	Wheat germ	Enriched oat flour	Required
Niacin	2 milligrams	3 milligrams	7 milligrams
Iron	3 milligrams	3 milligrams	9 milligrams
Thiamin	.5 milligram	.25 milligram	1 milligram
Cost	6 cents	8 cents	

Let x be the number of ounces of wheat germ, and let y be the number of ounces of enriched oat flour.

The required inequalities are:

$$\begin{cases} 2x+3y \geq 7 \\ 3x+3y \geq 9 \\ .5x+.25y \geq 1 \\ x \geq 0,\ y \geq 0 \end{cases} \text{ or } \begin{cases} y \geq -\dfrac{2}{3}x+\dfrac{7}{3} \\ y \geq -x+3 \\ y \geq -2x+4 \\ x \geq 0,\ y \geq 0 \end{cases}$$

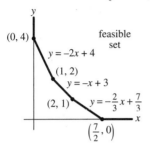

Vertex	Cost = $6x + 8y$
(0, 4)	32
(1, 2)	22
(2, 1)	20
$\left(\dfrac{7}{2}, 0\right)$	21

The minimum cost of 20 cents is achieved at (2, 1).
Use 2 ounces of wheat germ and 1 ounce of enriched oat flour.

3.

	Hardtops	Sports cars	Available
Assemble	8 labor-hours	18 labor-hours	360 labor-hours
Paint	2 labor-hours	2 labor-hours	50 labor-hours
Upholster	2 labor-hours	1 labor-hour	40 labor-hours
Profit	$90	$100	

Let x be the number of hardtops and let y be the number of Sports cars
The required inequalities are:

$$\begin{cases} 8x+18y \leq 360 \\ 2x+2y \leq 50 \\ 2x+y \leq 40 \\ x \geq 0,\ y \geq 0 \end{cases} \text{ or } \begin{cases} y \leq -\dfrac{4}{9}x+20 \\ y \leq -x+25 \\ y \leq -2x+40 \\ x \geq 0,\ y \geq 0 \end{cases}$$

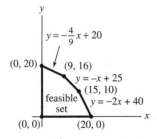

Vertex	Profit = 90x + 100y
(0, 0)	0
(0, 20)	2000
(9, 16)	2410
(15, 10)	2350
(20, 0)	1800

The maximum profit of $2410 is achieved at (9, 16).
Produce 9 hardtops and 16 sports cars.

4.

	Type A	Type B	Available
Peanuts	6 oz	12 oz	5400 oz
Raisins	1 oz	3 oz	1200 oz
Cashews	4 oz	2 oz	2400 oz
Revenue	$4.25	$6.55	

Let x be the amount of mixture A, and let y be the amount of mixture B.

Peanuts: $6x + 12y \le 5400$

Raisins: $x + 3y \le 1200$

Cashews: $4x + 2y \le 2400$

In standard form the inequalities are:

$$\begin{cases} y \le -\frac{1}{2}x + 450 \\ y \le -\frac{1}{3}x + 400 \\ y \le -2x + 1200 \\ x \ge 0,\ y \ge 0 \end{cases}$$

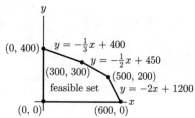

Find the vertices.
Lower left corner: (0, 0)

y-intercept of $y = -\frac{1}{3}x + 400$: (0, 400)

Intersection of $y = -\frac{1}{3}x + 400$ and $y = -\frac{1}{2}x + 450$:

$$-\frac{1}{3}x + 400 = -\frac{1}{2}x + 450$$

$$\frac{1}{6}x = 50$$

$$x = 300$$

$$y = -\frac{1}{3}x + 400 = -\frac{1}{3}(300) + 400 = 300$$

(300, 300)

Intersection of $y = -\frac{1}{2}x + 450$ and $y = -2x + 1200$:

$$-\frac{1}{2}x + 450 = -2x + 1200$$

$$\frac{3}{2}x = 750$$

$$x = 500$$

$$y = -\frac{1}{2}x + 450 = -\frac{1}{2}(500) + 450 = 200$$

(500, 200)

x-intercept of $y = -2x + 1200$: (600, 0)

Vertex	Revenue = 4.25x + 6.55y
(0, 0)	4.25(0) + 6.55(0) = 0
(0, 400)	4.25(0) + 6.55(400) = 2620
(300, 300)	4.25(300) + 6.55(300) = 3240
(500, 200)	4.25(500) + 6.55(200) = 3435
(600, 0)	4.25(600) + 6.55(0) = 2550

The maximum revenue is achieved at (500, 200).
Make 500 boxes of mixture A and 200 boxes of mixture B.

5.

	Elementary	Intermediate	Advanced
Profit	$8000	$7000	$1000
Variables	x	y	72 − x − y

The required inequalities are:

$$\begin{cases} 72 - x - y \geq 4 \\ x \geq 3y \\ y \geq 2(72 - x - y) \\ x \geq 0, y \geq 0 \end{cases} \text{ or } \begin{cases} y \leq -x + 68 \\ y \leq \frac{1}{3}x \\ y \leq -\frac{2}{3}x + 48 \\ x \geq 0, y \geq 0 \end{cases}$$

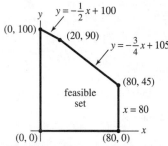

The objective function for the annual profit is $8000x + 7000y + 1000(72 - x - y)$, or $72{,}000 + 7000x + 6000y$.

Vertex	Profit = 72,000 + 7000x + 6000y
(48, 16)	504,000
(51, 17)	531,000
(60, 8)	540,000

The maximum annual profit of $540,000 is achieved at (60, 8).
Then $72 - x - y = 72 - 60 - 8 = 4$.
Publish 60 elementary books, 8 intermediate books, and 4 advanced books.

6.

	Rochester	Queens	Available
Transport time	15 hours	20 hours	2100 hours
Cost	$15	$30	$3000
Profit	$40	$30	

Let x be the number of computers sent from Rochester, and let y be the number of computers sent from Queens. The required inequalities are:

$$\begin{cases} 15x + 20y \leq 2100 \\ 15x + 30y \leq 3000 \\ x \leq 80 \\ y \leq 120 \\ x \geq 0, y \geq 0 \end{cases} \quad \text{or} \quad \begin{cases} y \leq -\dfrac{3}{4}x + 105 \\ y \leq -\dfrac{1}{2}x + 100 \\ x \leq 80 \\ y \leq 120 \\ x \geq 0, y \geq 0 \end{cases}$$

(The inequality $y \leq 120$ does not appear in the graph, as it is assured by the other inequalities.)

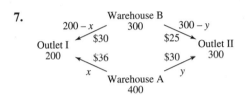

Vertex	Profit = 40x + 30y
(0, 0)	0
(0, 100)	3000
(20, 90)	3500
(80, 45)	4550
(80, 0)	3200

The maximum profit of $4550 is achieved at (80, 45).
Package 80 computers at Rochester and 45 computers at Queens.

7.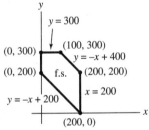

The required inequalities are:

$$\begin{cases} x \geq 0,\ y \geq 0 \\ 200 - x \geq 0 \\ 300 - y \geq 0 \\ x + y \leq 400 \\ (200 - x) + (300 - y) \leq 300 \end{cases} \quad \text{or} \quad \begin{cases} x \geq 0,\ y \geq 0 \\ x \leq 200 \\ y \leq 300 \\ y \leq -x + 400 \\ y \geq -x + 200 \end{cases}$$

The objective function for the cost is $36x + 30y + 30(200 - x) + 25(300 - y)$, or $13{,}500 + 6x + 5y$.

Vertex	Cost = 13,500 + 6x + 5y
(0, 200)	14,500
(0, 300)	15,000
(100, 300)	15,600
(200, 200)	15,700
(200, 0)	14,700

The minimum cost of $14,500 is achieved at (0, 200). Then $200 - x = 200 - 0 = 200$, and $300 - y = 300 - 200 = 100$.
Transport no refrigerators from warehouse A to outlet I, 200 refrigerators from warehouse A to outlet II, 200 refrigerators from warehouse B to outlet I, and 100 refrigerators from warehouse B to outlet II.

8.

	CD	Mutual fund	Stocks
Yield	.05	.07	.09
Variables	x	y	$10{,}000 - (x+y)$

The required inequalities are:

$$\begin{cases} y \le x + 10{,}000 - (x+y) \\ y + 10{,}000 - (x+y) \le 8000 \\ x \ge 0, y \ge 0 \\ x \le 10{,}000, y \le 10{,}000 \end{cases} \text{ or } \begin{cases} y \le 5000 \\ x \ge 2000 \\ y \ge 0 \\ x \le 10{,}000 \end{cases}$$

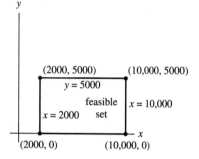

The objective function is [return] = $.05x + .07y + .09[10{,}000 - (x+y)]$ or $900 - .04x - .02y$.

Vertex	Return = $900 - .04x - .02y$
(2000, 0)	820
(2000, 5000)	720
(10,000, 5000)	400
(10,000, 0)	500

The maximum return of $820 is achieved at (2000, 0). Invest $2000 in the CD, $0 in mutual funds; and $8000 in stocks.

9. **a.** Yes; The added constraint may add to the original cost therefore it may increase the optimal cost.

   b. No; The added constraint will not decrease the original cost.

10. **a.** Yes; The removed constraint may increase the profit therefore it may increase the optimal profit.

    b. No; The removed constraint will not decrease profit.

11. Answers will vary.

12. One of the boundary points will always be a maximum or a minimum of the problem.

13. Every point on a line (or line segment) always has the same objective function value, provided the line (or line segment) has the same slope as a line with constant objective function.

Chapter 4

Exercises 4.1

1. $\begin{cases} 20x + 30y + u = 3500 \\ 50x + 10y + v = 5000 \\ -8x - 13y + M = 0 \end{cases}$

   Maximize M given $x \geq 0, y \geq 0, u \geq 0, v \geq 0$.

3. $\begin{cases} x + y + z + u = 100 \\ 3x + z + v = 200 \\ 5x + 10y + w = 100 \\ -x - 2y + 3z + M = 0 \end{cases}$

   Maximize M given $x \geq 0, y \geq 0, z \geq 0, u \geq 0, v \geq 0, w \geq 0$.

5. $\begin{cases} 4x + 6y - 7z + u = 16 \\ 3x + 2y + v = 11 \\ 9y + 3z + w = 21 \\ -3x - 5y - 12z + M = 0 \end{cases}$

   Maximize M given $x \geq 0, y \geq 0, z \geq 0, u \geq 0, v \geq 0, w \geq 0$.

7. a.
$$\begin{array}{cccccc} x & y & u & v & M & \\ \end{array}$$
$$\begin{bmatrix} 20 & 30 & 1 & 0 & 0 & | & 3500 \\ 50 & 10 & 0 & 1 & 0 & | & 5000 \\ -8 & -13 & 0 & 0 & 1 & | & 0 \end{bmatrix}$$

   b. $x = 0, y = 0, u = 3500, v = 5000, M = 0$

9. a.
$$\begin{array}{ccccccc} x & y & z & u & v & w & M \\ \end{array}$$
$$\begin{bmatrix} 1 & 1 & 1 & 1 & 0 & 0 & 0 & | & 100 \\ 3 & 0 & 1 & 0 & 1 & 0 & 0 & | & 200 \\ 5 & 10 & 0 & 0 & 0 & 1 & 0 & | & 100 \\ -1 & -2 & 3 & 0 & 0 & 0 & 1 & | & 0 \end{bmatrix}$$

   b. $x = 0, y = 0, z = 0, u = 100, v = 200, w = 100, M = 0$

11. a.
$$\begin{array}{ccccccc} x & y & z & u & v & w & M \\ \end{array}$$
$$\begin{bmatrix} 4 & 6 & -7 & 1 & 0 & 0 & 0 & | & 16 \\ 3 & 2 & 0 & 0 & 1 & 0 & 0 & | & 11 \\ 0 & 9 & 3 & 0 & 0 & 1 & 0 & | & 21 \\ -3 & -5 & -12 & 0 & 0 & 0 & 1 & | & 0 \end{bmatrix}$$

   b. $x = 0, y = 0, z = 0, u = 16, v = 11, w = 21, M = 0$

13. $x = 15, y = 0, u = 10, v = 0, M = 20$

15. $x = 10, y = 0, z = 15, u = 23, v = 0, w = 0,$
$M = -11$

17. a. Divide the first row by 2, then use matrix operations to change the remaining values in column 1 to zeros.

$$\begin{array}{c}\\x\\v\\M\end{array}\begin{array}{c}x \quad y \quad u \quad v \quad M\\ \left[\begin{array}{ccccc|c}1 & \frac{3}{2} & \frac{1}{2} & 0 & 0 & 6\\ 0 & -\frac{1}{2} & -\frac{1}{2} & 1 & 0 & 4\\ 0 & -5 & 5 & 0 & 1 & 60\end{array}\right]\end{array}$$

$x = 6, y = 0, u = 0, v = 4, M = 60$

b. Divide the first row by 3, then use matrix operations to change the remaining values in column 2 to zeros.

$$\begin{array}{c}\\y\\v\\M\end{array}\begin{array}{c}x \quad y \quad u \quad v \quad M\\ \left[\begin{array}{ccccc|c}\frac{2}{3} & 1 & \frac{1}{3} & 0 & 0 & 4\\ \frac{1}{3} & 0 & -\frac{1}{3} & 1 & 0 & 6\\ \frac{10}{3} & 0 & \frac{20}{3} & 0 & 1 & 80\end{array}\right]\end{array}$$

$x = 0, y = 4, u = 0, v = 6, M = 80$

c. Use matrix operations to change the remaining values in column 1 to zeros.

$$\begin{array}{c}\\u\\x\\M\end{array}\begin{array}{c}x \quad y \quad u \quad v \quad M\\ \left[\begin{array}{ccccc|c}0 & 1 & 1 & -2 & 0 & -8\\ 1 & 1 & 0 & 1 & 0 & 10\\ 0 & -10 & 0 & 10 & 1 & 100\end{array}\right]\end{array}$$

$x = 10, y = 0, u = -8, v = 0, M = 100$

d. Use matrix operations to change the remaining values in column 2 to zeros.

$$\begin{array}{c}\\u\\y\\M\end{array}\begin{array}{c}x \quad y \quad u \quad v \quad M\\ \left[\begin{array}{ccccc|c}-1 & 0 & 1 & -3 & 0 & -18\\ 1 & 1 & 0 & 1 & 0 & 10\\ 10 & 0 & 0 & 20 & 1 & 200\end{array}\right]\end{array}$$

$x = 0, y = 10, u = -18, v = 0, M = 200$

19. M becomes greatest after the pivot operation in part (d).

21. a. Group I variables are x and y.

Group II variables are u, v, and M.

b. (i) Divide the first row by 2, then use matrix operations to change the remaining values in column 1 to zeros.

$$\begin{array}{c}x \quad y \quad u \quad v \quad M\\ \left[\begin{array}{ccccc|c}1 & \frac{5}{2} & \frac{1}{2} & 0 & 0 & 50\\ 0 & -\frac{13}{2} & -\frac{3}{2} & 1 & 0 & 150\\ 0 & 18 & 5 & 0 & 1 & 500\end{array}\right]\end{array}$$ Feasible:

Group I variables: u and y Group II variables: x, v, M

(ii) Divide the first row by 5, then use matrix operations to change the remaining values in column 2 to zeros.

$$\begin{array}{c}x \quad y \quad u \quad v \quad M\\ \left[\begin{array}{ccccc|c}\frac{2}{5} & 1 & \frac{1}{5} & 0 & 0 & 20\\ \frac{13}{5} & 0 & -\frac{1}{5} & 1 & 0 & 280\\ -\frac{36}{5} & 0 & \frac{7}{5} & 0 & 1 & 140\end{array}\right]\end{array}$$ Feasible:

Group I variables: x and u Group II variables: y, v, M

(iii) Divide the second row by 3, then use matrix operations to change the remaining values in column 1 to zeros.

$$\begin{array}{c}x \quad y \quad u \quad v \quad M\\ \left[\begin{array}{ccccc|c}0 & \frac{13}{3} & 1 & -\frac{2}{3} & 0 & -100\\ 1 & \frac{1}{3} & 0 & \frac{1}{3} & 0 & 100\\ 0 & -\frac{11}{3} & 0 & \frac{10}{3} & 1 & 1000\end{array}\right]\end{array}$$

Not Feasible: Group I variables: y and v Group II variables: x, u, M

(iv) Use matrix operations to change the remaining value in column 2 to zeros.

$$\begin{array}{c}x \quad y \quad u \quad v \quad M\\ \left[\begin{array}{ccccc|c}-13 & 0 & 1 & -5 & 0 & -1400\\ 3 & 1 & 0 & 1 & 0 & 300\\ 11 & 0 & 0 & 7 & 1 & 2100\end{array}\right]\end{array}$$

Not Feasible: Group I variables: x and v Group II variables: y, u, M

c. M becomes greatest after the pivot operation in part (i).

Exercises 4.2

1. **a.** -12 is the most negative entry in the last row, and $\dfrac{6}{3} < \dfrac{10}{2}$. Pivot about the 3.

   b.
   $$\begin{array}{c} \\ u \\ y \\ M \end{array} \begin{array}{c} xyuvM \\ \left[\begin{array}{ccccc|c} \frac{16}{3} & 0 & 1 & -\frac{2}{3} & 0 & 6 \\ \frac{1}{3} & 1 & 0 & \frac{1}{3} & 0 & 2 \\ \hline 0 & 0 & 0 & 4 & 1 & 24 \end{array}\right] \end{array}$$

   c. $x = 0, y = 2, u = 6, v = 0, M = 24$

3. **a.** -2 is the only negative entry in the last row, and $\dfrac{5}{10} < \dfrac{12}{12}$. Pivot about the 10.

   b.
   $$\begin{array}{c} \\ u \\ y \\ M \end{array} \begin{array}{c} xyuvM \\ \left[\begin{array}{ccccc|c} -13 & 0 & 1 & -\frac{6}{5} & 0 & 6 \\ \frac{3}{2} & 1 & 0 & \frac{1}{10} & 0 & \frac{1}{2} \\ \hline 7 & 0 & 0 & \frac{1}{5} & 1 & 1 \end{array}\right] \end{array}$$

   c. $x = 0, y = \dfrac{1}{2}, u = 6, v = 0, M = 1$

In Exercises 5–25, the pivot elements are underlined.

5.
$$\begin{array}{c} \\ u \\ v \\ M \end{array} \begin{array}{c} xyuvM \\ \left[\begin{array}{ccccc|c} 1 & 1 & 1 & 0 & 0 & 7 \\ 1 & \underline{2} & 0 & 1 & 0 & 10 \\ \hline -1 & -3 & 0 & 0 & 1 & 0 \end{array}\right] \end{array}$$

$$\begin{array}{c} \\ u \\ y \\ M \end{array} \begin{array}{c} xyuvM \\ \left[\begin{array}{ccccc|c} \frac{1}{2} & 0 & 1 & -\frac{1}{2} & 0 & 2 \\ \frac{1}{2} & 1 & 0 & \frac{1}{2} & 0 & 5 \\ \hline \frac{1}{2} & 0 & 0 & \frac{3}{2} & 1 & 15 \end{array}\right] \end{array}$$

$x = 0, y = 5; M = 15$

7.
$$\begin{array}{c} \\ u \\ v \\ M \end{array} \begin{array}{c} xyuvM \\ \left[\begin{array}{ccccc|c} \underline{5} & 1 & 1 & 0 & 0 & 80 \\ 3 & 2 & 0 & 1 & 0 & 76 \\ \hline -4 & -2 & 0 & 0 & 1 & 0 \end{array}\right] \end{array}$$

$$\begin{array}{c} \\ x \\ v \\ M \end{array} \begin{array}{c} xyuvM \\ \left[\begin{array}{ccccc|c} 1 & \frac{1}{5} & \frac{1}{5} & 0 & 0 & 16 \\ 0 & \underline{\frac{7}{5}} & -\frac{3}{5} & 1 & 0 & 28 \\ \hline 0 & -\frac{6}{5} & \frac{4}{5} & 0 & 1 & 64 \end{array}\right] \end{array}$$

$$\begin{array}{c} \\ x \\ y \\ M \end{array} \begin{array}{c} xyuvM \\ \left[\begin{array}{ccccc|c} 1 & 0 & \frac{2}{7} & -\frac{1}{7} & 0 & 12 \\ 0 & 1 & -\frac{3}{7} & \frac{5}{7} & 0 & 20 \\ \hline 0 & 0 & \frac{2}{7} & \frac{6}{7} & 1 & 88 \end{array}\right] \end{array}$$

$x = 12, y = 20; M = 88$

9.
$$\begin{array}{c} \\ u \\ v \\ M \end{array} \begin{array}{c} xyzuvM \\ \left[\begin{array}{cccccc|c} 1 & 0 & \underline{2} & 1 & 0 & 0 & 10 \\ 0 & 3 & 1 & 0 & 1 & 0 & 24 \\ \hline -1 & -3 & -5 & 0 & 0 & 1 & 0 \end{array}\right] \end{array}$$

$$\begin{array}{c} \\ z \\ v \\ M \end{array} \begin{array}{c} xyzuvM \\ \left[\begin{array}{cccccc|c} \frac{1}{2} & 0 & 1 & \frac{1}{2} & 0 & 0 & 5 \\ -\frac{1}{2} & \underline{3} & 0 & -\frac{1}{2} & 1 & 0 & 19 \\ \hline \frac{3}{2} & -3 & 0 & \frac{5}{2} & 0 & 1 & 25 \end{array}\right] \end{array}$$

$$\begin{array}{c} \\ z \\ y \\ M \end{array} \begin{array}{c} xyzuvM \\ \left[\begin{array}{cccccc|c} \frac{1}{2} & 0 & 1 & \frac{1}{2} & 0 & 0 & 5 \\ -\frac{1}{6} & 1 & 0 & -\frac{1}{6} & \frac{1}{3} & 0 & \frac{19}{3} \\ \hline 1 & 0 & 0 & 2 & 1 & 1 & 44 \end{array}\right] \end{array}$$

$x = 0, y = \dfrac{19}{3}, z = 5; M = 44$

11.

$$\begin{array}{c|ccccc|c} & x & y & u & v & w & M & \\ \hline u & 5 & \underline{1} & 1 & 0 & 0 & 0 & 30 \\ v & 3 & 2 & 0 & 1 & 0 & 0 & 60 \\ w & 1 & 1 & 0 & 0 & 1 & 0 & 50 \\ \hline M & -2 & -3 & 0 & 0 & 0 & 1 & 0 \end{array}$$

$$\begin{array}{c|ccccc|c} & x & y & u & v & w & M & \\ \hline y & 5 & 1 & 1 & 0 & 0 & 0 & 30 \\ v & -7 & 0 & -2 & 1 & 0 & 0 & 0 \\ w & -4 & 0 & -1 & 0 & 1 & 0 & 20 \\ \hline M & 13 & 0 & 3 & 0 & 0 & 1 & 90 \end{array}$$

$x = 0, y = 30; M = 90$

Pivoting about the 2 instead gives the same solution although the tableau is different.

13.

$$\begin{array}{c|cccc|c} & x & y & u & v & M & \\ \hline u & 2 & \underline{3} & 1 & 0 & 0 & 400 \\ v & 1 & 1 & 0 & 1 & 0 & 150 \\ \hline M & -6 & -7 & 0 & 0 & 1 & 300 \end{array}$$

$$\begin{array}{c|cccc|c} & x & y & u & v & M & \\ \hline y & \frac{2}{3} & 1 & \frac{1}{3} & 0 & 0 & \frac{400}{3} \\ v & \frac{1}{3} & 0 & -\frac{1}{3} & 1 & 0 & \frac{50}{3} \\ \hline M & -\frac{4}{3} & 0 & \frac{7}{3} & 0 & 1 & \frac{3700}{3} \end{array}$$

$$\begin{array}{c|cccc|c} & x & y & u & v & M & \\ \hline y & 0 & 1 & 1 & -2 & 0 & 100 \\ x & 1 & 0 & -1 & 3 & 0 & 50 \\ \hline M & 0 & 0 & 1 & 4 & 1 & 1300 \end{array}$$

$x = 50, y = 100; M = 1300$

15. Let b be the number of large basketballs and f be the number of footballs manufactured. Maximize $2.50b + 2.00f$ subject to the constraints:

$$\begin{cases} 4b + 3f \leq 768 \\ 20b + 30f \leq 7200 \\ b \geq 0, f \geq 0 \end{cases}$$

(Note that 48 pounds was converted to ounces and 120 hours to minutes)

$$\begin{array}{c|cccc|c} & b & f & u & v & M & \\ \hline u & 4 & 3 & 1 & 0 & 0 & 768 \\ v & 20 & 30 & 0 & 1 & 0 & 7200 \\ \hline M & -2.5 & -2 & 0 & 0 & 1 & 0 \end{array}$$

$$\begin{array}{c|cccc|c} & b & f & u & v & M & \\ \hline b & 1 & \frac{3}{4} & \frac{1}{4} & 0 & 0 & 192 \\ v & 0 & \underline{15} & -5 & 1 & 0 & 3360 \\ \hline M & 0 & -\frac{1}{8} & \frac{5}{8} & 0 & 1 & 480 \end{array}$$

$$\begin{array}{c|cccc|c} & b & f & u & v & M & \\ \hline b & 1 & 0 & \frac{1}{2} & -\frac{1}{20} & 0 & 24 \\ f & 0 & 1 & -\frac{1}{3} & \frac{1}{15} & 0 & 224 \\ \hline M & 0 & 0 & \frac{7}{12} & \frac{1}{120} & 1 & 508 \end{array}$$

Therefore, 24 basketballs and 224 footballs should be manufactured.

17. Let c be the number of chairs, s be the number of sofas, and t be the number of tables manufactured each day. Maximize $80c + 70s + 120t$ subject to the constraints

$$\begin{cases} 6c + 3s + 8t \leq 768 \\ c + s + 2t \leq 144 \\ 2c + 5s \leq 216 \\ c \geq 0, s \geq 0, t \geq 0 \end{cases}$$

$$\begin{array}{c} \quad c \quad s \quad t \quad u \quad v \quad w \quad M \\ \begin{array}{c} u \\ v \\ w \\ M \end{array} \left[\begin{array}{ccccccc|c} 6 & 3 & 8 & 1 & 0 & 0 & 0 & 768 \\ 1 & 1 & 2 & 0 & 1 & 0 & 0 & 144 \\ 2 & 5 & 0 & 0 & 0 & 1 & 0 & 216 \\ -80 & -70 & -120 & 0 & 0 & 0 & 1 & 0 \end{array}\right] \end{array}$$

$$\begin{array}{c} \quad c \quad s \quad t \quad u \quad v \quad w \quad M \\ \begin{array}{c} u \\ t \\ w \\ M \end{array} \left[\begin{array}{ccccccc|c} 2 & -1 & 0 & 1 & -4 & 0 & 0 & 192 \\ \frac{1}{2} & \frac{1}{2} & 1 & 0 & \frac{1}{2} & 0 & 0 & 72 \\ 2 & 5 & 0 & 0 & 0 & 1 & 0 & 216 \\ -20 & -10 & 0 & 0 & 60 & 0 & 1 & 8640 \end{array}\right] \end{array}$$

$$\begin{array}{c} \quad c \quad s \quad t \quad u \quad v \quad w \quad M \\ \begin{array}{c} c \\ t \\ w \\ M \end{array} \left[\begin{array}{ccccccc|c} 1 & -\frac{1}{2} & 0 & \frac{1}{2} & -2 & 0 & 0 & 96 \\ 0 & \frac{3}{4} & 1 & -\frac{1}{4} & \frac{3}{2} & 0 & 0 & 24 \\ 0 & 6 & 0 & -1 & 4 & 1 & 0 & 24 \\ 0 & -20 & 0 & 10 & 20 & 0 & 1 & 10{,}560 \end{array}\right] \end{array}$$

$$\begin{array}{c} \quad c \quad s \quad t \quad u \quad v \quad w \quad M \\ \begin{array}{c} c \\ t \\ s \\ M \end{array} \left[\begin{array}{ccccccc|c} 1 & 0 & 0 & \frac{5}{12} & -\frac{5}{3} & \frac{1}{12} & 0 & 98 \\ 0 & 0 & 1 & -\frac{1}{8} & 1 & -\frac{1}{8} & 0 & 21 \\ 0 & 1 & 0 & -\frac{1}{6} & \frac{2}{3} & \frac{1}{6} & 0 & 4 \\ 0 & 0 & 0 & \frac{20}{3} & \frac{100}{3} & \frac{10}{3} & 1 & 10{,}640 \end{array}\right] \end{array}$$

98 chairs, 21 tables, 4 sofas.

19. Let b be the number of hours spent bicycling, j the number spent jogging, and s the number spent swimming each month. Maximize $200b + 475j + 275s$ subject to the constraints:

$$\begin{cases} b+j+s \leq 30 \\ s \leq 4 \\ -b+j-s \leq 0 \\ b \geq 0, j \geq 0, s \geq 0 \end{cases}$$

$$\begin{array}{c} \begin{array}{ccccccc} b & j & s & u & v & w & M \end{array} \\ \begin{array}{c} u \\ v \\ w \\ \hline M \end{array} \left[\begin{array}{ccccccc|c} 1 & 1 & 1 & 1 & 0 & 0 & 0 & 30 \\ 0 & 0 & 1 & 0 & 1 & 0 & 0 & 4 \\ -1 & \underline{1} & -1 & 0 & 0 & 1 & 0 & 0 \\ -200 & -475 & -275 & 0 & 0 & 0 & 1 & 0 \end{array} \right] \end{array}$$

$$\begin{array}{c} \begin{array}{ccccccc} b & j & s & u & v & w & M \end{array} \\ \begin{array}{c} u \\ v \\ j \\ \hline M \end{array} \left[\begin{array}{ccccccc|c} 2 & 0 & 2 & 1 & 0 & -1 & 0 & 30 \\ 0 & 0 & \underline{1} & 0 & 1 & 0 & 0 & 4 \\ -1 & 1 & -1 & 0 & 0 & 1 & 0 & 0 \\ -675 & 0 & -750 & 0 & 0 & 475 & 1 & 0 \end{array} \right] \end{array}$$

$$\begin{array}{c} \begin{array}{ccccccc} b & j & s & u & v & w & M \end{array} \\ \begin{array}{c} u \\ s \\ j \\ \hline M \end{array} \left[\begin{array}{ccccccc|c} \underline{2} & 0 & 0 & 1 & -2 & -1 & 0 & 22 \\ 0 & 0 & 1 & 0 & 1 & 0 & 0 & 4 \\ -1 & 1 & 0 & 0 & 1 & 1 & 0 & 4 \\ -675 & 0 & 0 & 0 & 750 & 475 & 1 & 3000 \end{array} \right] \end{array}$$

$$\begin{array}{c} \begin{array}{ccccccc} b & j & s & u & v & w & M \end{array} \\ \begin{array}{c} b \\ s \\ j \\ \hline M \end{array} \left[\begin{array}{ccccccc|c} 1 & 0 & 0 & \frac{1}{2} & -1 & -\frac{1}{2} & 0 & 11 \\ 0 & 0 & 1 & 0 & 1 & 0 & 0 & 4 \\ 0 & 1 & 0 & \frac{1}{2} & 0 & \frac{1}{2} & 0 & 15 \\ 0 & 0 & 0 & \frac{675}{2} & 75 & \frac{275}{2} & 1 & 10{,}425 \end{array} \right] \end{array}$$

11 hours bicycling, 4 hours swimming, 15 hours jogging. He will lose $\frac{10{,}425}{3500} \approx 3$ pounds.

21. Let a be the number of type A restaurants, b the number of type B restaurants, and c the number of type C restaurants. Maximize $40a + 30b + 25c$ subject to the constraints

$$\begin{cases} 600a + 400b + 300c \leq 48{,}000 \\ 15a + 9b + 5c \leq 1000 \\ a + b + c \leq 70 \\ a \geq 0,\ b \geq 0,\ c \geq 0 \end{cases}$$ (dollars in thousands)

$$\begin{array}{c} \begin{array}{ccccccc} a & b & c & u & v & w & M \end{array} \\ \begin{array}{c} u \\ v \\ w \\ M \end{array}\left[\begin{array}{ccccccc|c} 600 & 400 & 300 & 1 & 0 & 0 & 0 & 48{,}000 \\ 15 & 9 & 5 & 0 & 1 & 0 & 0 & 1000 \\ 1 & 1 & 1 & 0 & 0 & 1 & 0 & 70 \\ \hline -40 & -30 & -25 & 0 & 0 & 0 & 1 & 0 \end{array}\right] \end{array}$$

$$\begin{array}{c} \begin{array}{ccccccc} a & b & c & u & v & w & M \end{array} \\ \begin{array}{c} u \\ a \\ w \\ M \end{array}\left[\begin{array}{ccccccc|c} 0 & 40 & 100 & 1 & -40 & 0 & 0 & 8000 \\ 1 & \frac{3}{5} & \frac{1}{3} & 0 & \frac{1}{15} & 0 & 0 & \frac{200}{3} \\ 0 & \frac{2}{5} & \frac{2}{3} & 0 & -\frac{1}{15} & 1 & 0 & \frac{10}{3} \\ \hline 0 & -6 & -\frac{35}{3} & 0 & \frac{8}{3} & 0 & 1 & \frac{8000}{3} \end{array}\right] \end{array}$$

$$\begin{array}{c} \begin{array}{ccccccc} a & b & c & u & v & w & M \end{array} \\ \begin{array}{c} u \\ a \\ c \\ M \end{array}\left[\begin{array}{ccccccc|c} 0 & -20 & 0 & 1 & -30 & -150 & 0 & 7500 \\ 1 & \frac{2}{5} & 0 & 0 & \frac{1}{10} & -\frac{1}{2} & 0 & 65 \\ 0 & \frac{3}{5} & 1 & 0 & -\frac{1}{10} & \frac{3}{2} & 0 & 5 \\ \hline 0 & 1 & 0 & 0 & \frac{3}{2} & \frac{35}{2} & 1 & 2725 \end{array}\right] \end{array}$$

65 type A restaurants and 5 type C restaurants.

23. Let a be the number of bags of mix A, b the number of bags of mix B, and c the number of bags of mix C. Maximize $3a + 5b + 6c$ subject to the constraints

$$\begin{cases} 12a + 10b + 8c \leq 1200 \\ 5a + 6b + 8c \leq 800 \\ 3a + 4b + 4c \leq 600 \\ a \geq 0, b \geq 0, c \geq 0 \end{cases}$$

$$\begin{array}{c} \\ u \\ v \\ w \\ M \end{array} \begin{bmatrix} a & b & c & u & v & w & M & \\ 12 & 10 & 8 & 1 & 0 & 0 & 0 & 1200 \\ 5 & 6 & \underline{8} & 0 & 1 & 0 & 0 & 800 \\ 3 & 4 & 4 & 0 & 0 & 1 & 0 & 600 \\ -3 & -5 & -6 & 0 & 0 & 0 & 1 & 0 \end{bmatrix}$$

$$\begin{array}{c} \\ u \\ c \\ w \\ M \end{array} \begin{bmatrix} a & b & c & u & v & w & M & \\ 7 & 4 & 0 & 1 & -1 & 0 & 0 & 400 \\ \frac{5}{8} & \frac{3}{4} & 1 & 0 & \frac{1}{8} & 0 & 0 & 100 \\ \frac{1}{2} & 1 & 0 & 0 & -\frac{1}{2} & 1 & 0 & 200 \\ \frac{3}{4} & -\frac{1}{2} & 0 & 0 & \frac{3}{4} & 0 & 1 & 600 \end{bmatrix}$$

$$\begin{array}{c} \\ b \\ c \\ w \\ M \end{array} \begin{bmatrix} a & b & c & u & v & w & M & \\ \frac{7}{4} & 1 & 0 & \frac{1}{4} & -\frac{1}{4} & 0 & 0 & 100 \\ -\frac{11}{16} & 0 & 1 & -\frac{3}{16} & \frac{5}{16} & 0 & 0 & 25 \\ -\frac{5}{4} & 0 & 0 & -\frac{1}{4} & -\frac{1}{4} & 1 & 0 & 100 \\ \frac{13}{8} & 0 & 0 & \frac{1}{8} & \frac{5}{8} & 0 & 1 & 650 \end{bmatrix}$$

No bags of Mix A, 100 bags of Mix B, and 25 bags of Mix C.

25. $$\begin{array}{c} \\ u \\ v \\ w \\ M \end{array} \begin{bmatrix} x & y & z & u & v & w & M & \\ 1 & 1 & \underline{1} & 1 & 0 & 0 & 0 & 600 \\ 1 & 3 & 0 & 0 & 1 & 0 & 0 & 600 \\ 2 & 0 & 1 & 0 & 0 & 1 & 0 & 900 \\ -60 & -90 & -300 & 0 & 0 & 0 & 1 & 0 \end{bmatrix}$$

$$\begin{array}{c} \\ z \\ v \\ w \\ M \end{array} \begin{bmatrix} x & y & z & u & v & w & M & \\ 1 & 1 & 1 & 1 & 0 & 0 & 0 & 600 \\ 1 & 3 & 0 & 0 & 1 & 0 & 0 & 600 \\ 1 & -1 & 0 & -1 & 0 & 1 & 0 & 300 \\ 240 & 210 & 0 & 300 & 0 & 0 & 1 & 180{,}000 \end{bmatrix}$$

$x = 0, y = 0, z = 600; M = 180{,}000$

27.

$$\begin{array}{c} & x & y & u & v & M & \\ u & [5 & 1 & 1 & 0 & 0 & | & 8 \\ v & | 1 & 2 & 0 & 1 & 0 & | & 10 \\ M & [-2 & -4 & 0 & 0 & 1 & | & 0 \end{array}$$

$$\begin{array}{c} & x & y & u & v & M & \\ u & [\frac{9}{2} & 0 & 1 & -\frac{1}{2} & 0 & | & 3 \\ y & | \frac{1}{2} & 1 & 0 & \frac{1}{2} & 0 & | & 5 \\ M & [0 & 0 & 0 & 2 & 1 & | & 20 \end{array}$$

$x = 0, y = 5; M = 20$

29.

$$\begin{array}{c} & x & y & z & u & v & w & M & \\ u & [1 & 1 & 2 & 1 & 0 & 0 & 0 & | & 6 \\ v & | 1 & 5 & 2 & 0 & 1 & 0 & 0 & | & 20 \\ w & | 2 & 1 & 1 & 0 & 0 & 1 & 0 & | & 4 \\ M & [-3 & -2 & -2 & 0 & 0 & 0 & 1 & | & 0 \end{array}$$

$$\begin{array}{c} & x & y & z & u & v & w & M & \\ u & [0 & \frac{1}{2} & \frac{3}{2} & 1 & 0 & -\frac{1}{2} & 0 & | & 4 \\ v & | 0 & \frac{9}{2} & \frac{3}{2} & 0 & 1 & -\frac{1}{2} & 0 & | & 18 \\ x & | 1 & \frac{1}{2} & \frac{1}{2} & 0 & 0 & \frac{1}{2} & 0 & | & 2 \\ M & [0 & -\frac{1}{2} & -\frac{1}{2} & 0 & 0 & \frac{3}{2} & 1 & | & 6 \end{array}$$

$$\begin{array}{c} & x & y & z & u & v & w & M & \\ u & [-1 & 0 & 1 & 1 & 0 & -1 & 0 & | & 2 \\ v & | -9 & 0 & -3 & 0 & 1 & -5 & 0 & | & 0 \\ y & | 2 & 1 & 1 & 0 & 0 & 1 & 0 & | & 4 \\ M & [1 & 0 & 0 & 0 & 0 & 2 & 1 & | & 8 \end{array}$$

$x = 0, y = 4, z = 0; M = 8$

Pivoting about the $\frac{9}{2}$ in step 2 leads to the same solution. Pivoting about the $\frac{3}{2}$, first row, third column in step 2 leads to a different solution: $x = 0, y = 2, z = 2$. M is still 8.

Exercises 4.3

1.

$$\begin{array}{c} & x & y & u & v & M & \\ u & [1 & 1 & 1 & 0 & 0 & | & 5 \\ v & | 2 & -3 & 0 & 1 & 0 & | & -12 \\ M & [-40 & -30 & 0 & 0 & 1 & | & 0 \end{array}$$

$$\begin{array}{c} & x & y & u & v & M & \\ u & [\frac{5}{3} & 0 & 1 & \frac{1}{3} & 0 & | & 1 \\ y & | -\frac{2}{3} & 1 & 0 & -\frac{1}{3} & 0 & | & 4 \\ M & [-60 & 0 & 0 & -10 & 1 & | & 120 \end{array}$$

$$\begin{array}{c} & x & y & u & v & M & \\ x & [1 & 0 & \frac{3}{5} & \frac{1}{5} & 0 & | & \frac{3}{5} \\ y & | 0 & 1 & \frac{2}{5} & -\frac{1}{5} & 0 & | & \frac{22}{5} \\ M & [0 & 0 & 36 & 2 & 1 & | & 156 \end{array}$$

$x = \frac{3}{5}, y = \frac{22}{5}; M = 156$

3.

$$\begin{array}{c} & x & y & u & v & M & \\ u & [-1 & -1 & 1 & 0 & 0 & | & -3 \\ v & | -2 & 0 & 0 & 1 & 0 & | & -5 \\ M & [3 & 1 & 0 & 0 & 1 & | & 0 \end{array}$$

$$\begin{array}{c} & x & y & u & v & M & \\ y & [1 & 1 & -1 & 0 & 0 & | & 3 \\ v & | -2 & 0 & 0 & 1 & 0 & | & -5 \\ M & [2 & 0 & 1 & 0 & 1 & | & -3 \end{array}$$

$$\begin{array}{c} & x & y & u & v & M & \\ y & [0 & 1 & -1 & \frac{1}{2} & 0 & | & \frac{1}{2} \\ x & | 1 & 0 & 0 & -\frac{1}{2} & 0 & | & \frac{5}{2} \\ M & [0 & 0 & 1 & 1 & 1 & | & -8 \end{array}$$

$x = \frac{5}{2}, y = \frac{1}{2}$ Since $M = -8$, the minimum is 8. Pivoting about the -2 at the start leads to the same final matrix.

5.

$$\begin{array}{c} \\ u \\ v \\ w \\ t \\ M \end{array} \begin{array}{c} x \quad y \quad u \quad v \quad w \quad t \quad M \\ \left[\begin{array}{ccccccc|c} -2 & -1 & 1 & 0 & 0 & 0 & 0 & -11 \\ 1 & 1 & 0 & 1 & 0 & 0 & 0 & 10 \\ \frac{1}{3} & 1 & 0 & 0 & 1 & 0 & 0 & 6 \\ -\frac{1}{4} & \underline{-1} & 0 & 0 & 0 & 1 & 0 & -4 \\ \hline 13 & 4 & 0 & 0 & 0 & 0 & 1 & 0 \end{array}\right] \end{array}$$

$$\begin{array}{c} \\ u \\ v \\ w \\ y \\ M \end{array} \begin{array}{c} x \quad y \quad u \quad v \quad w \quad t \quad M \\ \left[\begin{array}{ccccccc|c} -\frac{7}{4} & 0 & 1 & 0 & 0 & -1 & 0 & -7 \\ \frac{3}{4} & 0 & 0 & 1 & 0 & 1 & 0 & 6 \\ \frac{1}{12} & 0 & 0 & 0 & 1 & \underline{1} & 0 & 2 \\ \frac{1}{4} & 1 & 0 & 0 & 0 & -1 & 0 & 4 \\ \hline 12 & 0 & 0 & 0 & 0 & 4 & 1 & -16 \end{array}\right] \end{array}$$

$$\begin{array}{c} \\ u \\ v \\ t \\ y \\ M \end{array} \begin{array}{c} x \quad y \quad u \quad v \quad w \quad t \quad M \\ \left[\begin{array}{ccccccc|c} -\frac{5}{3} & 0 & 1 & 0 & 1 & 0 & 0 & -5 \\ \frac{2}{3} & 0 & 0 & 1 & -1 & 0 & 0 & 4 \\ \frac{1}{12} & 0 & 0 & 0 & 1 & 1 & 0 & 2 \\ \frac{1}{3} & 1 & 0 & 0 & 1 & 0 & 0 & 6 \\ \hline \frac{35}{3} & 0 & 0 & 0 & -4 & 0 & 1 & -24 \end{array}\right] \end{array}$$

$$\begin{array}{c} \\ x \\ v \\ t \\ y \\ M \end{array} \begin{array}{c} x \quad y \quad u \quad v \quad w \quad t \quad M \\ \left[\begin{array}{ccccccc|c} 1 & 0 & -\frac{3}{5} & 0 & -\frac{3}{5} & 0 & 0 & 3 \\ 0 & 0 & \frac{2}{5} & 1 & -\frac{3}{5} & 0 & 0 & 2 \\ 0 & 0 & \frac{1}{20} & 0 & \frac{21}{20} & 1 & 0 & \frac{7}{4} \\ 0 & 1 & \frac{1}{5} & 0 & \frac{6}{5} & 0 & 0 & 5 \\ \hline 0 & 0 & 7 & 0 & 3 & 0 & 1 & -59 \end{array}\right] \end{array}$$

$x = 3, y = 5$; since $M = -59$, the minimum is 59. Other choices of pivot entries lead to the same final matrix.

7.

$$\begin{array}{c} \\ u \\ v \\ w \\ t \\ M \end{array} \begin{array}{c} x \quad y \quad u \quad v \quad w \quad t \quad M \\ \left[\begin{array}{ccccccc|c} -2 & -5 & 1 & 0 & 0 & 0 & 0 & -30 \\ 3 & \underline{-5} & 0 & 1 & 0 & 0 & 0 & -5 \\ 8 & 3 & 0 & 0 & 1 & 0 & 0 & 101 \\ -9 & 7 & 0 & 0 & 0 & 1 & 0 & 42 \\ \hline 2 & 7 & 0 & 0 & 0 & 0 & 1 & 0 \end{array}\right] \end{array}$$

$$\begin{array}{c} \\ u \\ y \\ w \\ t \\ M \end{array} \begin{array}{c} x \quad y \quad u \quad v \quad w \quad t \quad M \\ \left[\begin{array}{ccccccc|c} \underline{-5} & 0 & 1 & -1 & 0 & 0 & 0 & -25 \\ -\frac{3}{5} & 1 & 0 & -\frac{1}{5} & 0 & 0 & 0 & 1 \\ \frac{49}{5} & 0 & 0 & \frac{3}{5} & 1 & 0 & 0 & 98 \\ -\frac{24}{5} & 0 & 0 & \frac{7}{5} & 0 & 1 & 0 & 35 \\ \hline \frac{31}{5} & 0 & 0 & \frac{7}{5} & 0 & 0 & 1 & -7 \end{array}\right] \end{array}$$

$$\begin{array}{c} \\ x \\ y \\ w \\ t \\ M \end{array} \begin{array}{c} x \quad y \quad u \quad v \quad w \quad t \quad M \\ \left[\begin{array}{ccccccc|c} 1 & 0 & -\frac{1}{5} & \frac{1}{5} & 0 & 0 & 0 & 5 \\ 0 & 1 & -\frac{3}{25} & -\frac{2}{25} & 0 & 0 & 0 & 4 \\ 0 & 0 & \frac{49}{25} & -\frac{34}{25} & 1 & 0 & 0 & 49 \\ 0 & 0 & -\frac{24}{25} & \frac{59}{25} & 0 & 1 & 0 & 59 \\ \hline 0 & 0 & \frac{31}{25} & \frac{4}{25} & 0 & 0 & 1 & -38 \end{array}\right] \end{array}$$

$x = 5, y = 4$; since $M = -38$, the minimum is 38.

9. Minimize $3a + 1.5b$ subject to the constraints:
$$\begin{cases} 30a + 10b \geq 60 \\ 10a + 10b \geq 40 \\ 20a + 60b \geq 120 \\ a \geq 0, b \geq 0 \end{cases}$$

$$\begin{array}{c} \\ u \\ v \\ w \\ M \end{array} \begin{array}{c} a \quad\quad b \quad\quad u \quad v \quad w \quad M \\ \left[\begin{array}{cccccc|c} -30 & -10 & 1 & 0 & 0 & 0 & -60 \\ -10 & -10 & 0 & 1 & 0 & 0 & -40 \\ -20 & -60 & 0 & 0 & 1 & 0 & -120 \\ \hline 3 & \frac{3}{2} & 0 & 0 & 0 & 1 & 0 \end{array}\right] \end{array}$$

$$\begin{array}{c} \\ a \\ v \\ w \\ M \end{array} \begin{bmatrix} a & b & u & v & w & M & \\ 1 & \frac{1}{3} & -\frac{1}{30} & 0 & 0 & 0 & 2 \\ 0 & -\frac{20}{3} & -\frac{1}{3} & 1 & 0 & 0 & -20 \\ 0 & -\frac{160}{3} & -\frac{2}{3} & 0 & 1 & 0 & -80 \\ \hline 0 & \frac{1}{2} & \frac{1}{10} & 0 & 0 & 1 & -6 \end{bmatrix}$$

$$\begin{array}{c} \\ a \\ v \\ b \\ M \end{array} \begin{bmatrix} a & b & u & v & w & M & \\ 1 & 0 & -\frac{3}{80} & 0 & \frac{1}{160} & 0 & \frac{3}{2} \\ 0 & 0 & -\frac{1}{4} & 1 & -\frac{1}{8} & 0 & -10 \\ 0 & 1 & \frac{1}{80} & 0 & -\frac{3}{160} & 0 & \frac{3}{2} \\ \hline 0 & 0 & \frac{3}{32} & 0 & \frac{3}{320} & 1 & -\frac{27}{4} \end{bmatrix}$$

$$\begin{array}{c} \\ a \\ w \\ b \\ M \end{array} \begin{bmatrix} a & b & u & v & w & M & \\ 1 & 0 & -\frac{1}{20} & \frac{1}{20} & 0 & 0 & 1 \\ 0 & 0 & 2 & -8 & 1 & 0 & 80 \\ 0 & 1 & \frac{1}{20} & -\frac{3}{20} & 0 & 0 & 3 \\ \hline 0 & 0 & \frac{3}{40} & \frac{3}{40} & 0 & 1 & -\frac{15}{2} \end{bmatrix}$$

1 serving of food A, 3 servings of food B

11. Maximize $30a + 50b + 60c$ subject to the constraints:
$$\begin{cases} a+b+c \leq 600 \\ a \geq 100 \\ b \geq 50 \\ b+c \geq 200 \\ a \geq 0, b \geq 0, c \geq 0 \end{cases}$$

$$\begin{array}{c} \\ u \\ v \\ w \\ t \\ M \end{array} \begin{bmatrix} a & b & c & u & v & w & t & M & \\ 1 & 1 & 1 & 1 & 0 & 0 & 0 & 0 & 600 \\ -1 & 0 & 0 & 0 & 1 & 0 & 0 & 0 & -100 \\ 0 & -1 & 0 & 0 & 0 & 1 & 0 & 0 & -50 \\ 0 & -1 & -1 & 0 & 0 & 0 & 1 & 0 & -200 \\ \hline -30 & -50 & -60 & 0 & 0 & 0 & 0 & 1 & 0 \end{bmatrix}$$

$$\begin{array}{c} \\ u \\ a \\ w \\ t \\ M \end{array} \begin{array}{cccccccc} a & b & c & u & v & w & t & M \\ \left[\begin{array}{cccccccc|c} 0 & 1 & 1 & 1 & 1 & 0 & 0 & 0 & 500 \\ 1 & 0 & 0 & 0 & -1 & 0 & 0 & 0 & 100 \\ 0 & \underline{-1} & 0 & 0 & 0 & 1 & 0 & 0 & -50 \\ 0 & -1 & -1 & 0 & 0 & 0 & 1 & 0 & -200 \\ \hline 0 & -50 & -60 & 0 & -30 & 0 & 0 & 1 & 3000 \end{array}\right] \end{array}$$

$$\begin{array}{c} \\ u \\ a \\ b \\ t \\ M \end{array} \begin{array}{cccccccc} a & b & c & u & v & w & t & M \\ \left[\begin{array}{cccccccc|c} 0 & 0 & 1 & 1 & 1 & 1 & 0 & 0 & 450 \\ 1 & 0 & 0 & 0 & -1 & 0 & 0 & 0 & 100 \\ 0 & 1 & 0 & 0 & 0 & -1 & 0 & 0 & 50 \\ 0 & 0 & \underline{-1} & 0 & 0 & -1 & 1 & 0 & -150 \\ \hline 0 & 0 & -60 & 0 & -30 & -50 & 0 & 1 & 5500 \end{array}\right] \end{array}$$

$$\begin{array}{c} \\ u \\ a \\ b \\ c \\ M \end{array} \begin{array}{cccccccc} a & b & c & u & v & w & t & M \\ \left[\begin{array}{cccccccc|c} 0 & 0 & 0 & 1 & 1 & 0 & 1 & 0 & 300 \\ 1 & 0 & 0 & 0 & -1 & 0 & 0 & 0 & 100 \\ 0 & 1 & 0 & 0 & 0 & -1 & 0 & 0 & 50 \\ 0 & 0 & 1 & 0 & 0 & 1 & -1 & 0 & 150 \\ \hline 0 & 0 & 0 & 0 & -30 & 10 & -60 & 1 & 14{,}500 \end{array}\right] \end{array}$$

$$\begin{array}{c} \\ t \\ a \\ b \\ c \\ M \end{array} \begin{array}{cccccccc} a & b & c & u & v & w & t & M \\ \left[\begin{array}{cccccccc|c} 0 & 0 & 0 & 1 & 1 & 0 & 1 & 0 & 300 \\ 1 & 0 & 0 & 0 & -1 & 0 & 0 & 0 & 100 \\ 0 & 1 & 0 & 0 & 0 & -1 & 0 & 0 & 50 \\ 0 & 0 & 1 & 1 & 1 & 1 & 0 & 0 & 450 \\ \hline 0 & 0 & 0 & 60 & 30 & 10 & 0 & 1 & 32{,}500 \end{array}\right] \end{array}$$

Stock 100 of brand A, 50 of brand B, and 450 of brand C.

13. Let x = number of computers shipped from Chicago to Detroit. Let y = number of computers shipped from Chicago to Fletcher. Then (40 – x) equals the number of computers shipped from Boston to Detroit and (30 – y) represents the number of computers shipped from Boston to Fletcher.
    Minimize:
    $125(40-x)+100x+180(30-y)+160y$
    or $10400-25x-20y$

    Subject to:
    $$\begin{cases} x+y \le 80 \\ x+y \ge 20 \\ x \le 40 \\ y \le 30 \\ x \ge 0, y \ge 0 \end{cases}$$

$$\begin{array}{c} \\ u \\ v \\ w \\ t \\ M \end{array} \begin{array}{ccccccc} x & y & u & v & w & t & M \\ \left[\begin{array}{ccccccc|c} 1 & 1 & 1 & 0 & 0 & 0 & 0 & 80 \\ \underline{-1} & -1 & 0 & 1 & 0 & 0 & 0 & -20 \\ 1 & 0 & 0 & 0 & 1 & 0 & 0 & 40 \\ 0 & 1 & 0 & 0 & 0 & 1 & 0 & 30 \\ \hline -25 & -20 & 0 & 0 & 0 & 0 & 1 & -10400 \end{array}\right] \end{array}$$

	x	y	u	v	w	t	M	
u	0	0	1	1	0	0	0	60
x	1	1	0	−1	0	0	0	20
w	0	−1	0	1	1	0	0	20
t	0	1	0	0	0	1	0	30
M	0	5	0	−25	0	0	1	−9900

	x	y	u	v	w	t	M	
u	0	1	1	0	−1	0	0	40
x	1	0	0	0	1	0	0	40
v	0	−1	0	1	1	0	0	20
t	0	1	0	0	0	1	0	30
M	0	−20	0	0	25	0	1	−9400

	x	y	u	v	w	t	M	
u	0	0	1	0	−1	−1	0	10
x	1	0	0	0	1	0	0	40
v	0	0	0	1	1	1	0	50
y	0	1	0	0	0	1	0	30
M	0	0	0	0	25	20	1	−8800

Therefore, the minimum cost is $8800 when you ship all the computers from Chicago.

15.

	x	y	u	v	M	
u	4	1	1	0	0	5
v	−1	−3	0	1	0	−4
M	−1	2	0	0	1	0

	x	y	u	v	M	
x	1	$\frac{1}{4}$	$\frac{1}{4}$	0	0	$\frac{5}{4}$
v	0	$-\frac{11}{4}$	$\frac{1}{4}$	1	0	$-\frac{11}{4}$
M	0	$\frac{9}{4}$	$\frac{1}{4}$	0	1	$\frac{5}{4}$

	x	y	u	v	M	
x	1	0	$\frac{3}{11}$	$\frac{1}{11}$	0	1
y	0	1	$-\frac{1}{11}$	$-\frac{4}{11}$	0	1
M	0	0	$\frac{5}{11}$	$\frac{9}{11}$	1	−1

$x = 1, y = 1; M = −1$

Exercises 4.4

1. x (paring knives) goes from 14 to $14 + 54\left(-\frac{5}{27}\right) = 4$, y (pocket knives) goes from 8 to $8 + 54\left(\frac{7}{27}\right) = 22$, and the profit goes from $82 to $82 + 54\left(\frac{20}{27}\right) = \$122$.

3. The slack variable involved is w. x (sets to Rockville) goes from 20 to $20 + (50 - 45)(1) = 25$, y (sets to Annapolis) goes from 25 to $25 + (50 - 45)(0) = 25$, and the cost goes from $260 to $-[-260 + (50 - 45)(2)] = \$250$.

5. $100 + h\left(\frac{1}{4}\right) \geq 0$, $25 + h\left(-\frac{3}{16}\right) \geq 0$, and $100 + h\left(-\frac{1}{4}\right) \geq 0$, so

$h \geq -400$, $h \leq \frac{400}{3}$ and $h \leq 400$ or

$-400 \leq h \leq \frac{400}{3}$.

7. $\begin{bmatrix} 9 & 1 & 1 \\ 4 & 8 & -3 \end{bmatrix}$

9. $\begin{bmatrix} 7 \\ 6 \\ 5 \\ 1 \end{bmatrix}$

11. Yes

13. Minimize $[7\ 5\ 4]\begin{bmatrix} x \\ y \\ z \end{bmatrix}$ subject to the

constraints $\begin{bmatrix} 3 & 8 & 9 \\ 1 & 2 & 5 \\ 4 & 1 & 7 \end{bmatrix}\begin{bmatrix} x \\ y \\ z \end{bmatrix} \geq \begin{bmatrix} 75 \\ 80 \\ 67 \end{bmatrix}$ and

$\begin{bmatrix} x \\ y \\ z \end{bmatrix} \geq \begin{bmatrix} 0 \\ 0 \\ 0 \end{bmatrix}$.

15. Maximize $[3 \ 5]\begin{bmatrix} x \\ y \end{bmatrix}$ subject to the

constraints $\begin{bmatrix} 3 & 6 \\ 7 & 5 \\ 4 & 3 \end{bmatrix}\begin{bmatrix} x \\ y \end{bmatrix} \leq \begin{bmatrix} 90 \\ 138 \\ 120 \end{bmatrix}$ and

$\begin{bmatrix} x \\ y \end{bmatrix} \geq \begin{bmatrix} 0 \\ 0 \end{bmatrix}$.

17. Minimize $2x + 3y$ subject to the

constraints $\begin{cases} 7x + 4y \geq 33 \\ 5x + 8y \geq 44 \\ x + 3y \geq 55 \\ x \geq 0, \ y \geq 0 \end{cases}$.

19. Constraints

Cell	Name	Final Value	Shadow Price	Constraint R.H. Side	Allowable Increase	Allowable Decrease
B1		45	−2	45	10	20
B2		45	0	15	30	1E + 30
B3		20	0	30	1E + 30	10
B4		25	−1	25	20	10
B5		20	0	0	20	1E + 30
B6		25	0	0	25	1E + 30

The first line of the table corresponds to the constraint for the number of sets shipped from College Park. The shadow price is –2, and the range feasibility is [45 – 20, 45 + 10] or [25, 55].

Exercises 4.5

1. Primal: Maximize $[4 \ 2]\begin{bmatrix} x \\ y \end{bmatrix}$ subject to

$\begin{bmatrix} 5 & 1 \\ 3 & 2 \end{bmatrix}\begin{bmatrix} x \\ y \end{bmatrix} \leq \begin{bmatrix} 80 \\ 76 \end{bmatrix}$.

Dual: Minimize $[80 \ 76]\begin{bmatrix} u \\ v \end{bmatrix}$ subject to

$\begin{bmatrix} 5 & 3 \\ 1 & 2 \end{bmatrix}\begin{bmatrix} u \\ v \end{bmatrix} \geq \begin{bmatrix} 4 \\ 2 \end{bmatrix}$ and $\begin{bmatrix} u \\ v \end{bmatrix} \geq \begin{bmatrix} 0 \\ 0 \end{bmatrix}$.

Minimize $80u + 76v$ subject to the

constraints $\begin{cases} 5u + 3v \geq 4 \\ u + 2v \geq 2 \\ u \geq 0, \ v \geq 0 \end{cases}$.

3. Primal: Minimize $[10 \ 12]\begin{bmatrix} x \\ y \end{bmatrix}$ subject to

$\begin{bmatrix} 1 & 2 \\ -1 & 1 \\ 2 & 3 \end{bmatrix}\begin{bmatrix} x \\ y \end{bmatrix} \geq \begin{bmatrix} 1 \\ 2 \\ 1 \end{bmatrix}$ and $\begin{bmatrix} x \\ y \end{bmatrix} \geq \begin{bmatrix} 0 \\ 0 \end{bmatrix}$.

Dual: Maximize $[1 \ 2 \ 1]\begin{bmatrix} u \\ v \\ w \end{bmatrix}$ subject to

$\begin{bmatrix} 1 & -1 & 2 \\ 2 & 1 & 3 \end{bmatrix}\begin{bmatrix} u \\ v \\ w \end{bmatrix} \leq \begin{bmatrix} 10 \\ 12 \end{bmatrix}$ and $\begin{bmatrix} u \\ v \\ w \end{bmatrix} \geq \begin{bmatrix} 0 \\ 0 \\ 0 \end{bmatrix}$.

Maximize $u + 2v + w$ subject to the

constraints $\begin{cases} u - v + 2w \leq 10 \\ 2u + v + 3w \leq 12 \\ u \geq 0, \ v \geq 0, \ w \geq 0 \end{cases}$.

5. Primal: Minimize $\begin{bmatrix} 3 & 5 & 1 \end{bmatrix} \begin{bmatrix} x \\ y \\ z \end{bmatrix}$ subject

   to $\begin{bmatrix} -2 & 4 & 6 \\ 8 & 1 & 9 \end{bmatrix} \begin{bmatrix} x \\ y \\ z \end{bmatrix} \geq \begin{bmatrix} -7 \\ 10 \end{bmatrix}$ and $\begin{bmatrix} x \\ y \\ z \end{bmatrix} \geq \begin{bmatrix} 0 \\ 0 \\ 0 \end{bmatrix}$.

   Dual: Maximize $\begin{bmatrix} -7 & 10 \end{bmatrix} \begin{bmatrix} u \\ v \end{bmatrix}$ subject to

   $\begin{bmatrix} -2 & 8 \\ 4 & 1 \\ 6 & 9 \end{bmatrix} \begin{bmatrix} u \\ v \end{bmatrix} \leq \begin{bmatrix} 3 \\ 5 \\ 1 \end{bmatrix}$ and $\begin{bmatrix} u \\ v \end{bmatrix} \geq \begin{bmatrix} 0 \\ 0 \end{bmatrix}$.

   Maximize $-7u + 10v$ subject to the

   constraints $\begin{cases} -2u + 8v \leq 3 \\ 4u + v \leq 5 \\ 6u + 9v \leq 1 \\ u \geq 0, v \geq 0 \end{cases}$

7. $x = 12, y = 20, M = 88, u = \frac{2}{7}, v = \frac{6}{7}$.

9. $x = 0, y = 2, M = 24; u = 0, v = 12, w = 0$, $M = 24$

11. Maximize $3u + 5v$ subject to the

    constraints $\begin{cases} u + 2v \leq 3 \\ u \leq 1 \\ u \geq 0, v \geq 0 \end{cases}$.

    Solve the dual.

    $\begin{array}{c} \\ x \\ y \\ M \end{array} \begin{array}{c} u \quad v \quad x \quad y \quad M \\ \begin{bmatrix} 1 & 2 & 1 & 0 & 0 & 3 \\ 1 & 0 & 0 & 1 & 0 & 1 \\ -3 & -5 & 0 & 0 & 1 & 0 \end{bmatrix} \end{array}$

    $\begin{array}{c} \\ v \\ y \\ M \end{array} \begin{array}{c} u \quad v \quad x \quad y \quad M \\ \begin{bmatrix} \frac{1}{2} & 1 & \frac{1}{2} & 0 & 0 & \frac{3}{2} \\ 1 & 0 & 0 & 1 & 0 & 1 \\ -\frac{1}{2} & 0 & \frac{5}{2} & 0 & 1 & \frac{15}{2} \end{bmatrix} \end{array}$

    $\begin{array}{c} \\ v \\ u \\ M \end{array} \begin{array}{c} u \quad v \quad x \quad y \quad M \\ \begin{bmatrix} 0 & 1 & \frac{1}{2} & -\frac{1}{2} & 0 & 1 \\ 1 & 0 & 0 & 1 & 0 & 1 \\ 0 & 0 & \frac{5}{2} & \frac{1}{2} & 1 & 8 \end{bmatrix} \end{array}$

    $x = \frac{5}{2}, y = \frac{1}{2}$, minimum = 8; $u = 1, v = 1$, maximum = 8

13. Minimize $6u + 9v + 12w$ subject to the

    constraints $\begin{cases} u + 3v \geq 10 \\ -2u + w \geq 12 \\ v + 3w \geq 10 \\ u \geq 0, v \geq 0, w \geq 0 \end{cases}$.

    Solve the primal.

    $\begin{array}{c} \\ u \\ v \\ w \\ M \end{array} \begin{array}{c} x \quad y \quad z \quad u \quad v \quad w \quad M \\ \begin{bmatrix} 1 & -2 & 0 & 1 & 0 & 0 & 0 & 6 \\ 3 & 0 & 1 & 0 & 1 & 0 & 0 & 9 \\ 0 & 1 & 3 & 0 & 0 & 1 & 0 & 12 \\ -10 & -12 & -10 & 0 & 0 & 0 & 1 & 0 \end{bmatrix} \end{array}$

    $\begin{array}{c} \\ u \\ v \\ y \\ M \end{array} \begin{array}{c} x \quad y \quad z \quad u \quad v \quad w \quad M \\ \begin{bmatrix} 1 & 0 & 6 & 1 & 0 & 2 & 0 & 30 \\ 3 & 0 & 1 & 0 & 1 & 0 & 0 & 9 \\ 0 & 1 & 3 & 0 & 0 & 1 & 0 & 12 \\ -10 & 0 & 26 & 0 & 0 & 12 & 1 & 144 \end{bmatrix} \end{array}$

    $\begin{array}{c} \\ u \\ x \\ y \\ M \end{array} \begin{array}{c} x \quad y \quad z \quad u \quad v \quad w \quad M \\ \begin{bmatrix} 0 & 0 & \frac{17}{3} & 1 & -\frac{1}{3} & 2 & 0 & 27 \\ 1 & 0 & \frac{1}{3} & 0 & \frac{1}{3} & 0 & 0 & 3 \\ 0 & 1 & 3 & 0 & 0 & 1 & 0 & 12 \\ 0 & 0 & \frac{88}{3} & 0 & \frac{10}{3} & 12 & 1 & 174 \end{bmatrix} \end{array}$

    $x = 3, y = 12, z = 0$, maximum = 174; $u = 0, v = \frac{10}{3}, w = 12$, minimum = 174

15. Suppose we can hire workers out at a profit of u dollars per hour, sell the steel at a profit of v dollars per unit, and sell the wood at a profit of w dollars per unit. To find the minimum profit at which that should be done, minimize $90u + 138v + 120w$ subject to the constraints

    $\begin{cases} 3u + 7v + 4w \geq 3 \\ 6u + 5v + 3w \geq 5 \\ u \geq 0, v \geq 0, w \geq 0 \end{cases}$.

17. Suppose we can buy anthracite at u dollars per ton, ordinary coal at v dollars per ton, and bituminous coal at w dollars per ton. To find the maximum cost at which this should be done, maximize $80u + 60v + 75w$ subject to the constraints
$$\begin{cases} 4u + 4v + 7w \leq 150 \\ 10u + 5v + 5w \leq 200 \\ u \geq 0, v \geq 0, w \geq 0 \end{cases}$$

19. The new primal problem is to maximize $3x + 5y + pz$, where p is the profit per table knife and z is the number of table knives produced, subject to the constraints
$$\begin{cases} 3x + 6y + 4z \leq 90 \\ 7x + 5y + 6z \leq 138 \\ 4x + 3y + 2z \leq 120 \\ x \geq 0, y \geq 0, z \geq 0 \end{cases}$$
The dual is to minimize $90u + 138v + 120w$ subject to the constraints
$$\begin{cases} 3u + 7v + 4w \geq 3 \\ 6u + 5v + 3w \geq 5 \\ 4u + 6v + 2w \geq p \\ u \geq 0, v \geq 0, w \geq 0 \end{cases}$$
The original solution, with $u = \dfrac{20}{27}, v = \dfrac{1}{9}, w = 0$, will still be optimal if
$$4\left(\dfrac{20}{27}\right) + 6\left(\dfrac{1}{9}\right) + 2(0) = 3.63 \geq p.$$

The minimum profit per table knife that needs to be realized to warrant adding table knives to the product line is $3.63.

21. Dual: Maximize $5u + 8v$ subject to the constraints
$$\begin{cases} u + 2v \leq 16 \\ 3u + 4v \leq 42 \\ u \geq 0, v \geq 0 \end{cases}$$

	u	v	x	y	M	
x	1	2	1	0	0	16
y	3	4	0	1	0	42
M	−5	−8	0	0	1	0

	u	v	x	y	M	
v	$\frac{1}{2}$	1	$\frac{1}{2}$	0	0	8
y	1	0	−2	1	0	10
M	−1	0	4	0	1	64

	u	v	x	y	M	
v	0	1	$\frac{3}{2}$	$-\frac{1}{2}$	0	3
u	1	0	−2	1	0	10
M	0	0	2	1	1	74

$x = 2, y = 1, M = 74$

Chapter 4 Review of Fundamental Concepts

1. There are 3 things to check. First, the problem must be a maximization problem. Second, each variable must be greater than or equal to zero. Third, all other (nontrivial) constraints can be written in the form: [linear expression] ≤ [nonnegative number].

2. A slack variable is a variable that, when added to the left-hand side of the constraint makes the inequality into an equation. It is a nonnegative quantity and thus "picks up the slack". A group I variable is a variable that is set equal to zero in the particular solution. A group II variable is a variable whose particular value is read directly from the right-hand side of the equation.

3. First convert the linear programming problem into a system of equations by introducing slack variables to the nontrivial constraints and the objective function. Then form the augmented matrix corresponding to this system of equations including labeling each column with its corresponding variable.

4. Refer to the shaded box titled "The Simplex Method for Problems in Standard Form" on page 155.

5. First multiply the objective function by negative one and change "minimize" to "maximize" so that now you are maximizing the opposite of the original objective function. Apply the appropriate simplex method to obtain the solution. Multiply the value of M by negative one to find the minimum value of the objective function.

6. Refer to the shaded box titled "The Simplex Method for Problems in Nonstandard Form" on page 161.

7. Refer to the shaded box titled "The Dual of a Linear Programming Problem" on page 175.

8. Refer to the shaded box titled "The Fundamental Theorem of Duality" on page 176.

9. Let the matrix A be the coefficient matrix of the nontrivial inequalities, X be the column matrix of the variables, B be the column matrix of the right hand sides of the nontrivial constraints, and C be the row matrix of the coefficients of the objective function. Then a maximization problem would look like: Maximize CX subject to the constraints $AX \leq B, X \geq 0$ (where 0 is a matrix of all zeros).

10. "Sensitivity analysis" means using the final simplex tableau to look at what happens if small changes are made in the available resources.

11. Introducing a new product means introducing a new variable to the original problem (and thus a new column). In the dual, this becomes a new constraint. If the new product were to have no effect on the optimal solution, then the solution to the dual of the original problem would still be optimal and can be used to determine the value that would need to be assigned to the new variable in the objective function to alter the solution.

Chapter 4 Review Exercises

1.
$$\begin{array}{c} \\ u \\ v \\ M \end{array} \begin{array}{|ccccc|c|} x & y & u & v & M & \\ \hline 2 & 1 & 1 & 0 & 0 & 7 \\ -1 & 1 & 0 & 1 & 0 & 1 \\ -3 & -4 & 0 & 0 & 1 & 0 \end{array}$$

$$\begin{array}{c} \\ u \\ y \\ M \end{array} \begin{array}{|ccccc|c|} x & y & u & v & M & \\ \hline 3 & 0 & 1 & -1 & 0 & 6 \\ -1 & 1 & 0 & 1 & 0 & 1 \\ -7 & 0 & 0 & 4 & 1 & 4 \end{array}$$

$$\begin{array}{c} \\ x \\ y \\ M \end{array} \begin{array}{|ccccc|c|} x & y & u & v & M & \\ \hline 1 & 0 & \tfrac{1}{3} & -\tfrac{1}{3} & 0 & 2 \\ 0 & 1 & \tfrac{1}{3} & \tfrac{2}{3} & 0 & 3 \\ 0 & 0 & \tfrac{7}{3} & \tfrac{5}{3} & 1 & 18 \end{array}$$

$x = 2, y = 3, M = 18$

2.
$$\begin{array}{c} \\ u \\ v \\ M \end{array} \begin{array}{|ccccc|c|} x & y & u & v & M & \\ \hline 1 & 1 & 1 & 0 & 0 & 7 \\ 4 & 3 & 0 & 1 & 0 & 24 \\ -2 & -5 & 0 & 0 & 1 & 0 \end{array}$$

$$\begin{array}{c} \\ y \\ v \\ M \end{array} \begin{array}{|ccccc|c|} x & y & u & v & M & \\ \hline 1 & 1 & 1 & 0 & 0 & 7 \\ 1 & 0 & -3 & 1 & 0 & 3 \\ 3 & 0 & 5 & 0 & 1 & 35 \end{array}$$

$x = 0, y = 7, M = 35$

3.
$$\begin{array}{c} \\ u \\ v \\ w \\ M \end{array} \begin{array}{|cccccc|c|} x & y & u & v & w & M & \\ \hline 1 & 2 & 1 & 0 & 0 & 0 & 14 \\ 1 & 1 & 0 & 1 & 0 & 0 & 9 \\ 3 & 2 & 0 & 0 & 1 & 0 & 24 \\ -2 & -3 & 0 & 0 & 0 & 1 & 0 \end{array}$$

$$\begin{array}{c} \\ y \\ v \\ w \\ M \end{array} \begin{array}{|cccccc|c|} x & y & u & v & w & M & \\ \hline \tfrac{1}{2} & 1 & \tfrac{1}{2} & 0 & 0 & 0 & 7 \\ \tfrac{1}{2} & 0 & -\tfrac{1}{2} & 1 & 0 & 0 & 2 \\ 2 & 0 & -1 & 0 & 1 & 0 & 10 \\ -\tfrac{1}{2} & 0 & \tfrac{3}{2} & 0 & 0 & 1 & 21 \end{array}$$

$$\begin{array}{c} \begin{array}{cccccc} x & y & u & v & w & M \end{array} \\ \begin{array}{c} y \\ x \\ w \\ \hline M \end{array} \left[\begin{array}{cccccc|c} 0 & 1 & 1 & -1 & 0 & 0 & 5 \\ 1 & 0 & -1 & 2 & 0 & 0 & 4 \\ 0 & 0 & 1 & -4 & 1 & 0 & 2 \\ \hline 0 & 0 & 1 & 1 & 0 & 1 & 23 \end{array} \right] \end{array}$$

$x = 4,\ y = 5,\ M = 23$

4.
$$\begin{array}{c} \begin{array}{cccccc} x & y & u & v & w & M \end{array} \\ \begin{array}{c} u \\ v \\ w \\ \hline M \end{array} \left[\begin{array}{cccccc|c} 1 & 2 & 1 & 0 & 0 & 0 & 10 \\ 4 & 3 & 0 & 1 & 0 & 0 & 30 \\ -2 & \underline{1} & 0 & 0 & 1 & 0 & 0 \\ \hline -3 & -7 & 0 & 0 & 0 & 1 & 0 \end{array} \right] \end{array}$$

$$\begin{array}{c} \begin{array}{cccccc} x & y & u & v & w & M \end{array} \\ \begin{array}{c} u \\ v \\ y \\ \hline M \end{array} \left[\begin{array}{cccccc|c} \underline{5} & 0 & 1 & 0 & -2 & 0 & 10 \\ 10 & 0 & 0 & 1 & -3 & 0 & 30 \\ -2 & 1 & 0 & 0 & 1 & 0 & 0 \\ \hline -17 & 0 & 0 & 0 & 7 & 1 & 0 \end{array} \right] \end{array}$$

$$\begin{array}{c} \begin{array}{cccccc} x & y & u & v & w & M \end{array} \\ \begin{array}{c} x \\ v \\ y \\ \hline M \end{array} \left[\begin{array}{cccccc|c} 1 & 0 & \frac{1}{5} & 0 & -\frac{2}{5} & 0 & 2 \\ 0 & 0 & -2 & 1 & 1 & 0 & 10 \\ 0 & 1 & \frac{2}{5} & 0 & \frac{1}{5} & 0 & 4 \\ \hline 0 & 0 & \frac{17}{5} & 0 & \frac{1}{5} & 1 & 34 \end{array} \right] \end{array}$$

$x = 2,\ y = 4,\ M = 34$

5.
$$\begin{array}{c} \begin{array}{cccc} x & y & u & v & M \end{array} \\ \begin{array}{c} u \\ v \\ \hline M \end{array} \left[\begin{array}{cccc|c} \underline{-7} & -5 & 1 & 0 & 0 & -40 \\ -1 & -4 & 0 & 1 & 0 & -9 \\ \hline 1 & 1 & 0 & 0 & 1 & 0 \end{array} \right] \end{array}$$

$$\begin{array}{c} \begin{array}{cccc} x & y & u & v & M \end{array} \\ \begin{array}{c} x \\ v \\ \hline M \end{array} \left[\begin{array}{cccc|c} 1 & \frac{5}{7} & -\frac{1}{7} & 0 & 0 & \frac{40}{7} \\ 0 & -\frac{23}{7} & -\frac{1}{7} & 1 & 0 & -\frac{23}{7} \\ \hline 0 & \frac{2}{7} & \frac{1}{7} & 0 & 1 & -\frac{40}{7} \end{array} \right] \end{array}$$

$$\begin{array}{c} \begin{array}{ccccc} x & y & u & v & M \end{array} \\ \begin{array}{c} x \\ y \\ \hline M \end{array} \left[\begin{array}{ccccc|c} 1 & 0 & -\frac{4}{23} & \frac{5}{23} & 0 & 5 \\ 0 & 1 & \frac{1}{23} & -\frac{7}{23} & 0 & 1 \\ \hline 0 & 0 & \frac{3}{23} & \frac{2}{23} & 1 & -6 \end{array} \right] \end{array}$$

$x = 5,\ y = 1,\ M = -6$; the minimum is 6.

6.
$$\begin{array}{c} \begin{array}{ccccc} x & y & u & v & M \end{array} \\ \begin{array}{c} u \\ v \\ \hline M \end{array} \left[\begin{array}{ccccc|c} \underline{-1} & -1 & 1 & 0 & 0 & -6 \\ -1 & -2 & 0 & 1 & 0 & 0 \\ \hline 3 & 2 & 0 & 0 & 1 & 0 \end{array} \right] \end{array}$$

$$\begin{array}{c} \begin{array}{ccccc} x & y & u & v & M \end{array} \\ \begin{array}{c} x \\ v \\ \hline M \end{array} \left[\begin{array}{ccccc|c} 1 & \underline{1} & -1 & 0 & 0 & 6 \\ 0 & -1 & -1 & 1 & 0 & 6 \\ \hline 0 & -1 & 3 & 0 & 1 & -18 \end{array} \right] \end{array}$$

$$\begin{array}{c} \begin{array}{ccccc} x & y & u & v & M \end{array} \\ \begin{array}{c} y \\ v \\ \hline M \end{array} \left[\begin{array}{ccccc|c} 1 & 1 & -1 & 0 & 0 & 6 \\ 1 & 0 & -2 & 1 & 0 & 12 \\ \hline 1 & 0 & 2 & 0 & 1 & -12 \end{array} \right] \end{array}$$

$x = 0,\ y = 6,\ M = -12$; the minimum is 12.

7.
$$\begin{array}{c} \begin{array}{cccccc} x & y & u & v & w & M \end{array} \\ \begin{array}{c} u \\ v \\ w \\ \hline M \end{array} \left[\begin{array}{cccccc|c} -1 & -4 & 1 & 0 & 0 & 0 & -8 \\ -1 & -1 & 0 & 1 & 0 & 0 & -5 \\ \underline{-2} & -1 & 0 & 0 & 1 & 0 & -7 \\ \hline 20 & 30 & 0 & 0 & 0 & 1 & 0 \end{array} \right] \end{array}$$

$$\begin{array}{c} \begin{array}{cccccc} x & y & u & v & w & M \end{array} \\ \begin{array}{c} u \\ v \\ x \\ \hline M \end{array} \left[\begin{array}{cccccc|c} 0 & -\frac{7}{2} & 1 & 0 & -\frac{1}{2} & 0 & -\frac{9}{2} \\ 0 & -\frac{1}{2} & 0 & 1 & -\frac{1}{2} & 0 & -\frac{3}{2} \\ 1 & \frac{1}{2} & 0 & 0 & -\frac{1}{2} & 0 & \frac{7}{2} \\ \hline 0 & 20 & 0 & 0 & 10 & 1 & -70 \end{array} \right] \end{array}$$

$$\begin{array}{c} \begin{array}{cccccc} x & y & u & v & w & M \end{array} \\ \begin{array}{c} u \\ w \\ x \\ \hline M \end{array} \left[\begin{array}{cccccc|c} 0 & \underline{-3} & 1 & -1 & 0 & 0 & -3 \\ 0 & 1 & 0 & -2 & 1 & 0 & 3 \\ 1 & 1 & 0 & -1 & 0 & 0 & 5 \\ \hline 0 & 10 & 0 & 20 & 0 & 1 & 100 \end{array} \right] \end{array}$$

$$\begin{array}{c} \\ y \\ w \\ x \\ M \end{array} \begin{bmatrix} x & y & u & v & w & M & \\ 0 & 1 & -\frac{1}{3} & \frac{1}{3} & 0 & 0 & 1 \\ 0 & 0 & \frac{1}{3} & -\frac{7}{3} & 1 & 0 & 2 \\ 1 & 0 & \frac{1}{3} & -\frac{4}{3} & 0 & 0 & 4 \\ \hline 0 & 0 & \frac{10}{3} & \frac{50}{3} & 0 & 1 & -110 \end{bmatrix}$$

$x = 4$, $y = 1$, $M = -110$; the minimum is 110.

8.

$$\begin{array}{c} \\ u \\ v \\ w \\ M \end{array} \begin{bmatrix} x & y & u & v & w & M & \\ -2 & -1 & 1 & 0 & 0 & 0 & -10 \\ -3 & -2 & 0 & 1 & 0 & 0 & -18 \\ -1 & \underline{-2} & 0 & 0 & 1 & 0 & -10 \\ \hline 5 & 7 & 0 & 0 & 0 & 1 & 0 \end{bmatrix}$$

$$\begin{array}{c} \\ u \\ v \\ y \\ M \end{array} \begin{bmatrix} x & y & u & v & w & M & \\ -\frac{3}{2} & 0 & 1 & 0 & -\frac{1}{2} & 0 & -5 \\ -2 & 0 & 0 & 1 & \underline{-1} & 0 & -8 \\ \frac{1}{2} & 1 & 0 & 0 & -\frac{1}{2} & 0 & 5 \\ \hline \frac{3}{2} & 0 & 0 & 0 & \frac{7}{2} & 1 & -35 \end{bmatrix}$$

$$\begin{array}{c} \\ u \\ w \\ y \\ M \end{array} \begin{bmatrix} x & y & u & v & w & M & \\ -\frac{1}{2} & 0 & 1 & -\frac{1}{2} & 0 & 0 & -1 \\ 2 & 0 & 0 & -1 & 1 & 0 & 8 \\ \frac{3}{2} & 1 & 0 & -\frac{1}{2} & 0 & 0 & 9 \\ \hline -\frac{11}{2} & 0 & 0 & \frac{7}{2} & 0 & 1 & -63 \end{bmatrix}$$

$$\begin{array}{c} \\ x \\ w \\ y \\ M \end{array} \begin{bmatrix} x & y & u & v & w & M & \\ 1 & 0 & -2 & 1 & 0 & 0 & 2 \\ 0 & 0 & \underline{4} & -3 & 1 & 0 & 4 \\ 0 & 1 & 3 & -2 & 0 & 0 & 6 \\ \hline 0 & 0 & -11 & 9 & 0 & 1 & -52 \end{bmatrix}$$

$$\begin{array}{c} \\ x \\ u \\ y \\ M \end{array} \begin{bmatrix} x & y & u & v & w & M & \\ 1 & 0 & 0 & -\frac{1}{2} & \frac{1}{2} & 0 & 4 \\ 0 & 0 & 1 & -\frac{3}{4} & \frac{1}{4} & 0 & 1 \\ 0 & 1 & 0 & \frac{1}{4} & -\frac{3}{4} & 0 & 3 \\ \hline 0 & 0 & 0 & \frac{3}{4} & \frac{11}{4} & 1 & -41 \end{bmatrix}$$

$x = 4$, $y = 3$, $M = -41$; the minimum is 41.

9.

$$\begin{array}{c} \\ t \\ u \\ v \\ w \\ M \end{array} \begin{bmatrix} x & y & z & t & u & v & w & M & \\ 1 & 0 & 0 & 1 & 0 & 0 & 0 & 0 & 4 \\ 0 & 1 & 0 & 0 & 1 & 0 & 0 & 0 & 6 \\ 0 & 0 & 1 & 0 & 0 & 1 & 0 & 0 & 8 \\ 4 & 3 & 2 & 0 & 0 & 0 & 1 & 0 & 38 \\ \hline -36 & -48 & -70 & 0 & 0 & 0 & 0 & 1 & 0 \end{bmatrix}$$

$$\begin{array}{c} \\ t \\ u \\ z \\ w \\ M \end{array} \begin{bmatrix} x & y & z & t & u & v & w & M & \\ 1 & 0 & 0 & 1 & 0 & 0 & 0 & 0 & 4 \\ 0 & 1 & 0 & 0 & 1 & 0 & 0 & 0 & 6 \\ 0 & 0 & 1 & 0 & 0 & 1 & 0 & 0 & 8 \\ 4 & 3 & 0 & 0 & 0 & -2 & 1 & 0 & 22 \\ \hline -36 & -48 & 0 & 0 & 0 & 70 & 0 & 1 & 560 \end{bmatrix}$$

$$\begin{array}{c} \\ t \\ y \\ z \\ w \\ M \end{array} \begin{bmatrix} x & y & z & t & u & v & w & M & \\ 1 & 0 & 0 & 1 & 0 & 0 & 0 & 0 & 4 \\ 0 & 1 & 0 & 0 & 1 & 0 & 0 & 0 & 6 \\ 0 & 0 & 1 & 0 & 0 & 1 & 0 & 0 & 8 \\ \underline{4} & 0 & 0 & 0 & -3 & -2 & 1 & 0 & 4 \\ \hline -36 & 0 & 0 & 0 & 48 & 70 & 0 & 1 & 848 \end{bmatrix}$$

$$\begin{array}{c} \\ t \\ y \\ z \\ x \\ M \end{array} \begin{bmatrix} x & y & z & t & u & v & w & M & \\ 0 & 0 & 0 & 1 & \frac{3}{4} & \frac{1}{2} & -\frac{1}{4} & 0 & 3 \\ 0 & 1 & 0 & 0 & 1 & 0 & 0 & 0 & 6 \\ 0 & 0 & 1 & 0 & 0 & 1 & 0 & 0 & 8 \\ 1 & 0 & 0 & 0 & -\frac{3}{4} & -\frac{1}{2} & \frac{1}{4} & 0 & 1 \\ \hline 0 & 0 & 0 & 0 & 21 & 52 & 9 & 1 & 884 \end{bmatrix}$$

$x = 1$, $y = 6$, $z = 8$, $M = 884$

10.

	x	y	z	w	t	u	v	M	
t	6	9	12	15	1	0	0	0	672
u	1	−1	2	2	0	1	0	0	92
v	5	10	−5	4	0	0	1	0	280
M	−3	−4	−5	−4	0	0	0	1	0

	x	y	z	w	t	u	v	M	
t	0	15	0	3	1	−6	0	0	120
z	$\frac{1}{2}$	$-\frac{1}{2}$	1	1	0	$\frac{1}{2}$	0	0	46
v	$\frac{15}{2}$	$\frac{15}{2}$	0	9	0	$\frac{5}{2}$	1	0	510
M	$-\frac{1}{2}$	$-\frac{13}{2}$	0	1	0	$\frac{5}{2}$	0	1	230

	x	y	z	w	t	u	v	M	
y	0	1	0	$\frac{1}{5}$	$\frac{1}{15}$	$-\frac{2}{5}$	0	0	8
z	$\frac{1}{2}$	0	1	$\frac{11}{10}$	$\frac{1}{30}$	$\frac{3}{10}$	0	0	50
v	$\frac{15}{2}$	0	0	$\frac{15}{2}$	$-\frac{1}{2}$	$\frac{11}{2}$	1	0	450
M	$-\frac{1}{2}$	0	0	$\frac{23}{10}$	$\frac{13}{30}$	$-\frac{1}{10}$	0	1	282

	x	y	z	w	t	u	v	M	
y	0	1	0	$\frac{1}{5}$	$\frac{1}{15}$	$-\frac{2}{5}$	0	0	8
z	0	0	1	$\frac{3}{5}$	$\frac{1}{15}$	$-\frac{1}{15}$	$-\frac{1}{15}$	0	20
x	1	0	0	1	$-\frac{1}{15}$	$\frac{11}{15}$	$\frac{2}{15}$	0	60
M	0	0	0	$\frac{14}{5}$	$\frac{2}{5}$	$\frac{4}{15}$	$\frac{1}{15}$	1	312

$x = 60, y = 8, z = 20, w = 0, M = 312$

11. Minimize $14u + 9v + 24w$ subject to the constraints $\begin{cases} u + v + 3w \geq 2 \\ 2u + v + 2w \geq 3 \\ u \geq 0, v \geq 0, w \geq 0 \end{cases}$.

12. Maximize $8u + 5v + 7w$ subject to the constraints $\begin{cases} u + v + 2w \leq 20 \\ 4u + v + w \leq 30 \\ u \geq 0, v \geq 0, w \geq 0 \end{cases}$.

13. Primal: $x = 4, y = 5$, maximum = 23; Dual: $u = 1, v = 1, w = 0$, minimum = 23

14. Primal: $x = 4, y = 1$, minimum = 110;
    Dual: $u = \frac{10}{3}, v = \frac{50}{3}, w = 0$, maximum = 110

15. $A = \begin{bmatrix} 1 & 2 \\ 1 & 1 \\ 3 & 2 \end{bmatrix}, B = \begin{bmatrix} 14 \\ 9 \\ 24 \end{bmatrix},$
    $C = \begin{bmatrix} 2 & 3 \end{bmatrix}, X = \begin{bmatrix} x \\ y \end{bmatrix}$

    Primal: Maximize CX subject to $AX \leq B$, $X \geq 0$.

    Dual: $U = \begin{bmatrix} u \\ v \\ w \end{bmatrix}$

    Minimize $B^T U$ subject to $A^T U \geq C^T, U \geq 0$.

16. $A = \begin{bmatrix} 1 & 4 \\ 1 & 1 \\ 2 & 1 \end{bmatrix}, B = \begin{bmatrix} 8 \\ 5 \\ 7 \end{bmatrix},$
    $C = \begin{bmatrix} 20 & 30 \end{bmatrix}, X = \begin{bmatrix} x \\ y \end{bmatrix}$

    Primal: Minimize CX subject to $AX \geq B$, $X \geq 0$.

    Dual: $U = \begin{bmatrix} u \\ v \\ w \end{bmatrix}$

    Maximize $B^T U$ subject to $A^T U \leq C^T, U \geq 0$

17. a. Let a be the number of attack sticks and b the number of defense sticks. Maximize $16a + 20b$ subject to the

constraints $\begin{cases} 2a+b \leq 120 \\ a+3b \leq 150 \\ 2a+2b \leq 140 \\ a \geq 0, b \geq 0 \end{cases}$.

$$\begin{array}{c} \begin{array}{cccccc} a & b & u & v & w & M \end{array} \\ \begin{array}{c} u \\ v \\ w \\ M \end{array} \left[\begin{array}{cccccc|c} 2 & 1 & 1 & 0 & 0 & 0 & 120 \\ 1 & 3 & 0 & 1 & 0 & 0 & 150 \\ 2 & 2 & 0 & 0 & 1 & 0 & 140 \\ \hline -16 & -20 & 0 & 0 & 0 & 1 & 0 \end{array}\right] \end{array}$$

$$\begin{array}{c} \begin{array}{cccccc} a & b & u & v & w & M \end{array} \\ \begin{array}{c} u \\ b \\ w \\ M \end{array} \left[\begin{array}{cccccc|c} \frac{5}{3} & 0 & 1 & -\frac{1}{3} & 0 & 0 & 70 \\ \frac{1}{3} & 1 & 0 & \frac{1}{3} & 0 & 0 & 50 \\ \frac{4}{3} & 0 & 0 & -\frac{2}{3} & 1 & 0 & 40 \\ \hline -\frac{28}{3} & 0 & 0 & \frac{20}{3} & 0 & 1 & 1000 \end{array}\right] \end{array}$$

$$\begin{array}{c} \begin{array}{cccccc} a & b & u & v & w & M \end{array} \\ \begin{array}{c} u \\ b \\ a \\ M \end{array} \left[\begin{array}{cccccc|c} 0 & 0 & 1 & \frac{1}{2} & -\frac{5}{4} & 0 & 20 \\ 0 & 1 & 0 & \frac{1}{2} & -\frac{1}{4} & 0 & 40 \\ 1 & 0 & 0 & -\frac{1}{2} & \frac{3}{4} & 0 & 30 \\ \hline 0 & 0 & 0 & 2 & 7 & 1 & 1280 \end{array}\right] \end{array}$$

30 attack sticks, 40 defense sticks

b. The new problem is to maximize $16a + 20b + pc$, where c is the number of tennis rackets and p is the profit on each racket, subject to the

constraints: $\begin{cases} 2a+b+c \leq 120 \\ a+3b+4c \leq 150 \\ 2a+2b+2c \leq 140 \\ c \geq 0, b \geq 0, c \geq 0 \end{cases}$.

The dual problem is to minimize $120u + 150v + 140w$ subject to the

constraints: $\begin{cases} 2u+v+2w \geq 16 \\ u+3v+2w \geq 20 \\ u+4v+2w \geq p \\ u \geq 0, v \geq 0, w \geq 0 \end{cases}$.

The original solution, with $u = 0$, $v = 2$, $w = 7$ will be optimal if $0 + 4(2) + 2(7) \geq p$. Since $0 + 4(2) + 2(7) = \$22$, that is the profit per tennis racket that needs to be realized to justify the diversification.

18. The new problem is to maximize $70a + 210b + 140c + pd$, where d is the number of brand D stereo systems and p is the profit per system, subject to the constraints:

$\begin{cases} a+b+c+d \leq 100 \\ 5a+4b+4c+3d \leq 480 \\ 40a+20b+30c+30d \leq 3200 \\ a \geq 0, b \geq 0, c \geq 0, d \geq 0 \end{cases}$.

The dual problem is to minimize $100u + 480v + 3200w$, subject to the

constraints: $\begin{cases} u+5v+40w \geq 70 \\ u+4v+20w \geq 210 \\ u+4v+30w \geq 140 \\ u+3v+30w \geq p \\ u \geq 0, v \geq 0, w \geq 0 \end{cases}$.

The original solution, with $u = 210$, $v = 0$, $w = 0$, will still be optimal if $210 + 3(0) + 30(0) \geq p$. Since $210 + 3(0) + 30(0) = \$210$, that is the required profit per brand D system. The original solution was to sell 100 units of brand B at a profit of $210 each. Brand D units have to be at least this profitable. (Differences in storage space and commission turned out not to matter.)

4. The shadow price for Brie is $2.00.
The shadow price for Stilton is $1.00.

Chapter 5

Exercises 5.1

1. **a.** $S' = \{5, 6, 7\}$

   b. $S \cup T = \{1, 2, 3, 4, 5, 7\}$

   c. $S \cap T = \{1, 3\}$

   d. $S' \cap T = \{5, 7\}$

3. **a.** $R \cup S = \{a, b, c, d, e, f\}$

   b. $R \cap S = \{c\}$

   c. $S \cap T = \varnothing$

   d. $S' \cap R = \{a, b\}$

5. $\varnothing, \{1\}, \{2\}, \{1, 2\}$

7. **a.** $M \cap F = \{$all male college students who like football$\}$

   b. $M' = \{$college students who aren't males$\}$

   c. $M' \cap F' = \{$college students who aren't males who don't like football$\}$

   d. $M \cup F = \{$all male college students or all college students who like football$\}$

9. **a.** $S = \{1984, 1987, 1999, 2003, 2006, 2010\}$

   b. $T = \{1985, 1989, 1991, 1995, 1996, 1997, 1998, 1999, 2003, 2009\}$

   c. $S \cap T = \{1999, 2003\}$

   d. $S \cup T = \{1984, 1985, 1987, 1989, 1991, 1995, 1996, 1997, 1998, 1999, 2003, 2006, 2009, 2010\}$

   e. $S' \cap T = \{1985, 1989, 1991, 1995, 1996, 1997, 1998, 2009\}$

   f. $S \cap T' = \{1984, 1987, 2006, 2010\}$

11. From 1984 to 2011, during only four years did the Standard and Poor's Index increase by 2% or more during the first 5 days and not increase by 16% or more for that year.

13. **a.** $R \cup S = \{a, b, c, e\}$
    $(R \cup S)' = \{d, f\}$

    b. $R \cup S \cup T = \{a, b, c, e, f\}$

    c. $R \cap S = \{a, c\}$
    $R \cap S \cap T = (R \cap S) \cap T = \varnothing$

    d. $T' = \{a, b, c, d\}$
    $R \cap S \cap T' = (R \cap S) \cap T' = \{a, c\}$

    e. $R' = \{d, e, f\}; S \cap T = \{e\}$
    $R' \cap S \cap T = R' \cap (S \cap T) = \{e\}$

    f. $S \cup T = \{a, c, e, f\}$

    g. $R \cup S = \{a, b, c, e\};$
    $R \cup T = \{a, b, c, e, f\}$
    $(R \cup S) \cap (R \cup T) = \{a, b, c, e\}$

    h. $R \cap S = \{a, c\}; R \cap T = \varnothing$
    $(R \cap S) \cup (R \cap T) = \{a, c\}$

    i. $R' = \{d, e, f\}; T' = \{a, b, c, d\}$
    $R' \cap T' = \{d\}$

15. $(S')' = S$

17. $S \cup S' = U$

19. $T \cap S \cap T' = S \cap (T \cap T') = S \cap \varnothing = \varnothing$

21. $\{$divisions that had increases in labor costs or total revenue$\} = L \cup T$

23. $\{$divisions that made a profit despite an increase in labor costs$\} = L \cap P$

25. $\{$profitable divisions with increases in labor costs and total revenue$\} = P \cap L \cap T$

27. $\{$applicants who have not received speeding tickets$\} = S'$

29. $\{$applicants who have received speeding tickets, caused accidents, or were arrested for drunk

driving}
= $S \cup A \cup D$

31. {applicants who have not both caused accidents and received speeding tickets but who have been arrested for drunk driving}
= $(A \cap S)' \cap D$

33. $A \cap D$ = {male students at Mount College}

35. $A \cap B$ = {people who are both teachers and students at Mount College}

37. $A \cup C' = A \cup D$ = {Students at Mount College who aren't female}

39. $D' = C$ = {females at Mount College}

41. {people who don't like strawberry ice cream} = S'

43. {people who like vanilla or chocolate but not strawberry ice cream} = $(V \cup C) \cap S'$

45. {people who like neither chocolate nor vanilla ice cream} = $(V \cup C)'$

47. a. $R = \{B, C, D, E\}$
    b. $S = \{C, D, E, F\}$
    c. $T = \{A, D, E, F\}$
    d. $R' = \{A, F\}$
       $R' \cup S = \{A, C, D, E, F\}$
    e. $R' \cap T = \{A, F\}$
    f. $R \cap S = \{C, D, E\}$
       $R \cap S \cap T = (R \cap S) \cap T = \{D, E\}$

49. There are eight different ways. They are no toppings, mustard, relish, onions, mustard and relish, mustard and onions, relish and onions, all three toppings.

51. Any subset of T with 2 as an element is an example. Possible answer: {2}

53. If S is a subset of T, then $S \cup T = T$.

55. True; 5 is an element of the set {3, 5, 7}.

57. True; {b} is a subset of the set {b, c}.

59. False; 0 is not an element of the empty set $\varnothing$.

61. True; any set is a subset of itself.

Exercises 5.2

1. $n(S \cup T) = n(S) + n(T) - n(S \cap T)$
   $= 5 + 4 - 2 = 7$

3. $n(S \cup T) = n(S) + n(T) - n(S \cap T)$
   $15 = 7 + 8 - n(S \cap T)$
   $n(S \cap T) = 7 + 8 - 15 = 0$

5. $n(S \cup T) = n(S) + n(T) - n(S \cap T)$
   $13 = n(S) + 7 - 5$
   $n(S) = 13 - 7 + 5 = 11$

7. S is a subset of T.

9. Let P = {adults in South America fluent in Portuguese} and
   S = {adults in South America fluent in Spanish}.
   Then $P \cup S$ = {adults in South America fluent in Portuguese or Spanish} and
   $P \cap S$ = {adults in South America fluent in Portuguese and Spanish}.
   $n(P) = 160$, $n(S) = 155$,
   $n(P \cup S) = 293$ (numbers in millions)
   $n(P \cup S) = n(P) + n(S) - n(P \cap S)$
   $293 = 160 + 155 - n(P \cap S)$
   $n(P \cap S) = 160 + 155 - 293 = 22$
   22 million are fluent in both languages.

11. Let U = {all letters of the alphabet},
    let V = {letters with vertical symmetry}, and
    H = {letters with horizontal symmetry}.
    Then $V \cup H$ = {letters with vertical or horizontal symmetry} and
    $V \cap H$ = {letters with both vertical and horizontal symmetry}.
    $n(V) = 11$, $n(H) = 9$, $n(V \cap H) = 4$
    $n(V \cup H) = n(V) + n(H) - n(V \cap H)$
    $n(V \cup H) = 11 + 9 - 4 = 16$
    $n((V \cup H)') = n(U) - n(V \cup H) = 26 - 16 = 10$
    There are 10 letters with no symmetry.

13. Let A = {cars with automatic transmission} and P = {cars with power steering}.
Then $A \cup P$ = {cars with automatic transmission or power steering} and
$A \cap P$ = {cars with both automatic transmission and power steering},
$n(A) = 325$, $n(P) = 216$, $n(A \cap P) = 89$
$$n(A \cup P) = n(A) + n(P) - n(A \cap P)$$
$$= 325 + 216 - 89$$
$$= 452$$
452 cars were manufactured with at least one of the two options.

15. Consists of points not in S but in T.

17. Consists of points in S or not in T.

19. $(S' \cap T)' = S \cup T'$

Consists of points in S or not in T.

21. Consists of points in S but not in T or points in T but not in S.

23. $S \cup (S \cap T) = S$

Consists of points in S.

25. $S \cup S' = U$

Consists of all points.

27. Consists of points in R and S but not in T.

29. Consists of points in R or points in both S and T.

31. Consists of points in R but not in S or points in both R and T.

33. Consists of points in both R and T.

35. Consists of points not in R, S, or T.

37. Consists of points in R and T or points in S but not T.

39. $S' \cup (S \cap T)' = S' \cup S' \cup T' = S' \cup T'$

41. $(S' \cup T)' = S \cap T'$

43. $T \cup (S \cap T)' = T \cup S' \cup T'$
$$= (T \cup T') \cup S'$$
$$= U$$

45. S'

47. $R \cap T$

49. $R' \cap S \cap T$

51.

$T \cup (R \cap S')$

53. First draw a Venn diagram for $(R \cap S') \cup (S \cap T') \cup (T \cap R')$.

The set consists of the complement.
$(R \cap S \cap T) \cup (R' \cap S' \cap T')$

55. People who are not illegal aliens or everyone over the age of 18 who is employed

57. Everyone over the age of 18 who is unemployed

59. B is a subset of A', so $A' \cup B = A'$.
Noncitizens or legal aliens who are unemployed

Exercises 5.3

1. $5 + 6 = 11$

3. $6 + 5 + 15 + 20 = 46$

5. 11

7. $19 + 10 + 5 + 6 + 15 + 20 = 75$

9. $10 + 5 + 15 = 30$

11. $n(S \cap T') = n(S) - n(S \cap T) = 5 - 2 = 3$
$n(S' \cap T) = n(T) - n(S \cap T) = 6 - 2 = 4$
$n(S \cup T) = n(S) + n(T) - n(S \cap T)$
$\quad = 5 + 6 - 2 = 9$
$n(S' \cap T') = n(U) - n(S \cup T) = 14 - 9 = 5$

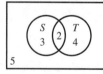

13. $n(S \cap T) = n(S) + n(T) - n(S \cup T)$
    $ = 12 + 14 - 18 = 8$
    $n(S \cap T') = n(S) - n(S \cap T) = 12 - 8 = 4$
    $n(S' \cap T) = n(T) - n(S \cap T) = 14 - 8 = 6$
    $n(S' \cap T') = n(U) - n(S \cup T) = 20 - 18 = 2$

    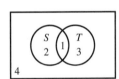

15. $n(S \cup T) = n(U) - n(S' \cap T') = 75 - 40 = 35$
    $n(S \cap T) = n(S) + n(T) - n(S \cup T) = 15 + 25 - 35 = 5$
    $n(S \cap T') = n(S) - n(S \cap T) = 15 - 5 = 10$
    $n(S' \cap T) = n(T) - n(S \cap T) = 25 - 5 = 20$

17. $n(S \cap T) = n(S) + n(T) - n(S \cup T) = 3 + 4 - 6 = 1$
    $n(U) = n(S' \cup T') + n(S \cap T) = 9 + 1 = 10$
    $n(S \cap T') = n(S) - n(S \cap T) = 3 - 1 = 2$
    $n(S' \cap T) = n(T) - n(S \cap T) = 4 - 1 = 3$
    $n(S' \cap T') = n(U) - n(S \cup T) = 10 - 6 = 4$

19. $n(R \cap S \cap T') = n(R \cap S) - n(R \cap S \cap T) = 7 - 2 = 5$
    $n(R \cap S' \cap T) = n(R \cap T) - n(R \cap S \cap T) = 6 - 2 = 4$
    $n(R' \cap S \cap T) = n(S \cap T) - n(R \cap S \cap T) = 5 - 2 = 3$
    $n(R \cap S' \cap T') = n(R) - n(R \cap S \cap T') - n(R \cap S' \cap T) - n(R \cap S \cap T) = 17 - 5 - 4 - 2 = 6$
    $n(R' \cap S \cap T') = n(S) - n(R \cap S \cap T') - n(R' \cap S \cap T) - n(R \cap S \cap T) = 17 - 5 - 3 - 2 = 7$
    $n(R' \cap S' \cap T) = n(T) - n(R \cap S' \cap T) - n(R' \cap S \cap T) - n(R \cap S \cap T) = 17 - 4 - 3 - 2 = 8$
    $n(R \cup S \cup T) = 6 + 7 + 8 + 5 + 4 + 3 + 2 = 35$
    $n(R' \cap S' \cap T') = n(U) - n(R \cup S \cup T) = 44 - 35 = 9$

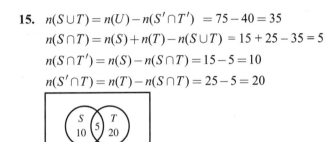

21. $n(R) = n(R \cup S) + n(R \cap S) - n(S) = 21 + 7 - 14 = 14$
$n(U) = n(R) + n(R') = 22 + 14 = 36$
$n(R \cap S \cap T') = n(R \cap S) - n(R \cap S \cap T) = 7 - 5 = 2$
$n(R \cap S' \cap T) = n(R \cap T) - n(R \cap S \cap T) = 11 - 5 = 6$
$n(R' \cap S \cap T) = n(S \cap T) - n(R \cap S \cap T) = 9 - 5 = 4$
$n(R \cap S' \cap T') = n(R) - n(R \cap S \cap T') - n(R \cap S' \cap T) - n(R \cap S \cap T) = 14 - 2 - 6 - 5 = 1$
$n(R' \cap S \cap T') = n(S) - n(R \cap S \cap T') - n(R' \cap S \cap T) - n(R \cap S \cap T) = 14 - 2 - 4 - 5 = 3$
$n(R' \cap S' \cap T) = n(T) - n(R \cap S' \cap T) - n(R' \cap S \cap T) - n(R \cap S \cap T) = 22 - 6 - 4 - 5 = 7$
$n(R \cup S \cup T) = 1 + 3 + 7 + 2 + 6 + 4 + 5 = 28$
$n(R' \cap S' \cap T') = n(U) - n(R \cup S \cup T) = 36 - 28 = 8$

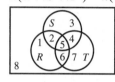

23. Let U = {high school students surveyed}, R = {students who like rock music}, and H = {students who like hip-hop music}.
$n(U) = 70$; $n(R) = 35$; $n(H) = 15$; $n(R \cap H) = 5$
$n(R \cup H) = n(R) + n(H) - n(R \cap H) = 35 + 15 - 5 = 45$
$n((R \cup H)') = n(U) - n(R \cup H) = 70 - 45 = 25$
25 students do not like either rock or hip-hop music.

25. Let U = {lines}, V = {lines with verbs}, A = {lines with adjectives}
$n(U) = 14$; $n(V) = 11$; $n(A) = 9$;
$n(V \cap A) = 7$
$n(V \cap A') = n(V) - n(V \cap A) = 11 - 7 = 4$
Four lines have a verb with no adjective.
$n(V' \cap A) = n(A) - n(V \cap A) = 9 - 7 = 2$
Two lines have an adjective but no verb.
$n(V \cup A) = n(V) + n(A) - n(V \cap A) = 11 + 9 - 7 = 13$
$n((V \cup A)') = n(U) - n(V \cap A) = 14 - 13 = 1$
One line has neither an adjective nor a verb.

For Exercises 26–30, let U = {students who took the exam}, F = {students who correctly answered the first question}, S = {students who correctly answered the second question}. Then $n(U) = 130$, $n(F) = 90$, $n(S) = 62$, $n(F \cap S) = 50$. Draw and complete the Venn diagram as follows.

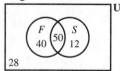

27. $n((F \cup S)') = 28$

29. $n(S \cap F') = 12$

31. Let U = {students in finite math}, M = {male students}, B = {students who are business majors}, and F = {first-year students}.
$n(U) = 35$; $n(M) = 22$; $n(B) = 19$; $n(F) = 27$;
$n(M \cap B) = 14$; $n(M \cap F) = 17$;
$n(B \cap F) = 15$; $n(M \cap B \cap F) = 11$

a. Draw a Venn diagram as shown.

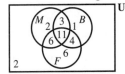

b. $n(F' \cap M' \cap B') = 2$
There are two upperclass women nonbusiness majors.

c. $n(M' \cap B) = 1 + 4 = 5$
There are five women business majors.

For Exercises 33–38, let $U = \{$UN members$\}$,
$R = \{$members with red in their flag$\}$,
$W = \{$members with white in their flag$\}$,
$B = \{$members with blue in their flag$\}$.
Then $n(U) = 193$, $n(R) = 145$, $n(W) = 132$, $n(B) = 104$,
$n(R \cap W) = 103$, $n(R \cap B) = 66$,
$n(W \cap B) = 73$, $n(R \cap W \cap B) = 52$. Draw and complete the Venn diagram as follows.

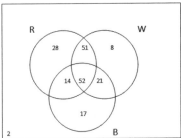

33. $n(R \cap (W \cup B)') = 28$

35. $n((R \cup W \cup B)') = 2$

37. $n(R \cap W \cap B') = 51$

For Exercises 39–44, let $U = \{$people surveyed$\}$,
$I = \{$people who learned from the Internet$\}$,
$T = \{$people who learned from television$\}$,
$N = \{$people who learned from newspapers$\}$.
Then $n(U) = 400$, $n(I) = 180$, $n(T) = 190$, $n(N) = 190$,
$n(I \cap T) = 80$, $n(I \cap N) = 90$,
$n(T \cap N) = 50$, $n(I \cap T \cap N) = 30$. Draw and complete the Venn diagram as follows.

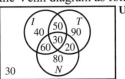

39. $n((I \cap N') \cup (I' \cap N)) = 40 + 50 + 80 + 20$
$= 190$

41. $n((I \cup T) \cap N') = 40 + 90 + 50 = 180$

43. $n((I \cap T' \cap N') \cup (I' \cap T \cap N') \cup (I' \cap T' \cap N))$
$= 40 + 90 + 80$
$= 210$

45. Draw and complete the Venn diagram as follows:

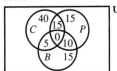

$40 + 15 + 15 + 15 + 5 + 10 + 0 = 100$
The correct answer is (d).

For Exercises 46–50, $n(U) = 4000$, $n(F) = 2000$,
$n(S) = 3000$, $n(L) = 500$,
$n(F \cap S) = 1500$, $n(F \cap L) = 300$,
$n(S \cap L) = 200$, $n(F \cap S \cap L) = 50$. Draw and complete the Venn diagram as follows.

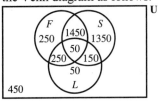

47. $(L \cup F \cup S)' = 450$

49. $L \cup S \cup F' = 4000 - 250 = 3750$

51. Let $U = \{$college students surveyed$\}$,
$F = \{$first-year students$\}$,
$D = \{$voted Democratic$\}$,
$n(U) = 100$; $n(F) = 50$; $n(D) = 55$
$n(F' \cap D') = n((F \cup D)') = 25$
$n(F \cup D) = n(U) - n((F \cup D)') = 100 - 25 = 75$
$n(F \cap D) = n(F) + n(D) - n(F \cup D)$
$= 50 + 55 - 75$
$= 30$
30 freshmen voted Democratic.

53. Let $U = \{$students$\}$, $D = \{$students who passed the diagnostic test$\}$, $C = \{$students who passed the course$\}$.
$n(U) = 30$; $n(D) = 21$; $n(C) = 23$; $n(D \cap C') = 2$

$n(D \cap C) = n(D) - n(D \cap C') = 21 - 2 = 19$

$n(D' \cap C) = n(C) - n(D \cap C) = 23 - 19 = 4$

Four students passed the course even though they failed the diagnostic test.

For Exercises 55–60, let U = {students}, M = {male students}, B = {biology majors}. Then $n(U) = 52$, $n(M \cap B) = 5$, $n(M' \cap B) = 15$, and $n(M' \cap B') = 12$. Therefore $n(B) = 5 + 15 = 20$, $n(M \cup B) = 52 - 12 = 40$, so $40 = n(M) + 20 - 5$ (inclusion-exclusion) and $n(M) = 40 - 20 + 5 = 25$.
Draw and complete the Venn diagram as follows.

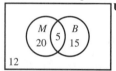

55. 40

57. 27

59. 20

For Exercises 61–68, let U = {surveyed students}, R = {students who like rock}, C = {students who like country}, J = {students who like jazz}. Then $n(U) = 190$, $n(R) = 114$, $n(C) = 50$, $n(R \cap J) = 15$, $n(C \cap J) = 11$, $n(R' \cap C' \cap J) = 20$, $n(R \cap C' \cap J) = 10$, $n(R \cap C \cap J') = 9$ and $n((R \cup C \cup J)') = 20$. Draw and complete the Venn diagram as follows.

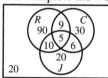

61. $n(R \cap C' \cap J') = 90$

63. $n(R' \cap C \cap J) = 6$

65. $n((R \cap C' \cap J') \cup (R' \cap C \cap J') \cup (R' \cap C' \cap J)) = 90 + 30 + 20 = 140$

67. $n((R \cap C) \cup (R \cap J) \cup (C \cap J)) = 9 + 10 + 6 + 5$
$= 30$

69. Let U = {executives}, F = {executives who read *Fortune*}, T = {executives who read *Time*}, M = {executives who read *Money*}.
$n(U) = 180$, $n(F) = 75$, $n(T) = 70$, $n(M) = 55$, $n(F \cap T) = 25$, $n(T \cap M) = 25$, $n(M \cap T \cap N) = 5$
Draw and complete the Venn diagram as follows.

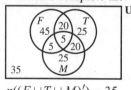

$n((F \cup T \cup M)') = 35$

71. Let U = {students}, P = {students who play piano}, V = {students who play violin} and C = {students who play clarinet}. Then let $x = n(P \cap V \cap C)$ and complete the Venn diagram as follows.

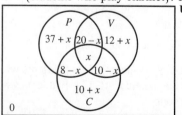

$37 + x + 12 + x + 10 + x + (20 - x) + (8 - x) + (10 - x) + x = 97 + x = 100$, so $x = 3$.

SSM: Finite Math Chapter 5: Sets and Counting

Exercises 5.4

1. $3 \cdot 2 = 6$ routes

3. $49 \cdot 48 \cdot 47 = 110{,}544$ possibilities

5. $20 \cdot 19 \cdot 18 = 6840$ possibilities

7. 30 because $30 \cdot 29 = 870$

9. **a.** $8 \cdot 7 \cdot 6 \cdot 5 \cdot 4 \cdot 3 \cdot 2 \cdot 1 = 40{,}320$ ways

    b. $5 \cdot 4 \cdot 3 \cdot 2 \cdot 1 \cdot 3 \cdot 2 \cdot 1 = 720$ ways

11. $26 \cdot 26 = 676$ words

13. $4 \cdot 3 \cdot 2 \cdot 1 = 24$ words

15. $2 \cdot 3 = 6$ outfits

17. $3 \cdot 12 \cdot 10 \cdot 10 \cdot 10 \cdot 10 = 360{,}000$ serial numbers

19. $10^9 - 1 = 999{,}999{,}999$ social security numbers

21. $8 \cdot 2 \cdot 10 = 160$ area codes

23. $9 \cdot 10 \cdot 10 \cdot 1 \cdot 1 = 900$ 5-digit palindromes

25. $26 \cdot 26 \cdot 1 \cdot 1 = 676$ 4-letter palindromes

27. $15 \cdot 15 = 225$ matchups

29. $3200 \cdot 2 \cdot 24 \cdot 52 = 7{,}987{,}200$ deals per year

31. $26 \cdot 26 \cdot 26 = 17{,}576$ sets of unique initials. Since there are 20,000 students, at least two students have the same set of initials

33. $3 \cdot 6 = 18$ different sweaters

35. $5 \cdot 4 = 20$ different mismatched sets

37. $2^6 = 64$ possible sequences

39. $2^5 = 32$ possible ways

41. $4^{10} = 1{,}048{,}576$ possible ways

43. $10^5 = 100{,}000$ possible zip codes

45. $8 \cdot 7 \cdot 6 \cdot 5 \cdot 4 \cdot 3 \cdot 2 \cdot 1 = 40{,}320$ ways

    $40320 \cdot 15 = 604{,}800$ seconds

    $\dfrac{604800}{60} = 10{,}080$ minutes

    $\dfrac{10080}{60} = 168$ hours

    $\dfrac{168}{24} = 7$ days

47. $6 \cdot 7 \cdot 4 = 168$ days or 24 weeks

49. $5 \cdot 11 \cdot (7 \cdot 2 + 1) \cdot 10 = 8250$ different ways

51. $2^4 = 16$ possible ways

53. $5 \cdot 2 \cdot 3 \cdot 2 \cdot 2 \cdot 8 = 960$ different cars

55. **a.** $9 \cdot 8 \cdot 7 \cdot 6 \cdot 5 \cdot 4 \cdot 3 \cdot 2 \cdot 1 = 362{,}880$

    b. $8 \cdot 7 \cdot 6 \cdot 5 \cdot 4 \cdot 3 \cdot 2 \cdot 1 \cdot 1 = 40{,}320$

    c. $1 \cdot 6 \cdot 5 \cdot 4 \cdot 3 \cdot 2 \cdot 1 \cdot 1 \cdot 1 = 720$

57. $\dfrac{10 \cdot 9}{2} + 10 \cdot 10 = 145$ handshakes

59. $4 \cdot 3 \cdot 3 \cdot 3 \cdot 3 \cdot 3 = 972$ ways

61. $7 \cdot 4 \cdot 2^6 = 1792$ different ballots

    $8 \cdot 5 \cdot 3^6 = 29{,}160$ different ballots

63. $2^4 = 16$ possible ways

Exercises 5.5

1. $P(4, 2) = 4 \cdot 3 = 12$

3. $P(6, 3) = 6 \cdot 5 \cdot 4 = 120$

5. $C(10, 3) = \dfrac{P(10, 3)}{3!} = \dfrac{10 \cdot 9 \cdot 8}{3 \cdot 2 \cdot 1} = 120$

7. $C(5, 4) = \dfrac{P(5, 4)}{4!} = \dfrac{5 \cdot 4 \cdot 3 \cdot 2}{4 \cdot 3 \cdot 2 \cdot 1} = 5$

9. $P(7, 1) = 7$

11. $P(n, 1) = n$

13. $C(4, 4) = \dfrac{P(4, 4)}{4!} = \dfrac{4 \cdot 3 \cdot 2 \cdot 1}{4 \cdot 3 \cdot 2 \cdot 1} = 1$

15. $C(n, n-2) = \dfrac{P(n, n-2)}{(n-2)!}$
    $= \dfrac{n \cdot (n-1) \cdots 4 \cdot 3}{(n-2) \cdot (n-3) \cdots 2 \cdot 1}$
    $= \dfrac{n \cdot (n-1)}{2 \cdot 1}$
    $= \dfrac{n(n-1)}{2}$

17. $6! = 6 \cdot 5 \cdot 4 \cdot 3 \cdot 2 \cdot 1 = 720$

19. $\dfrac{9!}{7!} = \dfrac{9 \cdot 8 \cdot 7 \cdot 6 \cdot 5 \cdot 4 \cdot 3 \cdot 2 \cdot 1}{7 \cdot 6 \cdot 5 \cdot 4 \cdot 3 \cdot 2 \cdot 1} = 9 \cdot 8 = 72$

21. Permutation; order matters

23. Combination; order does not matter

25. Permutation; order matters

27. $4! = 4 \cdot 3 \cdot 2 \cdot 1 = 24$ ways

29. $C(9, 2) = \dfrac{P(9,2)}{2!} = \dfrac{9 \cdot 8}{2 \cdot 1} = 36$ selections

31. $C(8, 4) = \dfrac{P(8,4)}{4!} = \dfrac{8 \cdot 7 \cdot 6 \cdot 5}{4 \cdot 3 \cdot 2 \cdot 1} = 70$ ways

33. $P(40, 5) = 40 \cdot 39 \cdot 38 \cdot 37 \cdot 36$
    $= 78,960,960$ ways

35. $C(10, 5) = \dfrac{P(10,5)}{5!}$
    $= \dfrac{10 \cdot 9 \cdot 8 \cdot 7 \cdot 6}{5 \cdot 4 \cdot 3 \cdot 2 \cdot 1}$
    $= 252$ ways

37. $C(100, 3) = \dfrac{P(100,3)}{3!}$
    $= \dfrac{100 \cdot 99 \cdot 98}{3 \cdot 2 \cdot 1}$
    $= 161,700$ possible samples

    $C(7, 3) = \dfrac{P(7,3)}{3!}$
    $= \dfrac{7 \cdot 6 \cdot 5}{3 \cdot 2 \cdot 1}$
    $= 35$ defective samples

39. $P(200, 3) = 200 \cdot 199 \cdot 198$
    $= 7,880,400$ ways

41. $C(52, 5) = \dfrac{P(52,5)}{5!}$
    $= \dfrac{52 \cdot 51 \cdot 50 \cdot 49 \cdot 48}{5 \cdot 4 \cdot 3 \cdot 2 \cdot 1}$
    $= 2,598,960$ hands

43. $C(13, 5) = \dfrac{P(13, 5)}{5!}$
    $= \dfrac{13 \cdot 12 \cdot 11 \cdot 10 \cdot 9}{5 \cdot 4 \cdot 3 \cdot 2 \cdot 1}$
    $= 1287$ hands

45. $5! = 5 \cdot 4 \cdot 3 \cdot 2 \cdot 1 = 120$ ways

47. **a.** $C(10, 4) = \dfrac{P(10,4)}{4!}$
    $= \dfrac{10 \cdot 9 \cdot 8 \cdot 7}{4 \cdot 3 \cdot 2 \cdot 1}$
    $= 210$ ways

    b. $C(10, 6) = \dfrac{P(10,6)}{6!}$
    $= \dfrac{10 \cdot 9 \cdot 8 \cdot 7 \cdot 6 \cdot 5}{6 \cdot 5 \cdot 4 \cdot 3 \cdot 2 \cdot 1}$
    $= 210$ ways

    c. They are the same because taking four sweaters is the same as leaving 6 sweaters.

49. $C(8, 2) = \dfrac{P(8, 2)}{2!}$

$= \dfrac{8 \cdot 7}{2 \cdot 1}$

$= 28$ games

51. $2 \cdot 7! = 2 \cdot 7 \cdot 6 \cdot 5 \cdot 4 \cdot 3 \cdot 2 \cdot 1$

$= 10,080$ batting orders

53. $\dfrac{C(59,6)}{C(49,6)} = \dfrac{45,057,474}{13,983,816} \approx 3.22$ Choice (b)

55. Moe: $C(9, 2) = \dfrac{P(9, 2)}{2!}$

$= \dfrac{9 \cdot 8}{2 \cdot 1}$

$= 36$ choices

Joe: $C(7,3) = \dfrac{P(7,3)}{3!}$

$= \dfrac{7 \cdot 6 \cdot 5}{3 \cdot 2 \cdot 1}$

$= 35$ choices

Thus Joe is correct

57. $4! \cdot P(4,3) \cdot P(5,3) \cdot P(6,3) \cdot P(7,3)$

$24 \cdot 24 \cdot 60 \cdot 120 \cdot 210 = 870,912,000$ pictures

59. $3! \cdot 3! \cdot 3! \cdot 3!$

$6 \cdot 6 \cdot 6 \cdot 6 = 1296$ ways

61. Through trial are error, you will find that 10 people were at the party.

63. $C(15,3) + (15 \cdot 14) + 15 = 455 + 210 + 15$

$= 680$ side dish options

65. $720 - 3! - 5!$

$720 - 6 - 120 = 594$

67. a. $C(45,5) = 1,221,759$ possible lottery tickets

b. $C(100,4) = 3,921,225$ possible lottery tickets

c. The first lottery has a better chance of winning

Exercises 5.6

1. a. $2^8 = 256$ outcomes

b. $C(8, 3) = \dfrac{8 \cdot 7 \cdot 6}{3 \cdot 2 \cdot 1} = 56$ outcomes

3. a. $C(7,5) + C(7,6) + C(7,7) =$

$21 + 7 + 1 = 29$ outcomes

b. $2^7 - 29 = 128 - 29 = 99$ outcomes

5. a. $2C(6,3) = \dfrac{2 \cdot 6 \cdot 5 \cdot 4}{3 \cdot 2 \cdot 1} = 40$ ways

b. $2C(5,3) = \dfrac{2 \cdot 5 \cdot 4 \cdot 3}{3 \cdot 2 \cdot 1} = 20$ ways

7. $C(10,5) \cdot C(5,4) \cdot C(1,1) =$

$252 \cdot 5 \cdot 1 = 1260$ ways

9. $C(7,2) = \dfrac{7 \cdot 6}{2 \cdot 1} = 21$ ways

11. $C(5,2) \cdot C(4,2) =$

$10 \cdot 6 = 60$ ways

13. $C(6,2) = \dfrac{6 \cdot 5}{2 \cdot 1} = 15$ ways

15. c. The two points where the combinations stop are the two points that make up the intersection B. Therefore, the sum of these two points will be the single combination.

   d. $C(8,3) + C(8,4) = C(9,4)$
   $56 + 70 = 126$

17. $C(8,3) = 56$ outcomes

19. $C(6,2) = 15$ sequences

21. a. $C(12,4) = 495$ samples

   b. $C(8,4) = 70$ samples

   c. $C(8,2) \cdot C(4,2) = 28 \cdot 6 = 168$ samples

   d. $C(8,3) \cdot C(4,1) + 70 =$
   $56 \cdot 4 + 70 = 294$ samples

23. a. $C(10,3) = 120$ ways

   b. $C(8,3) = 56$ ways

   c. $120 - 56 = 64$

25. $C(4,2) \cdot C(5,2) = 6 \cdot 10 = 60$ ways

27. $C(4,3) \cdot C(4,2) = 4 \cdot 6 = 24$ ways

29. $13 C(4,3) \cdot 12 C(4,2) = 13 \cdot 4 \cdot 12 \cdot 6$
   $= 3744$ ways

31. $C(7,5) \cdot 5! \cdot 21 \cdot 20 = 21 \cdot 120 \cdot 21 \cdot 20$
   $= 1,058,400$ ways

33. $C(10,5) \cdot P(21,5) = 252 \cdot 2,441,880$
   $= 615,353,760$ ways

35. $7 \cdot 3! \cdot 6! = 7 \cdot 6 \cdot 720$
   $= 30,240$ ways

37. $C(9,5) = 126$ ways

39. $C(26,22) \cdot C(10,7) = 14,950 \cdot 120$
   $= 1,794,000$ ways

41. $C(12,6) = 924$ ways

43. $\dfrac{C(100,50)}{2^{100}} \approx 0.07959 = 7.96\%$

45. $\dfrac{C(50,10) \cdot C(50,10)}{C(100,20)} \approx 0.1969 = 19.7\%$

Exercises 5.7

1. $\binom{6}{2} = C(6,2) = \dfrac{6 \cdot 5}{2 \cdot 1} = 15$

3. $\binom{8}{1} = C(8,1) = \dfrac{8}{1} = 8$

5. $\binom{18}{16} = C(18,16) = C(18,2) = \dfrac{18 \cdot 17}{2 \cdot 1} = 153$

7. $\binom{7}{0} = C(7,0) = 1$

9. $\binom{8}{8} = C(8,8) = 1$

11. $\binom{n}{n-1} = C(n, n-1) = C(n,1) = \dfrac{n}{1} = n$

13. $0! = 1$

15. $n \cdot (n-1)! = n!$

17. $\binom{6}{0} + \binom{6}{1} + \binom{6}{2} + \binom{6}{3} + \binom{6}{4} + \binom{6}{5} + \binom{6}{6}$
   $= 2^6$
   $= 64$

19. The number of terms in a binomial expansion is the exponent plus 1, so there are 20 terms.

21. $\binom{10}{0} x^{10} + \binom{10}{1} x^9 y + \binom{10}{2} x^8 y^2$
   $= x^{10} + 10 x^9 y + 45 x^8 y^2$

23. $\binom{15}{13}x^2y^{13} + \binom{15}{14}xy^{14} + \binom{15}{15}y^{15}$
    $= 105x^2y^{13} + 15xy^{14} + y^{15}$

25. $\binom{20}{10}x^{10}y^{10} = 184,756x^{10}y^{10}$

27. $\binom{4}{2} = C(4,2) = 6$

29. $\binom{11}{7} = 330$

31. $\binom{9}{0}x^9 + \binom{9}{1}x^8(2y) + \binom{9}{2}x^7(2y)^2$
    $= x^9 + 9x^8(2y) + 36x^7(4y^2)$
    $= x^9 + 18x^8y + 144x^7y^2$

33. $\binom{12}{6}x^6(-3y)^6 = 924x^6(729y^6)$
    $= 673,596x^6y^6$

35. $\binom{7}{4}x^3(-3y)^4 = 35x^3(81y^4) = 2835x^3y^4$

37. $2^6 = 64$ subsets

39. $2^4 = 16$ tips

41. $2^5 = 32$ options (The number of subsets of any size taken from a set of five elements.)

43. $2^8 - 1 = 255$ ways

45. $2 \cdot 3 \cdot 2^{15} = 196,608$ types

47. $2^7 - C(7,6) - C(7,7) = 128 - 7 - 1$
    $= 120$ ways

49. $2^8 - C(8,0) - C(8,1) = 256 - 1 - 8$
    $= 247$ ways

51. $\left(2^9 - C(9,8) - C(9,9)\right)\left(2^{10} - C(10,9) - C(10,10)\right)$
    $= (512 - 9 - 1)(1024 - 10 - 1)$
    $= (502)(1013)$
    $= 508,526$ ways

53. No, the exponents on the variables add to 7, but the combination portion of the terms shows that they should add to 8.

55. $\binom{5}{0} + \binom{5}{2} + \binom{5}{4} = \binom{5}{1} + \binom{5}{3} + \binom{5}{5}$
    $1 + 10 + 5 = 5 + 10 + 1$
    $16 = 16$

Chapter 5: Sets and Counting

Exercises 5.8

1. $\dfrac{5!}{3!1!1!} = 20$

3. $\dfrac{6!}{2!1!2!1!} = 180$

5. $\dfrac{7!}{3!2!2!} = 210$

7. $\dfrac{12!}{4!4!4!} = 34,650$

9. $\dfrac{12!}{5!3!2!2!} = 166,320$

11. $\dfrac{1}{5!} \cdot \dfrac{15!}{(3!)^5} = 1,401,400$

13. $\dfrac{1}{3!} \cdot \dfrac{18!}{(6!)^3} = 2,858,856$

15. $\dbinom{20}{7,\,5,\,8} = \dfrac{20!}{7!5!8!} = 99,768,240$ reports

17. $\dbinom{8}{2,1,4,1} = \dfrac{8!}{2!1!4!1!} = 840$ words

19. $\dbinom{9}{3,2,4} = \dfrac{9!}{3!2!4!} = 1260$ ways

21. $\dfrac{1}{5!} \cdot \dfrac{20!}{(4!)^5} = 2,546,168,625$ ways

23. $\dbinom{30}{10,\,2,\,18} = \dfrac{30!}{10!2!18!}$
    $= 5,708,552,850$ ways

25. $\dbinom{4}{1,\,1,\,2} = \dfrac{4!}{1!1!2!} = 12$ ways

27. $\dfrac{1}{7!} \cdot \dfrac{14!}{(2!)^7} = 135,135$ ways

29. The number of ways ten students are to be divided into two five member teams for a basketball game is $\dfrac{10!}{(2)!(5!)^2} = 126$.

31. $\dbinom{n}{1,1,\cdots,1} = \dfrac{n!}{1!1!\cdots 1!} = n!$

33. $\dbinom{38}{10,\,12,\,10,\,6}$
    $= \dfrac{38!}{10!12!10!6!}$
    $= 115,166,175,166,136,334,240$ ways

35. $\dbinom{52}{13,13,13,13} = \dfrac{52!}{13!13!13!13!}$
    $= 5.4 \times 10^{28}$

    Therefore, there are more than one octillion.

Chapter 5 Fundamental Concept Check

1. A collection of objects

2. Set B is a subset of set A if every element of B is also an element of A.

3. an object in the set

4. The set of all elements under consideration; usually denoted by the capital letter U.

5. The set containing no elements; denoted by the symbol $\varnothing$.

6. The set consisting of those elements of U that are not in A.

7. The set consisting of those elements that are in both A and B.

8. The set consisting of those elements that are in A or B or both.

9. If a task consists of t choices performed consecutively, and the first choice can be performed in m_1 ways, for each of these the second choice can be performed in m_2 ways,

and so on, then the task can be performed in $m_1 \cdot m_2 \cdots m_t$ ways.

10. an arrangement of r of the n objects in a specific order.

11. $P(n,r) = \underbrace{n(n-1)(n-2)\cdots(n-r+1)}_{r \text{ terms}}$

12. A combination does not take order into account.

13. $C(n,r) = \dfrac{\overbrace{n(n-1)(n-2)\cdots(n-r+1)}^{r \text{ terms}}}{r!}$

14. $n! = n(n-1)(n-2)\cdots 1$

$\binom{n}{r} = C(n,r) = \dfrac{\overbrace{n(n-1)(n-2)\cdots(n-r+1)}^{r \text{ terms}}}{r!}$

$P(n,r) = \underbrace{n(n-1)(n-2)\cdots(n-r+1)}_{r \text{ terms}}$

15. $(x+y)^n = \binom{n}{0}x^n + \binom{n}{1}x^{n-1}y + \binom{n}{2}x^{n-2}y^2 + \cdots + \binom{n}{n-1}xy^{n-1} + \binom{n}{n}y^n$

16. 2^n

17. A decomposition of a set into an ordered sequence of disjoint subsets.

18. $\dfrac{n!}{n_1! n_2! \cdots n_m!}$, where the partition is of the type $(n_1, n_2, \cdots, n_m)$ and $n = n_1 + n_2 + \cdots + n_m$.

Chapter 5 Review Exercises

1. $\varnothing, \{a\}, \{b\}, \{a, b\}$

2. $(S \cup T')' = S' \cap T$

3. $C(16, 2) = \dfrac{16!}{2!14!} = 120$ possibilities

4. $2 \cdot 5! = 240$ ways

5.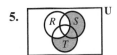

6. $\binom{12}{0}x^{12} + \binom{12}{1}x^{11}(-2y) + \binom{12}{2}x^{10}(-2y)^2$
$= x^{12} - 24x^{11}y + 264x^{10}y^2$

7. $C(8, 3) \cdot C(6, 2) = \dfrac{8!}{3!5!} \cdot \dfrac{6!}{2!4!} = 56 \cdot 15 = 840$

8. Let U = {people given pills}, P = {people who received placebos}, I = {people who showed improvement}.
$n(U) = 60; n(P) = 15; n(I) = 40; n(P' \cap I) = 30$
Draw and complete a Venn diagram as shown.

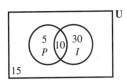

$n(P' \cap I') = 15$
Fifteen of the people who received the drug showed no improvement.

9. $7 \cdot 5 = 35$ combinations

10. $\binom{12}{2, 4, 6} = \dfrac{12!}{2!4!6!} = 13{,}860$

11. Let U = {applicants}, F = {applicants who speak French}, S = {applicants who speak Spanish}, and G = {applicants who speak German}.
    $n(U) = 115$; $n(F) = 70$; $n(S) = 65$; $n(G) = 65$; $n(F \cap S) = 45$; $n(S \cap G) = 35$; $n(F \cap G) = 40$; $n(F \cap S \cap G) = 35$. Draw and complete a Venn diagram.

    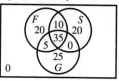

    $n((F \cup S \cup G)') = 0$
    None of the people speak none of the three languages.

12. $\binom{17}{15} = \binom{17}{2} = \frac{17 \cdot 16}{2 \cdot 1} = 136$

For Exercises 13–20, let U = {members of the Earth Club}, W = {members who thought the priority is clean water}, A = {members who thought the priority is clean air}, R = {members who thought the priority is recycling}. Then $n(U) = 100$, $n(W) = 45$, $n(A) = 30$, $n(R) = 42$, $n(W \cap A) = 13$, $n(A \cap R) = 20$, $n(W \cap R) = 16$, $n(W \cap A \cap R) = 9$. Draw and complete the Venn diagram as follows.

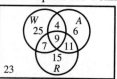

13. $n(A \cap W' \cap R') = 6$

14. $n((W \cap A') \cup (W' \cap A)) = (25 + 7) + (6 + 11) = 32 + 17 = 49$

15. $n((W \cup R) \cap A') = 25 + 15 + 7 = 47$

16. $n(A \cap R \cap W') = 11$

17. $n((W \cap A' \cap R') \cup (W' \cap A \cap R') \cup (W' \cap A' \cap R)) = 25 + 6 + 15 = 46$

18. $n(R') = 23 + 25 + 6 + 4 = 58$

19. $n(R \cap A') = 15 + 7 = 22$

20. $n((W \cup A \cup R)') = 23$

21. $C(9, 4) = C(9, 5) = 126$

22. $2^{20} = 1,048,576$ ways

23. Let S = {students who ski}, H = {students who play ice hockey}. Then
    $n(S \cup H) = n(S) + n(H) - n(S \cap H)$
    $= 400 + 300 - 150$
    $= 550$.

24. $6 \cdot 10 \cdot 8 = 480$ meals

25. $\binom{5}{1, 3, 1} = \frac{5!}{1! 3! 1!} = 20$ ways

26. The first digit can be anything but 0, the hundreds digit must be 3, and the last digit must be even.
$9 \cdot 1 \cdot 5 \cdot 10^4 = 450,000$

27. $9^2 \cdot 10^8 = 8,100,000,000$

28. $P(7,3) = 7 \cdot 6 \cdot 5 = 210$ ways

29. Strings of length 8 formed from the symbols a, b, c, d, e:
$5^8 = 390,625$ strings
Strings of length 8 formed from the symbols a, b, c, d:
$4^8 = 65,536$ strings
Strings with at least one e:
$390,625 - 65,536 = 325,089$ strings

30. $C(12, 5) = \dfrac{12!}{5!7!} = 792$

31. $C(30, 14) = \dfrac{30!}{14!16!} = 145,422,675$ groups

32. U = {households}
F = {households that get *Fancy Diet Magazine*}
C = {households that get *Clean Living Journal*}
$n(F \cup C) = n(F) + n(C) - n(F \cap C)$
$\quad = 4000 + 10,000 - 1500$
$\quad = 12,500$
$n((F \cup C)') = n(U) - n(F \cup C)$
$\quad = 40,000 - 12,500$
$\quad = 27,500$
27,500 households get neither.

33. $3^{10} = 59,049$ paths

34. $C(60, 10) = \dfrac{60!}{10!50!} = 75,394,027,566$ ways

35. $5^{10} = 9,765,625$ tests

36. $21^6 = 85,766,121$ strings

37. $C(10, 4) = \dfrac{10!}{4!6!} = 210$ ways

38. $\dfrac{1}{3!} \cdot \dfrac{21!}{(7!)^3} = 66,512,160$ ways

39. $14! = 87,178,291,200$ ways

40. $\dfrac{1}{4!} \cdot \dfrac{20!}{(5!)^4} = 488,864,376$ ways

41. $3 \cdot 5 \cdot 4 = 60$ ways

42. $3 \cdot C(10, 2) = 135$

43. A diagonal corresponds to a pair of vertices, except that adjacent pairs must be excluded. Hence there are
$C(n, 2) - n = \dfrac{n(n-1)}{2} - n = \dfrac{n(n-3)}{2}$ diagonals.

44. $8 \cdot 6 = 48$

45. $5! \cdot 4! \cdot 3! \cdot 2! \cdot 1! = 120 \cdot 24 \cdot 6 \cdot 2 \cdot 1 = 34,560$

46. $P(12, 5) = 12 \cdot 11 \cdot 10 \cdot 9 \cdot 8 = 95,040$

47. There are four suits. Once a suit is chosen, select 5 of the 13 cards.
Poker hands with cards of the same suit:
$4 \cdot C(13, 5) = 4 \cdot 1287 = 5148$ hands

48. $C(4, 3) \cdot C(48, 2) = 4 \cdot 1128 = 4512$ hands

49. Exactly the first two digits alike: $9 \cdot 1 \cdot 9 = 81$
First and last digit alike: $9 \cdot 9 \cdot 1 = 81$
Last two digits alike: $9 \cdot 9 \cdot 1 = 81$
Numbers with exactly two digits alike:
$81 + 81 + 81 = 243$

50. $9 \cdot 9 \cdot 8 = 648$ numbers

51. $3 \cdot 3 = 9$ pairings

52. $2 \cdot 3! \cdot 3! = 72$ ways

53. $24 \cdot 23 + 24 \cdot 23 \cdot 22 = 552 + 12,144$
$\quad = 12,696$ names

54. $3 \cdot 4! \cdot 2 = 3 \cdot 24 \cdot 2 = 144$ ways

55. Any two lines will intersect, so the number of intersections is $C(10, 2) = 45$.

56. First teacher: $\dfrac{1}{4!} \cdot \dfrac{24!}{(6!)^4} = 96{,}197{,}645{,}544$ ways

    Second teacher:
    $\dfrac{1}{6!} \cdot \dfrac{24!}{(4!)^6} = 4{,}509{,}264{,}634{,}875$ ways

    The second teacher has more options.

57. $\begin{pmatrix} 10 \\ 3, 4, 3 \end{pmatrix} = \dfrac{10!}{3!4!3!} = 4200$

58. Let n be the number of books.
    $n! = 120$, so $n = 5$. There are five books.

59. $7 \cdot 6 \cdot 5 \cdot 1 \cdot 4 \cdot 3 \cdot 2 \cdot 1 \cdot 1 = 5040$ orders

60. a. $C(12, 2) \cdot C(12, 3) = 66 \cdot 220 = 14{,}520$

    b. Select five out of twelve couples. Then determine the spouse from each couple.
    $C(12, 5) \cdot 2^5 = 25{,}344$

61. $C(7, 2) + C(7, 1) + C(7, 0) = 29$ ways

62. There are two choices for every row, so the number of paths from the A at the top to the final A is $2^6 = 64$ ways.

63. There are $2 \cdot 26^2$ 3-letter call letters and $2 \cdot 26^3$ 4-letter call letters, so $2 \cdot 26^2 + 2 \cdot 26^3 = 36{,}504$ call letters are possible.

64. Find n such that $n! = 479{,}001{,}600$.
    Testing numbers on computer, $n = 12$.

65. $\begin{pmatrix} 25 \\ 10, 9, 6 \end{pmatrix} = 16{,}360{,}143{,}800$ ways

66. First choose the three letters. There are $C(26, 3)$ ways to do this. Then choose the three numbers. There are $C(10, 3)$ ways to do this. Then order the six. There are 6! ways to do this.
    The number of license plates is
    $C(26, 3) \cdot C(10, 3) \cdot 6! = 224{,}640{,}000$.

Conceptual Exercises

67. If $A \cap B = \emptyset$ then A and B have no elements in common.

68. If $n(A \cup B) = n(A) + n(B)$, then A and B are disjoint sets and have no elements in common.

69. $n(n-1)! = n(n-1)(n-2)(n-3)\ldots 1 = n!$
    For example,
    $5 \cdot 4! = 5(4 \cdot 3 \cdot 2 \cdot 1) = 5 \cdot 4 \cdot 3 \cdot 2 \cdot 1 = 5!$.

70. Let $n = 1$. We have $1(1-1)! = 1(0)! = 1! = 1$ So, since 1 times any number is the number, we have $0! = 1! = 1$.

71. A permutation is an arrangement of items in which order is important. A combination is a subset of objects taken from a larger set in which the order of the objects does not make a difference.

72. The number of committees of size 6 is the same as the number of committees of size 4. For each committee of size 6, there is a committee consisting of the four people who were not chosen. For example, use letters A, B, C, D, E, F, G, H, I, J to represent the people. For the committee {A, B, C, D, E, F} the complementary committee is {G, H, I, J}.

73. $C(10, 3) = C(10, 7)$. The number of subsets of size 3 taken from a 10 member set is the same as the number of subsets of size 7. For example, consider the set $N = \{1, 2, 3, 4, 5, 6, 7, 8, 9, 10\}$. For the subset $A = \{1, 2, 3\}$, there is the corresponding subset $B = \{4, 5, 6, 7, 8, 9, 10\}$.

74. A committee of size 5 either includes or excludes John Doe. If the committee includes him, there are $C(10, 4)$ ways to choose the other 4 members. If the committee excludes him, there are $C(10, 5)$ ways to choose the 5 members. Hence $C(10, 4) + C(10, 5) = C(11, 5)$.

Chapter 6

Exercises 6.1

1. a. The set of all possible pairs: {RS, RT, RU, RV, ST, SU, SV, TU, TV, UV}

    b. The set of pairs containing R: {RS, RT, RU, RV}

    c. The set of pairs containing neither R nor S: {TU, TV, UV}

3. a. {HH, HT, TH, TT}

    b. {HH, HT}

5. a. {(I, red), (I, white), (II, red), (II, white)}

    b. All combinations with I: {(I, red), (I, white)}

7. a. S = {all positive numbers of minutes}

    b. $E \cap F$ = "more than 5 but less than 8 minutes"
    $E \cap G = \emptyset$ (There's no time longer than 5 minutes but less than 4 minutes.)
    E' = "5 minutes or less"
    F' = "8 minutes or more"
    $E' \cap F = E'$ = "5 minutes or less"
    $E' \cap F \cap G = G$ = "less than 4 minutes"
    $E \cup F = S$

9. a. Eight possible combinations:
    {(Fr, Lib), (Fr, Con), (So, Lib), (So, Con), (Jr, Lib), (Jr, Con), (Sr, Lib), (Sr, Con)}

    b. All combinations with Con: {(Fr, Con), (So, Con), (Jr, Con), (Sr, Con)}

    c. {(Jr, Lib)}

    d. All combinations with neither Fr nor Con: {(So, Lib), (Jr, Lib), (Sr, Lib)}

11. a. No; $E \cap F$ = {2}

    b. Yes; $F \cap G = \emptyset$

13. All combinations of members of S: $\emptyset, \{a\}, \{b\}, \{c\}, \{a, b\}, \{a, c\}, \{b, c\}, S$

15. Yes; $(E \cup F) \cap (E' \cap F')$ = {1, 2, 3} ∩ {4} = $\emptyset$

17. a. {0, 1, 2, 3, 4, 5, 6, 7, 8, 9, 10}

    b. More than half heads: {6, 7, 8, 9, 10}

19. a. No; there are blue-eyed males.

    b. Yes; a brown-eyed female doesn't have blue eyes.

    c. Yes; a brown-eyed female isn't male.

21. {0, 1, 2, 3, 4, 5, 6, 7, 8}

23. A possible outcome is (7, 4). There are $9 \cdot 9 = 81$ outcomes in the sample space.

25. A possible outcome is {2, 6, 9, 10}. There are $C(14, 4) - 1 = 1000$ possible combinations, so $\frac{17}{1000} = 0.017 = 1.7\%$ were assigned to the Chicago Bulls.

27. A sample space for the choice of murderers would be {Colonel Mustard, Miss Scarlet, Professor Plum, Mrs. White, Mr. Green, Mrs. Peacock}

    A sample space for the entire solution would be formed by placing each suspect, with each murder weapon, in each room.

    a. 6 suspects × 6 weapons × 9 rooms = 324 outcomes

    b. $E \cap F$ = "The murder occurred in the library with a gun."

    c. $E \cup F$ = "Either the murder occurred in the library, or it was done with a gun."

Exercises 6.2

1. Judgemental; the probability is an opinion.

3. Logical; the probability is based on theory.

5. Empirical; the probability is based on past data.

7. $\frac{1}{38} + \frac{1}{38} = \frac{2}{38} = \frac{1}{19}$

9. a. $\dfrac{191}{4487} \approx .04257$

    b. $\dfrac{191+81}{4487} = \dfrac{272}{4487} \approx .06062$

    c. $\dfrac{4487-272}{4487} = \dfrac{4215}{4487} \approx .9394$

11. a. $\dfrac{6}{26} = \dfrac{3}{13} \approx .2308$

    b. $\dfrac{5}{26} \approx .1923$

    c. $\dfrac{6+5-2}{26} = \dfrac{9}{26} \approx .3462$

13. a. E = "the numbers add up to 8" = {(2, 6), (3, 5), (4, 4), (5, 3), (6, 2)}

    $\Pr(E) = \dfrac{5}{36} \approx .1389$

    b. $\Pr(\text{sum is 2}) = \Pr((1,1)) = \dfrac{1}{36}$;

    $\Pr(\text{sum is 3}) = \Pr((1,2)) + \Pr((2,1)) = \dfrac{2}{36}$

    $\Pr(\text{sum is 4}) = \Pr((1,3)) + \Pr((2,2)) + \Pr((3,1)) = \dfrac{3}{36}$

    The probability that the sum is less than 5 is $\dfrac{1}{36} + \dfrac{2}{36} + \dfrac{3}{36} = \dfrac{1}{6} \approx .1667$.

15.

Kind of High School	Probability
Public	$\dfrac{168912}{204000} = .828$
Private	$\dfrac{34068}{204000} = .167$
Home School	$\dfrac{1020}{204000} = .005$

17. $\Pr(>3 \text{ and } \leq 15) = .2 + .26 + .25 = .71$

19. a. $\Pr(E) = .1 + .6 = .7$; $\Pr(F) = .6 + .1 = .7$

    b. $\Pr(E') = 1 - .7 = .3$

    c. $\Pr(E \cap F) = .6$

    d. $\Pr(E \cup F) = .1 + .6 + .1 = .8$

21. a.

Number of Colleges Applied to	Probability
1	.12
2	.21 − .12 = .09
3	.33 − .21 = .12
4	.47 − .33 = .14
≥ 5	1 − .47 = .53

    b. $.12 + .14 + .53 = .79$

23. a. No; The probabilities add to more than 1.

    b. No; One of the probabilities is a negative value.

    c. No; The probabilities do not add to 1.

25. $1 - \left(\dfrac{1}{3} + \dfrac{1}{2}\right) = \dfrac{1}{6}$

27. $\Pr(\text{Liberal}) = .28$; $\Pr(\text{middle}) = 2x$; $\Pr(\text{Conseravtive}) = x$

    $.28 + 2x + x = 1$
    $3x = .72$
    $x = .24$

29. The probability is 1 because the sum of a pair of dice must be odd or even.

31. $\Pr(E \cup F) = \Pr(E) + \Pr(F)$
    $= .4 + .5$
    $= .9$

33.

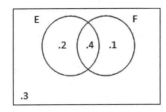

a. $\Pr(E \cup F) = .2 + .4 + .1 = .7$

b. $\Pr(E \cap F') = .2$

35.

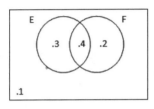

a. $\Pr(E \cup F) = .3 + .4 + .2 = .9$

b. $\Pr(E \cap F) = .4$

37. $\Pr(H \cap P) = \Pr(H) + \Pr(P) - \Pr(H \cup P)$
$= .7 + .8 - .9$
$= .6$

39. $10 \text{ to } 1 = \dfrac{10}{10+1} = \dfrac{10}{11}$

41. $.2 = \dfrac{1}{5} \Rightarrow 1 \text{ to } (5-1) = 1 \text{ to } 4$

43. $.3125 = \dfrac{5}{16} \Rightarrow 5 \text{ to } (16-5) = 5 \text{ to } 11$

45. $2 \text{ to } 9 = \dfrac{2}{2+9} = \dfrac{2}{11}$

47. a. Sparks: $5 \text{ to } 3 = \dfrac{3}{5+3} = \dfrac{3}{8}$

Meteors: $3 \text{ to } 1 = \dfrac{1}{3+1} = \dfrac{1}{4}$

Asteroids: $3 \text{ to } 2 = \dfrac{2}{3+2} = \dfrac{2}{5}$

Suns: $4 \text{ to } 1 = \dfrac{1}{4+1} = \dfrac{1}{5}$

b. $\dfrac{3}{8} + \dfrac{1}{4} + \dfrac{2}{5} + \dfrac{1}{5} = \dfrac{15}{40} + \dfrac{10}{40} + \dfrac{16}{40} + \dfrac{8}{40}$
$= \dfrac{49}{40}$

c. Bookies have to make a living. The payoffs are a little lower than they should be; thus allowing the bookie to make a profit.

49. There are more members (13) than Zodiac signs (12) so two or more members will always have the same Zodiac sign; thus the probability is 1.

51. This event never occurs; if 5 of the people receive the correct coat then so must the remaining person. Thus the probability is 0.

Exercises 6.3

1. a. $\dfrac{7}{13} \approx .5385$

b. $\dfrac{6}{13} \approx .4615$

c. $\Pr(\{3, 6, 9, 12\}) = \dfrac{4}{13} \approx .3077$

d. $\Pr(\{1, 3, 5, 6, 7, 9, 11, 12, 13\}) = \dfrac{9}{13}$
$\approx .6923$

3. a. $\dfrac{C(5, 2)}{C(9, 2)} = \dfrac{10}{36} = \dfrac{5}{18} \approx .2778$

b. $1 - \dfrac{C(5, 2)}{C(9, 2)} = 1 - \dfrac{10}{36} = \dfrac{13}{18} \approx .7222$

5. a. $\dfrac{C(6,4)+C(7,4)}{C(13,4)} = \dfrac{15+35}{715}$

   $= \dfrac{50}{715} = \dfrac{10}{143} \approx .0699$

   b. $\dfrac{C(6,4)+C(6,3)C(7,1)}{C(13,4)} = \dfrac{15+20\cdot 7}{715}$

   $= \dfrac{15+140}{715}$

   $= \dfrac{155}{715}$

   $= \dfrac{31}{143} \approx .2168$

7. Ways for exactly 2 to agree: $C(5, 2) \times C(4, 1)$
   Ways for all 3 to agree: $C(5, 3)$
   $\dfrac{C(5,2)\times C(4,1)+C(5,3)}{C(9,3)} = \dfrac{10\times 4+10}{84}$

   $= \dfrac{25}{42} \approx .5952$

9. $1 - \dfrac{C(6,3)}{C(10,3)} = 1 - \dfrac{20}{120} = \dfrac{5}{6} \approx .8333$

11. $1 - \dfrac{C(5,3)}{C(10,3)} = 1 - \dfrac{10}{120} = \dfrac{11}{12} \approx .9167$

13. $\dfrac{C(10,7)}{C(22,7)} = \dfrac{5}{7106} \approx .0007$

15. Ways for no girls to be chosen: $C(12, 7)$
    Ways for exactly 1 girl to be chosen:
    $C(12, 6) \times C(10, 1)$
    $1 - \dfrac{C(12,7)+C(12,6)\times C(10,1)}{C(22,7)}$

    $= 1 - \dfrac{792+924\times 10}{170{,}544} = \dfrac{16}{17} \approx .9412$

17. $1 - \dfrac{7\cdot 6\cdot 5}{7\cdot 7\cdot 7} = 1 - \dfrac{210}{343} = \dfrac{133}{343} \approx .3878$

19. $1 - \dfrac{30\times 29\times 28\times 27}{30^4} = \dfrac{47}{250} \approx .188$

21. $1 - \dfrac{P(20,8)}{20^8} \approx .8016$

23. Pr(at least one birthday on June 13)
    $= 1 - \left(\dfrac{364}{365}\right)^{25} \approx .06629$

    Because in Table 1 no particular date is being matched. Any two (or more) identical birthdays count as a success.

25. $\dfrac{C(6,2)\cdot 5^4}{6^6} = \dfrac{15\cdot 625}{46{,}656} = \dfrac{9375}{46{,}656} \approx .2009$

27. $\dfrac{3^4}{6^4} = \dfrac{81}{1296} = \dfrac{1}{16} \approx .0625$

29. $\dfrac{C(10,4)}{2^{10}} = \dfrac{210}{1024} = \dfrac{105}{512} \approx .2051$

31. $1 - \dfrac{6\times 5\times 4\times 3}{6^4} = \dfrac{13}{18} \approx .7222$

33. The tourist must travel 8 blocks of which 3 are south. Thus he has $C(8, 3) = 56$ ways to get to B from A.

    a. To get from A to B through C there are $C(3, 1) \cdot C(5, 2) = 30$ ways.
       The probability is $\dfrac{30}{56} = \dfrac{15}{28} \approx .5357$.

    b. To get from A to B through D there are $C(5, 1) \cdot C(3, 2) = 15$ ways.
       The probability is $\dfrac{15}{56} \approx .2679$.

    c. To get from A to B through C and D there are $C(3, 1) \cdot C(2, 0) \cdot C(3, 2) = 9$ ways.
       The probability is $\dfrac{9}{56} \approx .1607$.

    d. The number of ways to get from A to B through C or D is $30 + 15 - 9 = 36$.
       The probability is $\dfrac{36}{56} = \dfrac{9}{14} \approx .6429$.

35. $1 - \dfrac{4\cdot 4\cdot 4}{5\cdot 5\cdot 5} = 1 - \dfrac{64}{125}$

    $= \dfrac{61}{125} \approx .488$

37. $1 - \dfrac{12\cdot 11\cdot 10}{15\cdot 14\cdot 13} = 1 - \dfrac{1320}{2730}$

    $= \dfrac{1410}{2730} \approx .5165$

39. $\dfrac{5 \cdot 4 \cdot 3}{5^3} = \dfrac{60}{125} = \dfrac{12}{25} = .48$

41. There are 5! = 120 ways to arrange the family. First determine where the parents will stand. There are two ways with the man at one end. If the man doesn't stand at one end, there are $3 \cdot 2 = 6$ ways for the couple to stand together. Next determine the order of the children. There are 3! = 6 ways to order the children. Thus there are $(2 + 6) \cdot 6 = 48$ ways to stand with the parents together.

    The probability is $\dfrac{48}{120} = \dfrac{2}{5} = .4$.

43. There are $13 \cdot 12 = 156$ ways to choose the two denominations, $C(4, 3) = 4$ ways to choose the suits for the three of a kind and $C(4, 2) = 6$ ways to choose the suits for the pair; thus there are $156 \cdot 4 \cdot 6 = 3744$ possible full house hands, so the probability is $\dfrac{3744}{C(52, 5)} \approx .0014$.

45. There are $C(13, 2) = 78$ choices for the denominations of the two pairs, and $C(4, 2) = 6$ choices each for their suits; the remaining card has $52 - 8 = 44$ choices, for a total of $78 \cdot 6^2 \cdot 44 = 123{,}552$ possible two pair hands, so the probability is $\dfrac{123{,}552}{C(52, 5)} \approx .0475$.

47. **a.** There are 4 ways to select the suit of the 4 card group, $C(13, 4)$ ways to select their denominations, and $C(13, 3)^3$ ways to select 3 cards each from the remaining 3 suits. The probability of a 4-3-3-3 bridge hand is $\dfrac{4 \cdot C(13, 4) \cdot C(13, 3)^3}{C(52, 13)} \approx .1054$.

    b. There are $C(4, 2) = 6$ ways to choose the suits of the 4-card groups and 2 ways to choose the suit of the 3-card group (then the suit of the 2-card group is uniquely determined). The denominations can then be chosen in $C(13, 4)^2 \cdot C(13, 3) \cdot C(13, 2)$ ways, so the probability of a 4-4-3-2 bridge hand is $\dfrac{6 \cdot 2 \cdot C(13, 4)^2 \cdot C(13, 3) \cdot C(13, 2)}{C(52, 13)} \approx .2155$.

49. $\dfrac{2}{C(40, 6)} = \dfrac{1}{1{,}919{,}190} = .0000005211$

51. $\dfrac{C(6,3) \cdot C(43,3)}{C(49, 6)} = \dfrac{20 \cdot 12341}{13{,}983{,}816}$
    $= \dfrac{246{,}820}{13{,}983{,}816}$
    $\approx .01765$

53. $1 - \dfrac{P(100,15)}{100^{15}} \approx .6687$

55. $\dfrac{25-20}{80} = \dfrac{5}{80} = \dfrac{1}{16} = .0625 = 6.25\%$

    $\dfrac{25-20}{25} = \dfrac{5}{25} = \dfrac{1}{5} = .2 = 20\%$

    16 people will be needed because it helps 1 out of every 16 people.

57. $\dfrac{20-15}{100} = \dfrac{5}{100} = \dfrac{1}{20} = .05 = 5\%$

    $\dfrac{20-15}{20} = \dfrac{5}{20} = \dfrac{1}{4} = .25 = 25\%$

    20 people will be needed because it helps 1 out of every 20 people.

59. Let x represent the number of people in the control group that developed the condition.

    $\dfrac{x-12}{x} = \dfrac{.25}{1}$

    $.25x = x - 12$

    $-.75x = -12$

    $x = 16$

61. The probability that no two people choose the same card is $\dfrac{P(52, n)}{52^n}$, so the probability that at least two people pick the same card is $P_n = 1 - \dfrac{P(52, n)}{52^n}$.

    For $n = 5$, $P_5 \approx .1797$.

    $P_8 \approx .4324$ and $P_9 \approx .5197$.

    P_n increases as n increases, so $n = 9$ is the smallest value of n for which $P_n > 0.5$.

63. The probability that one or more people in a group of size n were born on a given specific day is $P_n = 1 - \left(\dfrac{364}{365}\right)^n$.

    For $n = 100$, $P_n \approx .240$.

    $P_{252} \approx .4991$ and $P_{253} \approx .5005$, so $n = 253$ is the smallest value of n for which $P_n > 0.5$.

Chapter 6: Probability

Exercises 6.4

1. **a.** $\Pr(E) = .3 + .1 = .4$

   b. $\Pr(F) = .1 + .4 = .5$

   c. $\Pr(E|F) = \dfrac{.1}{.5} = .2$

   d. $\Pr(F|E) = \dfrac{.1}{.4} = .25$

3. **a.** $\Pr(E|F) = \dfrac{.1}{.3} = \dfrac{1}{3}$

   b. $\Pr(F|E) = \dfrac{.1}{.5} = \dfrac{1}{5}$

   c. $\Pr(E|F') = \dfrac{.4}{.7} = \dfrac{4}{7}$

   d. $\Pr(E'|F') = \dfrac{.3}{.7} = \dfrac{3}{7}$

5. **a.** $\Pr(E \cap F) = \dfrac{1}{3} + \dfrac{5}{12} - \dfrac{2}{3} = \dfrac{1}{12}$

   b. $\Pr(E|F) = \dfrac{\frac{1}{12}}{\frac{5}{12}} = \dfrac{1}{5}$

   c. $\Pr(F|E) = \dfrac{\frac{1}{12}}{\frac{1}{3}} = \dfrac{1}{4}$

7. **a.** $\Pr(F|E) = \dfrac{\Pr(E \cap F)}{\Pr(E)}$

   $.25 = \dfrac{\Pr(E \cap F)}{.4}$

   $\Pr(E \cap F) = .1$

   b. $\Pr(E \cup F) = .4 + .3 - .1 = .6$

   c. $\Pr(E|F) = \dfrac{.1}{.3} = \dfrac{1}{3}$

   d. $\Pr(E' \cap F) = .3 - .1 = .2$

9. $\Pr(8 | \text{not } 7) = \dfrac{\Pr(8 \cap \text{not } 7)}{\Pr(\text{not } 7)}$

   $\Pr(8 | \text{not } 7) = \dfrac{\frac{5}{36}}{\frac{30}{36}}$

   $\Pr(8 | \text{not } 7) = \dfrac{5}{30} = \dfrac{1}{6}$

11. 0; because exactly one coin shows heads therefore there are two tails.

13. $\dfrac{[\text{number of outcomes that four are white}]}{[\text{number of outcomes that at least 1 is white}]}$

    $\dfrac{C(8,4)}{C(13,4) - C(5,4)} = \dfrac{70}{715 - 5}$

    $= \dfrac{70}{710}$

    $= \dfrac{7}{71} \approx .09859$

15. $\dfrac{[\text{number of outcomes that two are Democrat}]}{[\text{number of outcomes that at least 1 is Democrat}]}$

    $\dfrac{C(5,2)}{C(9,2) - C(4,2)} = \dfrac{10}{36 - 6}$

    $= \dfrac{10}{30}$

    $= \dfrac{1}{3}$

17. $\Pr(\text{grad} | \text{more } \$45{,}000) = \dfrac{\Pr(\text{grad and} > 45000)}{\Pr(> 45000)}$

    $= \dfrac{.10}{.25}$

    $= \dfrac{2}{5} = .4$

19. **a.** $\Pr(\text{Masters}) = \dfrac{735}{2602} \approx .2825$

    b. $\Pr(\text{Male}) = \dfrac{1137}{2602} \approx .4370$

    c. $\Pr(\text{Female}|\text{Masters}) = \dfrac{432}{735} \approx .5878$

    d. $\Pr(\text{Doctors}|\text{Female}) = \dfrac{90}{1465} \approx .0614$

21. a. $\Pr(\text{Officer}) = \dfrac{236.3}{1399.4} \approx .1689$

b. $\Pr(\text{Marine}) = \dfrac{199.0}{1399.4} \approx .1422$

c. $\Pr(\text{Officer and Marine}) = \dfrac{22.1}{1399.4} \approx .0158$

d. $\Pr(\text{Officer}|\text{Marine}) = \dfrac{22.1}{199.0} \approx .1111$

e. $\Pr(\text{Marine}|\text{Officer}) = \dfrac{22.1}{236.3} \approx .0935$

23. $\Pr(\$5|\$5) = \dfrac{\frac{1}{3}}{\frac{1}{2}} = \dfrac{2}{3}$

25. $\dfrac{4}{52} \cdot \dfrac{3}{51} = \dfrac{12}{2652} = \dfrac{1}{221} \approx .004525$

27. $\dfrac{1}{2}$; because the flip of the coin the fifth time is independent of the first four times.

29. $\Pr(\text{Health}|\text{Fem.}) = \dfrac{\Pr(\text{Health and Fem.})}{\Pr(\text{Fem.})}$

$.09 = \dfrac{\Pr(\text{Health and Fem.})}{.54}$

$\Pr(\text{Health and Fem.}) = .0486$

31. $\Pr(\text{Win with 2 point shot}) = .48 \cdot .5 = .24$

$\Pr(\text{Win with 3 point shot}) = .29$

Therefore, there is a better chance of winning if you take the 3 – point shot.

33. $\Pr(E \cap F) = .4 + .5 - .7 = .2$

$\Pr(E|F) = \dfrac{.2}{.5} = .4 = \Pr(E)$

$\Pr(F|E) = \dfrac{.2}{.4} = .5 = \Pr(F)$

Therefore, the two events are independent.

35. $\Pr(E \cap F) = (.5)(.6) = .3$

$\Pr(E \cup F) = .5 + .6 - .3 = .8$

37. Since the events are independent, $\Pr(F|E) = \Pr(F) = .6$

39. Since the events are independent,
$\Pr(F) = \dfrac{\Pr(E \cap F)}{1 - \Pr(E')} = \dfrac{.1}{1 - .6} = \dfrac{.1}{.4} = .25$

41. No; because the selection of the first ball affects the selection of the second ball.

43. Yes; because $\Pr(E \cap F) = \Pr(E) \cdot \Pr(F)$.

45. No; because $\Pr(E \cap F) \neq \Pr(E) \cdot \Pr(F)$.

47. No; because $\Pr(E \cap F) \neq \Pr(E) \cdot \Pr(F)$.

49. a. $\Pr(\text{pass all}) = (.80)(.75)(.60) = .36$

b. $\Pr(\text{pass first 2}) = (.80)(.75)(.40) = .24$

$\Pr(\text{pass first and third}) = (.80)(.25)(.60) = .12$

$\Pr(\text{pass last 2}) = (.2)(.75)(.60) = .09$

$\Pr(\text{pass at least 2}) = .36 + .24 + .12 + .09 = .81$

51. $(.99)^5 (.98)^5 (.975)^3 \approx .7967$

53. $(.7)^4 = .2401$

55. a. $1 - (.7)^4 = 1 - .2401 = .7599$

b. $(.7599)^{10} \approx .06420$

c. $1 - (.9358)^{20} \approx .7347$

57. Scoring 0: $1 - .6 = .4$

Scoring 1: $(.6)(1 - .6) = .24$

Scoring 2: $(.6)(.6) = .36$

59. Answer will vary.

61. $(.6)(.4) = .24$

   $(.4)(.6) = .24$

   Answers will vary

63. Answers will vary.

65. Answers will vary.

67. $26; \left(\frac{37}{38}\right)^{26} \approx .4999$

Exercises 6.5

1.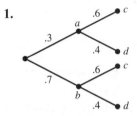

3.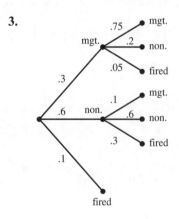

5. $.30 \times .75 + .60 \times .10 = .285$

7. $.10 + .30 \times .05 + .60 \times .30 = .295$

9. $\text{Pr(white, then red)} + \text{Pr(red, then red)}$
   $= \frac{2}{3} \times \frac{1}{2} + \frac{1}{3} \times \frac{3}{4} = \frac{7}{12}$

11.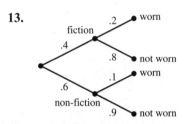

    $\text{Pr(king on 1st draw)} + \text{Pr(king on 2nd draw)}$
    $+ \text{Pr(king on 3rd draw)}$
    $= 1 - \text{Pr(not a king on 3rd draw)}$
    $= 1 - \frac{12}{13} \times \frac{47}{51} \times \frac{23}{25} = 1 - \frac{4324}{5525} = \frac{1201}{5525} \approx .22$

13.

    $.40 \times .20 + .60 \times .10 = .14$

15. $\Pr(\text{male}|\text{color-blind}) = \dfrac{\frac{1}{2} \times .05}{\frac{1}{2} \times .05 + \frac{1}{2} \times .004} = \dfrac{25}{27}$

17. $.5 \times .9 + .5 \times .7 = .8$

19. $\Pr(\text{fake}|\text{HH}) = \dfrac{\frac{1}{4} \times 1}{\frac{3}{4} \times \frac{1}{4} + \frac{1}{4} \times 1} = \dfrac{4}{7}$

21. **a.** $\Pr(\text{wins the point}) = .60 \times .75 + .40 \times .75 \times .50 = .60$

    b. $\Pr(\text{first serve good}|\text{wins service point}) = \dfrac{\Pr(\text{first serve good and wins service point})}{\Pr(\text{wins service point})} = \dfrac{.60 \times .75}{.60} = .75$

23.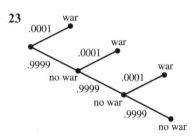

    $.0001 + .9999 \times .0001 + .9999^2 \times .0001$ or $1 - (.9999)^3 \approx .00029997$

25.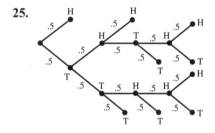

    $.5 + (.5)^3 + 2(.5)^5 = \dfrac{11}{16}$

27. **a.** $\Pr(\text{white}) = \dfrac{1}{2} \times \dfrac{1}{2} = \dfrac{1}{4}$

    $\Pr(\text{red}) = 1 - \dfrac{1}{4} = \dfrac{3}{4}$

    b. $\Pr(\text{red}) = .6 \times .5 + .4 \times 1 = .7$

29. $\Pr(\text{night}|\text{part-timer}) = \dfrac{.60 \times 2}{.25 \times 4 + .60 \times 2} = \dfrac{6}{11}$

31. $pr(W_2) = pr(R_1 \cap W_2) + pr(W_1 \cap W_2)$
$= pr(R_1)pr(W_2|R_1) + pr(W_1)p(W_2|W_1)$
$= \left(\frac{5}{10}\right)\cdot\left(\frac{12}{13}\right) + \left(\frac{5}{10}\right)\cdot 1$
$= \frac{25}{26}$

33. Pr(get same number of heads)
= Pr(2 heads) + Pr(1 head) + Pr(0 heads)
$= \frac{1}{4}\cdot\frac{1}{4} + \frac{1}{2}\cdot\frac{1}{2} + \frac{1}{4}\cdot\frac{1}{4} = \frac{3}{8}$

35. a. Since Stan always scores 400 points, Oliver wins .99 = 99% of the time.

   b. For Oliver to score the most points he must win all 400 games. The probability that this occurs is $.99^{400} \approx .0180$.

37. [number of ways two cards are left] = $C(52, 2)$ = 1326
   [number of ways two cards have same color] = $C(26, 2) + C(26, 2) = 650$
   [number of ways two cards have different color] = $26 \cdot 26 = 676$
   Pr(two cards have same color) = $\frac{650}{1326} = \frac{25}{51}$

   Pr(two cards have different colors) = $\frac{676}{1326} = \frac{26}{51}$

   Pr(guess color|same color) = 1
   Pr(guess color|different colors) = $\frac{1}{2}$
   Pr(guess color) = $\left(\frac{25}{51}\right)(1) + \left(\frac{26}{51}\right)\left(\frac{1}{2}\right) = \frac{38}{51}$

39. True; Sensitivity gives the percent of people that have a condition given the fact that they tested positive.

41. True; Based on the definition of specificity.

43. True; Sensitivity gives the percentage of people that have a given condition.

45. $\text{Pr}(\text{Hep}|\text{Positive}) = \frac{(.0005)(.95)}{(.0005)(.95) + (.9995)(.1)}$
$= \frac{.000475}{.000475 + .09995}$
$= \frac{.000475}{.100425} \approx .00473$

47. $\text{Pr}(\text{Cond.}|\text{Positive}) = \frac{9}{19} \approx .474$

49. $\text{Pr}(\text{Used}|\text{Positive}) = \frac{(.1)(1)}{(.1)(1) + (.9)(.01)}$
$= \frac{.1}{.1 + .009}$
$= \frac{.1}{.109} \approx .92$

Exercises 6.6

1. $\text{Pr}(\text{over 60}|\text{accident}) = \frac{.10 \times .04}{(.05 \times .06) + (.10 \times .04) + (.25 \times .02) + (.20 \times .015) + (.30 \times .025) + (.10 \times .04)}$
$= \frac{.004}{.0265}$
$= \frac{8}{53}$

3. $\Pr(\text{sophomore}|A) = \dfrac{.30 \times .4}{(.10 \times .2)+(.30 \times .4)+(.40 \times .3)+(.20 \times .1)} = \dfrac{.12}{.28} = \dfrac{3}{7}$

5. $\Pr(\geq \$75{,}000 | 2 \text{ or more cars}) = \dfrac{.05 \times .9}{(.10 \times .2)+(.20 \times .5)+(.35 \times .6)+(.30 \times .75)+(.05 \times .9)}$

    $= \dfrac{.045}{.6}$

    $= \dfrac{3}{40}$

    $= .075$

7. $\Pr(\text{passed exam}|A) = \dfrac{.80 \times .40}{.80 \times .40 + .20 \times .20}$

    $= \dfrac{.32}{.36}$

    $= \dfrac{8}{9}$

9. a. $(.20 \times .20) + (.15 \times .15) + (.25 \times .12) + (.30 \times .10) + (.10 \times .10) = .1325$

   b. $\Pr(\text{division C}|\text{bilingual}) = \dfrac{.25 \times .12}{.1325} = \dfrac{.03}{.1325} = \dfrac{12}{53} \approx .23$

11. $\Pr(\text{cancer}|\text{positive}) = \dfrac{\Pr(\text{cancer}) \times \Pr(\text{positive}|\text{cancer})}{\Pr(\text{cancer}) \times \Pr(\text{positive}|\text{cancer}) + \Pr(\text{no cancer}) \times \Pr(\text{positive}|\text{no cancer})}$

    $= \dfrac{.02 \times .75}{(.02 \times .75)+(.98 \times .30)} = \dfrac{5}{103} \approx .049$

13. a. $1 - .99 = .01$

    b. $\Pr(\text{pregnant}|\text{positive})$

    $= \dfrac{\Pr(\text{pregnant}) \times \Pr(\text{positive}|\text{pregnant})}{\Pr(\text{pregnant}) \times \Pr(\text{positive}|\text{pregnant}) + \Pr(\text{not pregnant}) \times \Pr(\text{positive}|\text{not pregnant})}$

    $= \dfrac{.40 \times .99}{(.40 \times .99)+(.60 \times .02)} = \dfrac{33}{34} \approx .971$

15. $\Pr(\text{steroids}|\text{positive}) = \dfrac{\Pr(\text{steroids}) \times \Pr(\text{positive}|\text{steroids})}{\Pr(\text{steroids}) \times \Pr(\text{positive}|\text{steroids}) + \Pr(\text{no steroids}) \times \Pr(\text{positive}|\text{no steroids})}$

    $= \dfrac{.10 \times .93}{(.10 \times .93)+(.90 \times .02)} = \dfrac{31}{37} \approx .838$

17. $\Pr(\text{Pat}|\text{singing}) = \dfrac{\Pr(\text{Pat}) \times \Pr(\text{singing}|\text{Pat})}{\Pr(\text{Pat}) \times \Pr(\text{singing}|\text{Pat}) + \Pr(\text{Alex}) \times \Pr(\text{singing}|\text{Alex})}$

$= \dfrac{.5 \times .9}{(.5 \times .9) + (.5 \times .5)} = \dfrac{.45}{.45 + .25} = \dfrac{9}{14} \approx .643$

19. **a.** $\Pr(\text{one is}) = \dfrac{13}{52} = \dfrac{1}{4}$

   b. $\Pr(\text{none is}|\text{random one isn't})$

   $= \dfrac{\Pr(\text{none is}) \times (\text{random one isn't}|\text{none is})}{\Pr(\text{none is}) \times \Pr(\text{random one isn't}|\text{none is}) + \Pr(\text{one is}) \times \Pr(\text{random one isn't}|\text{one is})}$

   $= \dfrac{\frac{3}{4} \times 1}{\left(\frac{3}{4} \times 1\right) + \left(\frac{1}{4} \times \frac{12}{13}\right)} = \dfrac{13}{17} \approx .765$

   c. $\Pr(\text{one is}|10 \text{ randoms aren't})$

   $= \dfrac{\Pr(\text{one is}) \times \Pr(10 \text{ randoms aren't}|\text{one is})}{\Pr(\text{one is}) \times \Pr(10 \text{ randoms aren't}|\text{one is}) + \Pr(\text{none is}) \times \Pr(10 \text{ randoms aren't}|\text{none is})}$

   $= \dfrac{\frac{1}{4} \times \left(\frac{12}{13}\right)^{10}}{\left[\frac{1}{4} \times \left(\frac{12}{13}\right)^{10}\right] + \left(\frac{3}{4} \times 1\right)}$

   $\approx .130$

21. **a.** $\Pr(\text{Lakeside}|\text{winner})$

   $= \dfrac{\Pr(\text{Lakeside}) \times \Pr(\text{winner}|\text{Lakeside})}{\Pr(\text{Lakeside}) \times \Pr(\text{winner}|\text{Lakeside}) + \Pr(\text{Pylesville}) \times \Pr(\text{winner}|\text{Pylesville}) + \Pr(\text{Millerville}) \times \Pr(\text{winner}|\text{Millerville})}$

   $= \dfrac{.40 \times .05}{(.40 \times .05) + (.20 \times .02) + (.40 \times .03)} = \dfrac{5}{9}$

   b. $\dfrac{.20 \times .02}{(.40 \times .05) \times (.20 \times .02) + (.40 \times .03)} = \dfrac{1}{9} \approx 11\%$

23.

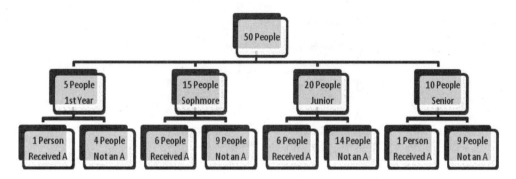

$$\Pr(\text{Sophmore} \mid A) = \frac{6}{1+6+6+1} = \frac{6}{14} = \frac{3}{7}$$

25.

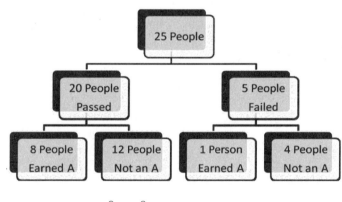

$$\Pr(\text{Passed} \mid A) = \frac{8}{8+1} = \frac{8}{9}$$

27.

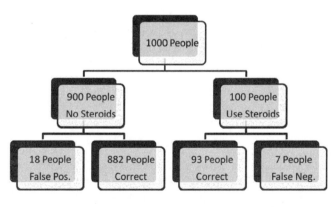

$$\Pr(\text{Used} \mid \text{Positive}) = \frac{93}{93+18} = \frac{93}{111} = \frac{31}{37} \approx .838$$

29.

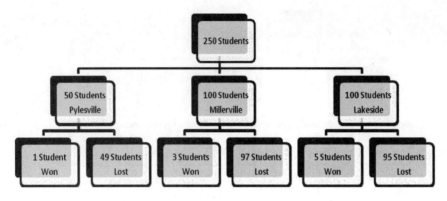

$$\Pr(\text{Lakeside} \mid \text{Winner}) = \frac{5}{1+3+5} = \frac{5}{9}$$

Exercises 6.7

1. Use seq(randInt(1,6),X,1,36,1)→☐ L_1.

   Theoretical probabilities: $\frac{1}{6}$ for each fall.

3. Use seq(randInt(1,1000),X,1,10,1)→L_1 where a number from 1–840 represents a successful freethrow and 841–1000 represents a miss.

5. Use seq(randInt(1,4),X,1,10,1)→L_1, where a = 1, b = 2, c = 3, and d = 4.

7. Answers will vary (see Example 4).

9. Answers will vary.

Chapter 6 Fundamental Concept Check

1. The set of all possible outcomes

2. $A \cup B$; $A \cap B$; A'

3. 0

4. The entire sample space.

5. $\Pr(A \cup B) = \Pr(A) + \Pr(B) - \Pr(A \cap B)$

6. Two events are independent if the occurrence of one has no effect on the occurrence of the other. The two events might have outcomes in common. Two events are mutually exclusive if they have no outcomes in common.

7. The probability of an event is the sum of the probabilities of the outcomes in the event.

8. k to $n-k$

SSM: Finite Math **Chapter 6:** Probability

9. $\dfrac{a}{a+b}$

10. $\Pr(A \cup B) = \Pr(A) + \Pr(B) - \Pr(A \cap B)$

11. $\dfrac{\Pr(E \cap F)}{\Pr(F)}$

12. Bayes' Theorem provides a way to calculate certain conditional probabilities. See page 289.

13. A tree diagram is a graphical device used to show all the possible outcomes of an experiment, and their probabilities, in a clear and uncomplicated manner. Each branch of the tree contains a probability.

Chapter 6 Review Exercises

1. **a.** The set of all possible pairs: {PN, PD, PQ, PH, ND, NQ, NH, DQ, DH, QH}

   b. The set of pairs containing an even number of cents: {PN, PQ, NQ, DH}

2. **a.** A male junior is elected

   b. A female junior is not elected

   c. A male or a junior is elected

3. $\Pr(E \cap F) = .4 + .3 - .5 = .2$

4. $\Pr(E \cup F) = .5 + .3 = .8$

5. $\dfrac{120 - (\text{speak Chinese or Spanish}) - (\text{speak French only})}{120} = \dfrac{120 - (30 + 50 - 12) - (75 - 30 - 15 + 7)}{120} = \dfrac{1}{8}$

6. There are $15 - 5 = 10$ who like only bicycling and $20 - 5 = 15$ who like only jogging, so the probability is $\dfrac{10 + 15}{50} = .5$.

7. $\Pr(\text{within 20 minutes}) = \dfrac{13}{13 + 12} = \dfrac{13}{25}$; $\Pr(\text{more than 20 minutes}) = 1 - \dfrac{13}{25} = \dfrac{12}{25}$

8. $\dfrac{1}{1 + 3708} = \dfrac{1}{3709}$

9. $26\% = \dfrac{13}{50}$; $a = 13$ and $a + b = 50$, so $b = 37$. The odds the person selected is under 18 are 13 to 37; the odds the person selected is 18 or older are 37 to 13.

10. $25\% = \dfrac{1}{4}$; $a = 1$ and $a + b = 4$, so $b = 3$. The odds are 1 to 3.

11. $\dfrac{5}{10} \times \dfrac{4}{9} \times \dfrac{3}{8} = \dfrac{1}{12}$

12. There are $C(4, 2) = 6$ ways to choose a "pair" of socks from the drawer; of these, 2 have the same color, so the probability is $\dfrac{2}{6} = \dfrac{1}{3}$.

13. $1 - \dfrac{C(5,1) \cdot C(95,3) + C(95,4)}{C(100,4)} = 1 - \dfrac{5 \cdot 138,415 + 3,183,545}{3,921,225}$
$= 1 - \dfrac{3,875,620}{3,921,225}$
$\approx .01163$

14. $\dfrac{C(5,1) \cdot C(4,2)}{C(9,3)} = \dfrac{5 \cdot 6}{84} = \dfrac{30}{84} = \dfrac{5}{14} \approx .3571$

15. a. Pr(prepared every question) $= \dfrac{C(8,6)}{C(10,6)} = \dfrac{28}{210} = \dfrac{2}{15}$

    b. Pr(not prepared on test) $= \dfrac{C(8,4)}{C(10,6)} = \dfrac{70}{210} = \dfrac{1}{3}$

16. $\dfrac{w+3}{(w+3)+(m+2)} = \dfrac{w+3}{w+m+5}$
    (e)

17. $1 - \left(\dfrac{1}{2}\right)^5 = \dfrac{31}{32}$

18. Pr(no tails) + Pr(1 tail each) + Pr(2 tails each) + Pr(3 tails each) $= \left(\dfrac{1}{8}\right)^2 + \left(\dfrac{3}{8}\right)^2 + \left(\dfrac{3}{8}\right)^2 + \left(\dfrac{1}{8}\right)^2 = \dfrac{5}{16}$

19. $\dfrac{2}{7} \times \dfrac{1}{6} = \dfrac{1}{21}$

20. Pr(select 4 winning teams) $= \left(\dfrac{1}{16}\right)^4 = \dfrac{1}{65,536}$

    Odds against $= 1 - \dfrac{1}{65,536} : \dfrac{1}{65,536} = 65,535 : 1$

21. a. $\left(\dfrac{1}{36}\right)^3$

    b. $\left(\dfrac{1}{10}\right)^4$

c. $\left(\dfrac{26}{36}\right)^3 \left(\dfrac{1}{2}\right)^4 = \left(\dfrac{17{,}576}{46{,}656}\right)\left(\dfrac{1}{16}\right) = \dfrac{2197}{93{,}312}$

22. a. Pr(all three cards are aces)
$= pr(A_1 \cap A_2 \cap A_3) = \dfrac{4}{52} \cdot \dfrac{4}{52} \cdot \dfrac{4}{52} = \dfrac{1}{2197}$

   b. Pr (at least one ace) = 1 − Pr(no aces)
$= 1 - \left(\dfrac{48}{52}\right)^3 = 1 - \left(\dfrac{12}{13}\right)^3 = \dfrac{469}{2197}$

23. Record the six consecutive outcomes; there are 6^6 possibilities, of which 6! contain each number exactly once. Hence the probability is $\dfrac{6!}{6^6} = \dfrac{5}{324}$.

24. Pr(4 different numbers) $= \dfrac{6 \cdot 5 \cdot 4 \cdot 3}{6 \cdot 6 \cdot 6 \cdot 6} = \dfrac{5}{18}$, so the odds in favor of getting four different numbers are $\dfrac{5}{18} : 1 - \dfrac{5}{18} = 5$ to 13.

25. $1 - \dfrac{365 \cdot 364 \cdot 363 \cdot 362 \cdot 361}{365^5} \approx .0271$.

26. $1 - \dfrac{6 \cdot 5 \cdot 4}{7 \cdot 7 \cdot 7} = 1 - \dfrac{120}{343} = \dfrac{223}{343} \approx .6501$

27. $\Pr(E \mid F) = \dfrac{.4 + .3 - .5}{.3} = \dfrac{.2}{.3} = \dfrac{2}{3}$

28. $\Pr(F) = \dfrac{\frac{1}{10}}{\frac{1}{7}} = \dfrac{7}{10}$

29. Pr(at least one tail appears in three coins given that at least one head appeared)
   1) Look at the sample space: HHH, HHT, HTH, HTT, THH, THT, TTH, TTT.
   2) The first seven outcomes have at least one head.
   3) Pr (one or more tails in the first seven outcomes) = 6/7

30. $\Pr(\text{one 3} \mid \text{no doubles}) = \dfrac{5+5}{30} = \dfrac{1}{3}$

31. a. Pr(employed) $= \dfrac{137.70}{148.93} \approx .9246$

   b. Pr(Male) $= \dfrac{79.34}{148.93} \approx .5327$

c. $\Pr(\text{Female}|\text{employed}) = \dfrac{64.41}{137.70} \approx .4678$

d. $\Pr(\text{employed}|\text{Female}) = \dfrac{64.41}{69.59} \approx .9256$

32. a. $\Pr(\text{eng}) = \dfrac{15}{50} = \dfrac{3}{10}$

b. $\Pr(\text{Eng}|\text{public}) = \dfrac{10}{25} = \dfrac{2}{5}$

c. $\Pr(\text{Private}|\text{eng}) = \dfrac{5}{15} = \dfrac{1}{3}$

d. $\Pr(\text{Public}|\text{eng}) = \dfrac{10}{15} = \dfrac{2}{3}$

33. $(.08)(.5) = .04$

34. $\Pr(R_1 \cap G_2 \cap G_3 \cap R_4) = \dfrac{10}{30} \cdot \dfrac{20}{29} \cdot \dfrac{19}{28} \cdot \dfrac{9}{27}$

$= \dfrac{95}{1827} \approx .0520$

35. No; $\Pr(F|E) = \dfrac{1}{6} > \Pr(F) = \dfrac{5}{36}$

36. No; $\Pr(F|E) = \dfrac{3}{4} \neq \Pr(F) = \dfrac{1}{2}$

37. Yes

38. Yes

39. a. $\dfrac{1}{4} \times \dfrac{1}{3} = \dfrac{1}{12}$

b. $\dfrac{1}{4} + \dfrac{1}{3} - \dfrac{1}{12} = \dfrac{1}{2}$

40. a. $(.4)(.75) = .3$

b. $.4 + .75 - .3 = .85$

SSM: Finite Math **Chapter 6:** Probability

41. $\Pr(A \cup B) = \dfrac{1}{2}$; $\Pr(A' \cap B) = \dfrac{1}{3}$

    $\Pr(A) + \Pr(A' \cap B) = \Pr(A \cup B)$

    $\Pr(A) = \dfrac{1}{2} - \dfrac{1}{3} = \dfrac{1}{6}$

42. $\Pr(A \text{ and } B) = (.4)(.3) = .12$

    $\Pr(A \text{ only}) = .3 - .12 = .18$

    $\Pr(B \text{ only}) = .4 - .12 = .28$

    $\Pr(\text{exactly } 1) = .18 + .28 = .46$

43. $\left(\dfrac{1}{3} \times \dfrac{2}{3}\right) + \left(\dfrac{2}{3} \times \dfrac{1}{3}\right) = \dfrac{4}{9}$

44. $\Pr(3 \text{ is drawn on 1st or 2nd draw}) = \dfrac{1}{3} + \left(\dfrac{1}{3}\right)\left(\dfrac{1}{2}\right) + \left(\dfrac{1}{3}\right)\left(\dfrac{1}{2}\right) = \dfrac{2}{3}$

45. You should switch. If you stay with your original choice your probability of winning remains $\dfrac{1}{3}$, whereas if you switch you lose only if your original choice was correct, so your probability of winning is $\dfrac{2}{3}$.

46. $(.60 \times .90) + (.40 \times .05) = .56 = 56\%$
    (a)

47. $\Pr(\text{both parents left-handed} \mid \text{child left-handed})$

    $= \dfrac{\Pr(\text{all three left-handed})}{\Pr(\text{child left-handed})}$

    $= \dfrac{.4 \times .25 \times .25}{(.4 \times .25 \times .25) + (.2 \times .25 \times .75) + (.2 \times .75 \times .25) + (.1 \times .75 \times .75)}$

    $= \dfrac{4}{25}$

48. $\Pr(\text{correct} \mid \text{rejected}) = \dfrac{\Pr(\text{correct}) \times \Pr(\text{rejected} \mid \text{correct})}{\Pr(\text{correct}) \times \Pr(\text{rejected} \mid \text{correct}) + \Pr(\text{incorrect}) \times \Pr(\text{rejected} \mid \text{incorrect})}$

    $= \dfrac{.80 \times .05}{(.80 \times .05) + (.20 \times .90)} = \dfrac{2}{11}$

49. $\Pr(C|\text{wrong}) = \dfrac{\Pr(C) \times \Pr(\text{wrong}|C)}{\Pr(C) \times \Pr(\text{wrong}|C) + \Pr(A) \times \Pr(\text{wrong}|A) + \Pr(B) \times \Pr(\text{wrong}|B)}$

$= \dfrac{.20 \times .05}{(.20 \times .05) + (.40 \times .02) + (.40 \times .03)}$

$= \dfrac{.01}{.03}$

$= \dfrac{1}{3}$

50. If n is the number of dragons, then $\dfrac{\text{\# of heads on 1-headed dragons}}{\text{\# of heads}} = \dfrac{\frac{n}{3}}{\frac{n}{3} + 2 \cdot \frac{n}{3} + 3 \cdot \frac{n}{3}} = \dfrac{1}{6}$

51. If E and F are independent events, then the outcome of F does not affect the outcome of E and vice versa. If the outcome of F does not affect the outcome of E, then neither would the outcome of F'.
Example: Let E = event you get a six on a die and F = event you get a H on a coin toss. Now, E and F are independent. F' is the event you get a T on a coin toss. The events E and F' are independent, so whether you get a H or T on the coin toss does not affect the probability of the six on a die.

52. If you know $\Pr(E \cup F)$, then you can compute
$\Pr(E \cap F) = \Pr(E) + \Pr(F) - \Pr(E \cup F)$. Alternatively, if you know that E and F are independent events, then you can compute
$\Pr(E \cap F) = \Pr(E) \cdot \Pr(F)$.

53. If two events are independent, then $\Pr(E \cap F) = \Pr(E) \cdot \Pr(F)$. Since the condition also states that they have nonzero probabilities, the right side of the equation has a nonzero value. Therefore, the probability of the intersection is nonzero, and events E and F must have at least one outcome in come, thus they are not mutually exclusive.

54. Suppose A and B are two mutually exclusive events with nonzero probabilities. Then $\Pr(A \cap B) = 0$ and $\Pr(A)\Pr(B) \neq 0$, so A and B are not independent.

55. True; The formula for a conditional probability is $\Pr(E|F) = \dfrac{\Pr(E \cap F)}{\Pr(F)}$.

Chapter 7

Exercises 7.1

1.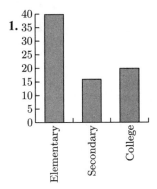

3.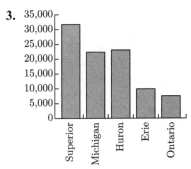

5.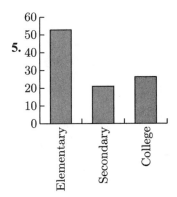

7. To find the central angle, multiply the percentage .10 by 360 to obtain 36°, choice (b).

9.
Lake	Percent	360° × Percent
Superior	33.5	120.6°
Michigan	23.6	85.0°
Huron	24.4	87.8°
Erie	10.5	37.8°
Ontario	8.0	28.8°

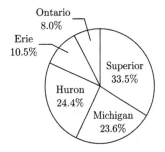

11. Pr(Master's or Doctorate) = .420 + .191 = .611

13.
0	⊞ ‖	7
1	‖	2
2	‖	2
3	‖	2
4	‖‖‖‖	4
5		0
6		0
7	‖	2
8	‖	1
9		0
10	‖	1

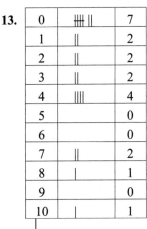

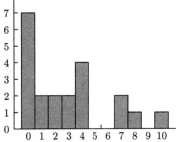

15. median of $\{2, 3, 5, 20\} = \dfrac{3+5}{2} = 4$

The median of $\{n, 1, 3, 5, 15\}$ equals 4 only if $n = 4$. Answer (c) is correct.

7-1
Copyright © 2014 Pearson Education, Inc.

17.
    2 5 7 10 15

19. min = 10, $Q_1 = 13$, $Q_2 = 17$, $Q_3 = 21$, max = 24;
    IQR = $Q_3 - Q_1 = 21 - 13 = 8$;

    10 13 17 21 24

21. min = 20, $Q_1 = 28$, $Q_2 = 42$, $Q_3 = 52.5$, max = 56;
    IQR = $Q_3 - Q_1 = 52.5 - 28 = 24.5$;

    20 28 42 52.5 56

23. (A) – (c), (B) – (d), (C) – (a), (D) – (b)

25. a. min = 200, $Q_1 = 400$, $Q_2 = 600$, $Q_3 = 700$, max = 800

    b. 25%

    c. 25%

    d. 50%

    e. 75%

27. Braves: .200, .227, .229, .243, .281, .286, .296, .317, .345, .350
    Brewers: .091, .150, .200, .250, .270, .280, .317, .333, .359

    .200 .229 .2835 .317 .350

    .091 .175 .270 .325 .359

    Possible prediction: The Braves seem more likely to win because their hitting is more consistent with a higher median.

Exercises 7.2

1. | Grade | Relative Frequency |
    |---|---|
    | 0 | $\frac{2}{25} = .08$ |
    | 1 | $\frac{3}{25} = .12$ |
    | 2 | $\frac{10}{25} = .40$ |
    | 3 | $\frac{6}{25} = .24$ |
    | 4 | $\frac{4}{25} = .16$ |

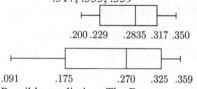

3. | Number of calls during minute | Relative Frequency |
    |---|---|
    | 20 | $\frac{3}{60} = .05$ |
    | 21 | $\frac{3}{60} = .05$ |
    | 22 | $\frac{0}{60} = 0$ |
    | 23 | $\frac{6}{60} = .10$ |
    | 24 | $\frac{18}{60} = .30$ |
    | 25 | $\frac{12}{60} = .20$ |
    | 26 | $\frac{0}{60} = 0$ |
    | 27 | $\frac{9}{60} = .15$ |
    | 28 | $\frac{6}{60} = .10$ |
    | 29 | $\frac{3}{60} = .05$ |

5. HHH, HHT, HTH, THH,
   HTT, THT, TTH, TTT

Number of Heads	Probability
0	$\dfrac{\binom{3}{0}}{2^3} = \dfrac{1}{8}$
1	$\dfrac{\binom{3}{1}}{2^3} = \dfrac{3}{8}$
2	$\dfrac{\binom{3}{2}}{2^3} = \dfrac{3}{8}$
3	$\dfrac{\binom{3}{3}}{2^3} = \dfrac{1}{8}$

7.

Number of Red Balls	Probability
0	$\dfrac{\binom{3}{0}\binom{4}{3}}{\binom{7}{3}} = \dfrac{4}{35}$
1	$\dfrac{\binom{3}{1}\binom{4}{2}}{\binom{7}{3}} = \dfrac{18}{35}$
2	$\dfrac{\binom{3}{2}\binom{4}{1}}{\binom{7}{3}} = \dfrac{12}{35}$
3	$\dfrac{\binom{3}{3}\binom{4}{0}}{\binom{7}{3}} = \dfrac{1}{35}$

9.

No. Red Balls	Player's Earnings	Probability
2	$5	$\dfrac{\binom{2}{2}\binom{4}{0}}{\binom{6}{2}} = \dfrac{1}{15}$
1	$1	$\dfrac{\binom{2}{1}\binom{4}{1}}{\binom{6}{2}} = \dfrac{8}{15}$
0	–$1	$\dfrac{\binom{2}{0}\binom{4}{2}}{\binom{6}{2}} = \dfrac{6}{15}$

11. $\Pr(5 \le X \le 7)$
$= \Pr(X=5) + \Pr(X=6) + \Pr(X=7)$
$= .2 + .1 + .3$
$= .6$

13.

k	$\Pr(X^2 = k)$
0	.1
1	.2
4	.3
9	.2
16	.2

15.

k	$\Pr(X - 1 = k)$
–1	.1
0	.2
1	.3
2	.2
3	.2

17.

k	$\Pr\left(\tfrac{1}{5}Y = k\right)$
1	.3
2	.4
3	.1
4	.1
5	.1

19.

$(X+1)^2 = k$	$\Pr((X+1)^2 = k)$
$(0+1)^2 = 1$	.1
$(1+1)^2 = 4$	.2
$(2+1)^2 = 9$	.3
$(3+1)^2 = 16$	.2
$(4+1)^2 = 25$	.2

21.

	Relative Frequency	
Grade	9 A.M. class	10 A.M. class
F	$\tfrac{10}{60} \approx .17$	$\tfrac{16}{100} = .16$
D	$\tfrac{15}{60} = .25$	$\tfrac{23}{100} = .23$
C	$\tfrac{20}{60} \approx .33$	$\tfrac{15}{100} = .15$
B	$\tfrac{10}{60} \approx .17$	$\tfrac{21}{100} = .21$
A	$\tfrac{5}{60} \approx .08$	$\tfrac{25}{100} = .25$

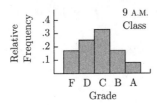

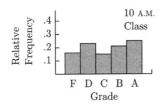

The 9 A.M. class has the distribution centered on the C grade with relatively few A's. The 10 A.M. class has a large percentage of A's and D's with fewer C's.

23. percentage with C or higher:
$$\left(\frac{10+6+4}{25}\right) \times 100\% = 80\%$$

25. a. less than 22: 3 + 3 = 6
more than 27: 6 + 3 = 9
combined: $\left(\frac{6+9}{60}\right) \times 100\% = 25\%$

b. between 23 and 25:
$\left(\frac{6+18+12}{60}\right) \times 100\% = 60\%$

c.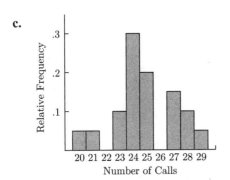

d. Estimated average number of calls would be 24 since that number has the highest frequency of occurrence.
It is actually ≈ 25.

27. a. $\Pr(U=4) = 1 - \left(\frac{3}{15} + \frac{2}{15} + \frac{4}{15} + \frac{5}{15}\right)$
$= 1 - \frac{14}{15}$
$= \frac{1}{15}$

b. $\Pr(U \geq 2) = \Pr(U=2) + \Pr(U=3) + \Pr(U=4)$
$= \frac{4}{15} + \frac{5}{15} + \frac{1}{15}$
$= \frac{10}{15}$
$= \frac{2}{3}$

c. $\Pr(U \leq 3) = 1 - \Pr(U=4)$
$= 1 - \frac{1}{15}$
$= \frac{14}{15}$

d.

$U+2 = K$	$\Pr(U+2 = K)$
$0+2 = 2$	$\frac{3}{15}$
$1+2 = 3$	$\frac{2}{15}$
$2+2 = 4$	$\frac{4}{15}$
$3+2 = 5$	$\frac{5}{15}$
$4+2 = 6$	$\frac{1}{15}$

$\Pr(U+2 < 4)$
$= \Pr(U+2 = 2) + \Pr(U+2 = 3)$
$= \frac{3}{15} + \frac{2}{15}$
$= \frac{5}{15}$
$= \frac{1}{3}$

e.

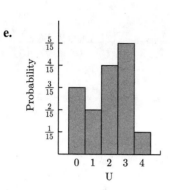

Exercises 7.3

1. $\binom{5}{2}(.3)^2(.7)^3 = (10)(.09)(.343) = .3087$

3. $\binom{4}{3}\left(\frac{1}{3}\right)^3\left(\frac{2}{3}\right)^1 = (4)\left(\frac{1}{27}\right)\left(\frac{2}{3}\right) = \frac{8}{81} \approx .0988$

5. $\Pr(X=4) = \binom{10}{4}\left(\frac{1}{2}\right)^4\left(\frac{1}{2}\right)^6 = \frac{105}{512} \approx .2051$

7. $\Pr(X=7) = \binom{10}{7}\left(\frac{1}{2}\right)^7\left(\frac{1}{2}\right)^3 = \frac{15}{128} \approx .1171875$

   $\Pr(X=8) = \binom{10}{8}\left(\frac{1}{2}\right)^8\left(\frac{1}{2}\right)^2 = \frac{45}{1024} \approx .043945$

   $\Pr(X=7 \text{ or } 8) = .1171875 + .043945 = .1611$

9. $\Pr(\text{At least } 1) = 1 - \Pr(X=0)$
   $= 1 - .0009766$
   $= .9990$

11. $\Pr(X=0) = \binom{4}{0}\left(\frac{1}{6}\right)^0\left(\frac{5}{6}\right)^4 = \frac{625}{1296} \approx .4823$

13. $\Pr(X=3) = \binom{4}{3}\left(\frac{1}{6}\right)^3\left(\frac{5}{6}\right)^1 = \frac{5}{324} \approx .015432$

    $\Pr(X=2 \text{ or } 3) = .115742 + .015432 = .1312$

15. $\Pr(X=2 \text{ or } 1 \text{ or } 0)$
    $= .1157 + .3858 + .4823 = .9838$

17. $\Pr(X=2) = \binom{8}{2}(.14)^2(.86)^6 \approx .2220$

19. $\Pr(X=4) = \binom{8}{4}(.14)^4(.86)^4 \approx .01471$

    $\Pr(X=5) = \binom{8}{5}(.14)^5(.86)^3 \approx .00192$

    $\Pr(X=4 \text{ or } 5) = .01471 + .00192 = .01663$

21. $\Pr(\text{At least } 3) = 1 - \Pr(X=0 \text{ or } 1 \text{ or } 2)$
    $= 1 - .299217 - .389679 - .222026$
    $= .08908$

23. $\Pr(X=7) = \binom{7}{7}(.764)^7(.236)^0 \approx .1519$

25. $\Pr(X=6) = \binom{7}{6}(.764)^6(.236)^1 \approx .32853$

    $\Pr(X=6 \text{ or } 7) = .32853 + .15193 = .4805$

27. **a.** $\Pr(X=2) = \binom{4}{2}\left(\frac{1}{2}\right)^2\left(\frac{1}{2}\right)^2 = \frac{3}{8} = .375$

    b. $\Pr(X=3) = 2\binom{4}{3}\left(\frac{1}{2}\right)^3\left(\frac{1}{2}\right)^1 = \frac{8}{16} = .5$

    c. $\Pr(X=4) = 2\binom{4}{4}\left(\frac{1}{2}\right)^4\left(\frac{1}{2}\right)^0 = \frac{2}{16} = .125$

29. $\Pr(X=3) = \binom{5}{3}(.76)^3(.24)^2 \approx .2529$

    $\Pr(X=5) = \binom{5}{5}(.76)^5(.24)^0 \approx .2536$

    The probability of getting 5 successes is higher.

31. Based on the histogram, the probability of getting 10 successes is higher.

33. **a.** The value with the highest probability is $X=2$.

    b. The value with a probability around .2 is $X=3$.

35. a. $1 - .4602 - .0796 = .4602$

b. $.4602 + .0796 = .5398$

37. The probability of success is .5 therefore the histogram will be symmetrical about the middle value, or 5.

39. 1; because the sum of all individual probabilities is 1.

41. Let "success" = "lives to 100." Then $p = .0217$, $q = .9783$, $n = 77$.
$$\Pr(X \geq 2) = 1 - \Pr(X = 0) - \Pr(X = 1)$$
$$= 1 - (.9783)^{77} - 77(.0217)^1 (.9783)^{76}$$
$$\approx .500$$

43. Let "success" = "adverse reaction." Then $p = .02$, $q = .98$, $n = 56$.
$$\Pr(X \geq 3) = 1 - \Pr(X = 0) - \Pr(X = 1) - \Pr(X = 2)$$
$$= 1 - (.98)^{56} - \binom{56}{1}(.02)^1(.98)^{55} - \binom{56}{2}(.02)^2(.98)^{54}$$
$$\approx .1018$$

45. Let "success" be "a vote for" the candidate. Then $p = .6$, $q = .4$, $n = 5$.
$$\Pr(X \leq 2)$$
$$= \Pr(X = 0) + \Pr(X = 1) + \Pr(X = 2)$$
$$= \binom{5}{0}(.6)^0(.4)^5 + \binom{5}{1}(.6)^1(.4)^4 + \binom{5}{2}(.6)^2(.4)^3$$
$$= .01024 + .0768 + .2304$$
$$\approx .3174$$

47. Let "success" = "defective." Then $p = .03$, $q = .97$, $n = 30$.
$$\Pr(X \geq 2) = 1 - \Pr(X = 0) - \Pr(X = 1)$$
$$= 1 - (.97)^{30} - 30(.03)^1(.97)^{29}$$
$$\approx .2269$$

49.

		Mother	
		A	a
Father	A	AA	Aa
	a	Aa	aa

Child's genes	Probability
AA	$\frac{1}{4}$
Aa	$\frac{2}{4}$
aa	$\frac{1}{4}$

Let "success" be "aa." Then
$p = \frac{1}{4}$, $q = \frac{3}{4}$, $n = 3$.
$$\Pr(X \geq 1) = 1 - \Pr(X = 0)$$
$$= 1 - \binom{3}{0}\left(\frac{1}{4}\right)^0 \left(\frac{3}{4}\right)^3$$
$$\approx 1 - .4219$$
$$= .5781$$

51. Let "success" be "gets a hit." Then $p = .3$, $q = .7$, $n = 4$.
$$\Pr(X = 0) = \binom{4}{0}(.3)^0(.7)^4 = .2401$$
$$\Pr(X = 3) = \binom{4}{3}(.3)^3(.7)^1 = .0756$$

53. Here $p = 0.82$. The expected value is
$\mu = np = 10(0.82) = 8.2$
$$\Pr(X = 8) = C(10,8)(0.82)^8(0.18)^2 = 0.2980$$
$$\Pr(X = 9) = C(10,9)(0.82)^9(0.18)^1 = 0.3017$$
So 9 is the most likely number.

55. **a.** For the underdog to win the series in five sets, the underdog must win two of the first four sets and win the fifth set. The probability that this occurs is $\binom{4}{2}(1-p)^2 p^2 (1-p) = \binom{4}{2}(1-p)^3 p^2$.

b. Using the same reasoning,
Pr(underdog wins in 3 sets) $= (1-p)^3$
Pr(underdog wins in 4 sets) $= \binom{3}{2}(1-p)^2 p(1-p)$
$= \binom{3}{2}(1-p)^3 p$

c.

Pr(underdog wins set) $= (1-p)^3 \left[1 + \binom{3}{1}p + \binom{4}{2}p^2\right]$
$= (1-p)^3(1+3p+6p^2)$
$= (1-.7)^3(1+3(.7)+6(.7)^2)$
$= .16308$

d. Suppose all five sets of the match are played even if one of the players wins three sets before the fifth set. Then the probability that the underdog wins at least three sets is
$\binom{5}{0}(1-p)^5 + \binom{5}{1}p(1-p)^4 + \binom{5}{2}p^2(1-p)^3$.

Substituting $p = .7$ gives a probability $\approx .16308$.

57. 17; because $1 - \binom{17}{0}\left(\frac{1}{6}\right)^0 \left(\frac{5}{6}\right)^{17} = .9549$

59. 114; because
$1 - \binom{114}{0}(.00045)^0 (.99955)^{114} = .0500$

61. Using a TI83 graphing calculator input the following steps:
binomcdf(100, .5, 60) − binomcdf(100, .5, 39) the answer will be approximately .9648.

63. Using a TI83 graphing calculator input the following steps:
$1 -$ binomcdf(100, .6, 59) $\approx .5433$

65. **a.** Substituting $p = .6$ in this formula gives a probability $\approx .2898$.

b. Suppose as in Problem 56(d) above that all $2n - 1$ games are played. Then the probability that the favorite wins at least n games is
$\binom{2n-1}{n}p^n(1-p)^{n-1} + \binom{2n-1}{n+1}p^{n+1}(1-p)^{n-2}$
$+ \cdots + \binom{2n-1}{2n-1}p^{2n-1}$.

Some experimenting shows that for $p = .6$, $n = 21$ is the smallest value of n for which this probability exceeds .9. Since the number of games is $2n - 1 = 2(21) - 1 = 41$.

Exercises 7.4

1. $E(X) = 0(.15) + 1(.2) + 2(.1) + 3(.25) + 4(.3) = 2.35$

3. **a.** GPA $= \dfrac{4+4+4+3+3+3+3+2+2+1}{10}$
$= \dfrac{29}{10}$
$= 2.9$

b.

Grade	Relative Frequency
4	$\frac{3}{10} = .3$
3	$\frac{4}{10} = .4$
2	$\frac{2}{10} = .2$

SSM: Finite Math   Chapter 7: Probability and Statistics

                    1                    $\frac{1}{10} = .1$

   c. $E(X) = 4(.3) + 3(.4) + 2(.2) + 1(.1)$
       $= 2.9$

5. $\bar{x}_A = 0(.3) + 1(.3) + 2(.2) + 3(.1) + 4(0) + 5(.1)$
   $= 1.5$
   $\bar{x}_B = 0(.2) + 1(.3) + 2(.3) + 3(.1) + 4(.1) + 5(0)$
   $= 1.6$
   Group A had fewer cavities.

7. $E(X) = 1(.2) + 2(.4) + 3(.3) + 4(.1)$
   $= 2.3$

9. $E(X) = 7(.25) + 8(.25) + 9(.25) + 10(.25)$
   $= 8.5$

11. | Earnings | Probability |
     |----------|-------------|
     | –$1 | $\frac{37}{38}$ |
     | $35 | $\frac{1}{38}$ |

   $E(X) = -1\left(\frac{37}{38}\right) + 35\left(\frac{1}{38}\right) \approx -\$.0526$

13. | Earnings | Probability |
     |----------|-------------|
     | –50¢ | $\frac{2}{6} = \frac{1}{3}$ |
     | 0¢ | $\left(\frac{4}{6}\right)\left(\frac{2}{5}\right) = \frac{4}{15}$ |
     | 50¢ | $\left(\frac{4}{6}\right)\left(\frac{3}{5}\right)\left(\frac{2}{4}\right) = \frac{1}{5}$ |
     | $1 | $\left(\frac{4}{6}\right)\left(\frac{3}{5}\right)\left(\frac{2}{4}\right)\left(\frac{2}{3}\right) = \frac{2}{15}$ |
     | $1.50 | $\left(\frac{4}{6}\right)\left(\frac{3}{5}\right)\left(\frac{2}{4}\right)\left(\frac{1}{3}\right)\left(\frac{2}{2}\right) = \frac{1}{15}$ |

   $E(X) = -.5\left(\frac{1}{3}\right) + 0\left(\frac{4}{15}\right) + .5\left(\frac{1}{5}\right) + 1\left(\frac{2}{15}\right) + 1.5\left(\frac{1}{15}\right)$
   $\approx \$.1667$

15. Let x be the cost of the policy.
    $E(X) = (-x)(.9) + (10{,}000 - x)(.1) = -x + 1000$
    The expected value is zero if $x = 1000$.
    He should be willing to pay up to $1000.

17. $\frac{5 \times .300 + 4 \times .350}{9} \approx .322$
    Answer (b) is correct.

19. | Recorded Value | Probability |
     |----------------|-------------|
     | 1 | $\frac{1}{36}$ |
     | 2 | $\frac{3}{36}$ |
     | 3 | $\frac{5}{36}$ |
     | 4 | $\frac{7}{36}$ |
     | 5 | $\frac{9}{36}$ |
     | 6 | $\frac{11}{36}$ |

$$E(X) = 1\left(\frac{1}{36}\right) + 2\left(\frac{3}{36}\right) + 3\left(\frac{5}{36}\right) + 4\left(\frac{7}{36}\right) + 5\left(\frac{9}{36}\right) + 6\left(\frac{11}{36}\right) \approx 4.47$$

21. $E(X) = -1\left(\frac{125}{216}\right) + 1\left(\frac{75}{216}\right) + 3\left(\frac{15}{216}\right) + 5\left(\frac{1}{216}\right) = \frac{0}{216} = 0$

23. $E(X) = 15(.217) = 3.255$

25. The expected number of times that a 5 or a 6 will appear is $\mu = np = 30\left(\frac{1}{3}\right) = 10$

27. The expected value of points when taking one three point shot is $\mu = 3(.40) = 1.2$ points. The expected value of taking three free-throws when the success probability is 60% is $\mu = 3(.60) = 1.8$, which is the greater expected value.

29.
Number of Republicans	Probability
0	$\frac{10}{84}$
1	$\frac{40}{84}$
2	$\frac{30}{84}$
3	$\frac{4}{84}$

$E(\text{Rep.}) = 0\left(\frac{10}{84}\right) + 1\left(\frac{40}{84}\right) + 2\left(\frac{30}{84}\right) + 3\left(\frac{4}{84}\right) = \frac{4}{3} \approx 1.333$

$E(\text{Dem.}) = 3 - E(\text{Rep.}) = 3 - \frac{4}{3} = \frac{5}{3} \approx 1.667$

31. No Replacing

Number of Red	Probability
0	$\frac{1}{35}$
1	$\frac{12}{35}$
2	$\frac{18}{35}$
3	$\frac{4}{35}$

$E(\text{Red}) = 0\left(\frac{1}{35}\right) + 1\left(\frac{12}{35}\right) + 2\left(\frac{18}{35}\right) + 3\left(\frac{4}{35}\right)$
$= \frac{12}{7} \approx 1.714$

Replacing

Number of Red	Probability
0	$\frac{27}{343}$
1	$\frac{108}{343}$
2	$\frac{144}{343}$
3	$\frac{64}{343}$

$$E(\text{Red}) = 0\left(\frac{27}{343}\right) + 1\left(\frac{108}{343}\right) + 2\left(\frac{144}{343}\right) + 3\left(\frac{64}{343}\right)$$
$$= \frac{12}{7} \approx 1.714$$

33. Let x = chance of rain.
$$-8000 + 40{,}000x = 0$$
$$x = .20 \to 20\%$$
Answer (a) is correct. (Attendance revenue is irrelevant, since it is not affected by the decision whether or not to purchase insurance.)

35. $\dfrac{7x + 4y}{x + y}$

   Answer (b) is correct.

37. Solve $\dfrac{16 \cdot 54 + 14x}{30} = 56.1$
$$14x + 864 = 1683$$
$$14x = 819$$
$$x = 58.5°$$

39. $\dfrac{5 + 6 + x}{3} = \dfrac{2 + 7 + 9}{3}$
$$11 + x = 18$$
$$x = 7$$
Answer (d) is correct.

41. Let d = size of Dick's card collection.
$$\dfrac{\frac{3}{2}d + d + \frac{1}{2}d}{3} = 120$$
$$d = 120$$
Tom has $\dfrac{3}{2}d = \dfrac{3}{2}(120) = 180$
Answer (d) is correct.

43. Let x = number of cases.
$$20\left(\frac{1}{2}x\right) + 30\left(\frac{1}{2}x\right) = 75{,}000$$
$$25x = 75{,}000$$
$$x = 3000 \text{ cases}$$
Answer (a) is correct.

Exercises 7.5

1. $m = 70(.5) + 71(.2) + 72(.1) + 73(.2) = 71$
   $\sigma^2 = (70-71)^2(.5) + (71-71)^2(.2) + (72-71)^2(.1) + (73-71)^2(.2)$
   $= .5 + 0 + .1 + .8$
   $= 1.4$

3. B

5. a. $\mu_A = -10\left(\frac{1}{5}\right) + 20\left(\frac{3}{5}\right) + 25\left(\frac{1}{5}\right) = 15$

   $\mu_B = 0(.3) + 10(.4) + 30(.3) = 13$

   $\sigma_A^2 = (-10-15)^2\left(\frac{1}{5}\right) + (20-15)^2\left(\frac{3}{5}\right) + (25-15)^2\left(\frac{1}{5}\right) = 125 + 15 + 20 = 160$

   $\sigma_B^2 = (0-13)^2(.3) + (10-13)^2(.4) + (30-13)^2(.3) = 50.7 + 3.6 + 86.7 = 141$

   b. Investment A

   c. Investment B

7. a. $\mu_A = 100(.1) + 101(.2) + 102(.3) + 103(0) + 104(0) + 105(.2) + 106(.2) = 103$
   $\sigma_A^2 = (100-103)^2(.1) + (101-103)^2(.2) + \cdots + (106-103)^2(.2) = 4.6$
   $\mu_B = 100(0) + 101(.2) + 102(0) + 103(.2) + 104(.1) + 105(.2) + 106(.3) = 104$
   $\sigma_B^2 = (100-104)^2(0) + (101-104)^2(.2) + \cdots + (106-104)^2(.3) = 3.4$

   b. Business B

   c. Business B

9. The number of heads is a binomial random variable X with $n = 12$, $p = .5$ so $\mu_X = 12 \times .5 = 6$, $\sigma_X = \sqrt{12 \times .5 \times .5} \approx 1.732$.

11. The number of defective widgets is a binomial random variable X with $n = 200$, $p = .015$ so $\mu_X = 200 \times .015 = 3$, $\sigma_X = \sqrt{200 \times .015 \times .985} \approx 1.719$.

13. a. $35 - c = 25$ and $35 + c = 45 \Rightarrow c = 10$.
    Probability $\geq 1 - \dfrac{5^2}{10^2} = 1 - \dfrac{25}{100} = .75$

    b. $35 - c = 20$ and $35 + c = 50 \Rightarrow c = 15$
    Probability $\geq 1 - \dfrac{5^2}{15^2} \approx .89$

    c. $35 - c = 29$ and $35 + c = 41 \Rightarrow c = 6$
    Probability $\geq 1 - \dfrac{5^2}{6^2} \approx .31$

15. $\mu = 3000, \sigma = 250$
$3000 - c = 2000$ and $3000 + c = 4000 \Rightarrow c = 1000$
$$\text{Probability} \geq 1 - \frac{250^2}{1000^2} = .9375$$
Number of bulbs to replace: $\geq 5000(.9375) \approx 4688$

17. $\text{Probability} = 1 - \frac{6^2}{c^2} = \frac{7}{16}$
$16c^2 - 576 = 7c^2$
$9c^2 = 576$
$c = \sqrt{64} = 8$

19. a. $\mu = 2\left(\frac{1}{36}\right) + 3\left(\frac{2}{36}\right) + \cdots + 12\left(\frac{1}{36}\right) = 7$
$\sigma^2 = (2-7)^2\left(\frac{1}{36}\right) + \cdots + (12-7)^2\left(\frac{1}{36}\right)$
$= \frac{210}{36}$
$= \frac{35}{6}$

b. $\Pr(4 \leq X \leq 10)$
$= \frac{3}{36} + \frac{4}{36} + \frac{5}{36} + \frac{6}{36} + \frac{5}{36} + \frac{4}{36} + \frac{3}{36}$
$= \frac{30}{36}$
$= \frac{5}{6}$

c. $7 - c = 4$ and $7 + c = 10 \Rightarrow c = 3$
$$\text{Probability} \geq 1 - \frac{\frac{35}{6}}{3^2} = \frac{19}{54}$$

21. $E(X) = (-2 - 1 + 0 + 1 + 2)(.2) = 0$
$E(X^2) = (4 + 1 + 0 + 1 + 4)(.2) = 2$
$\text{Var}(X) = E(X^2) - E(X)^2 = 2 - 0 = 2$

23.

$2X$	Probability
–2	$\frac{1}{8}$
–1	$\frac{3}{8}$
0	$\frac{1}{8}$
1	$\frac{1}{8}$
2	$\frac{2}{8}$

$$\mu = -2\left(\frac{1}{8}\right) - 1\left(\frac{3}{8}\right) + \cdots + 2\left(\frac{2}{8}\right) = 0$$

$$\sigma^2_{2X} = (-2-0)^2\left(\frac{1}{8}\right) + \cdots + (2-0)^2\left(\frac{2}{8}\right) = 2$$

$$\sigma^2_{2X} = 4\sigma^2_X = 4\left(\frac{1}{2}\right) = 2$$

25. $\mu = \dfrac{59{,}794 + 58{,}587 + \cdots + 47{,}800}{8} = 53{,}294.63$

$\sigma^2 = \dfrac{(59{,}794 - 53{,}294.63)^2 + \cdots + (47{,}800 - 53{,}294.63)^2}{8}$

$\sigma \approx 4224.40$

27. $\mu = 4.72$, $\sigma \approx 1.40$

Exercises 7.6

1. $A(1.25) = .8944$

3. $1 - A(.25) = 1 - .5987 = .4013$

5. $A(1.5) - A(.5) = .9332 - .6915 = .2417$

7. $A(-.5) + (1 - A(.5)) = .3085 + (1 - .6915)$
   $\qquad\qquad\qquad\qquad\qquad = .6170$

9. $\Pr(Z \geq z) = .0401$
   $A(z) = 1 - .0401 = .9599$
   $z = 1.75$

11. $\Pr(-z \leq Z \leq z) = .5468$

    $A(-z) = \dfrac{1 - .5468}{2} = .2266$

    $-z = -.75$
    $z = .75$

13. The 90th percentile of the standard normal distribution is 1.28 (use table or InvNorm(0.90) on TI 83)

15. $\mu = 6, \sigma \approx 2$

17. $\mu = 9, \sigma \approx 1$

19. $\dfrac{6-8}{\frac{3}{4}} = -\dfrac{2}{1} \cdot \dfrac{4}{3} = -\dfrac{8}{3}$

21. $\dfrac{x-8}{\frac{3}{4}} = 10$

    $x = \dfrac{30}{4} + 8 = \dfrac{62}{4} = 15\dfrac{1}{2}$

    Answer (a) is correct.

23. $\Pr(X \geq 9) = \Pr\left(Z \geq \dfrac{9-10}{\frac{1}{2}}\right)$

    $= \Pr(Z \geq -2)$
    $= 1 - \Pr(Z \leq -2)$
    $= 1 - .0228$
    $= .9772$

25. $\Pr(6 \leq X \leq 10) = \Pr\left(\dfrac{6-7}{2} \leq Z \leq \dfrac{10-7}{2}\right)$

    $= \Pr(-.50 \leq Z \leq 1.50)$
    $= .9332 - .3085$
    $= .6247$

27. $\Pr(-2 \leq Z \leq 2) = A(2) - A(-2)$

    $= .9772 - .0228$
    $= .9544$

29. From Table 2 we see that $\Pr(Z \leq 2) = .9772$ for standard normal Z. Solve $\dfrac{6-5}{\sigma} = 2$: $2\sigma = 1$,

    $\sigma = .5$.

31. $\mu = 3.3, \sigma \approx .2$

    $\Pr(X \geq 4) = \Pr\left(Z \geq \dfrac{4-3.3}{.2}\right)$

    $= \Pr(Z \geq 3.5)$
    $= 1 - \Pr(Z \leq 3.5)$
    $= 1 - .9998$
    $= .0002$

33. $\mu = 6, \sigma = .02$

    $\Pr(5.95 \leq X \leq 6.05)$

    $= \Pr\left(\dfrac{5.95-6}{.02} \leq Z \leq \dfrac{6.05-6}{.02}\right)$

    $= \Pr(-2.5 \leq Z \leq 2.5)$
    $= .9938 - .0062$
    $= .9876$

35. $\mu = 5.4, \sigma = .6$

    $\Pr(X > 5.832) \approx .2358$

37. $\mu = 30{,}000, \sigma = 4000$

    $\Pr(X > 39{,}000)$

    $= \Pr\left(Z > \dfrac{39{,}000 - 30{,}000}{4000}\right)$

    $= \Pr(Z > 2.25)$
    $= 1 - \Pr(Z \leq 2.25)$
    $= 1 - .9878$
    $= .0122$

39. $\mu = 520$, $\sigma = 75$

   a. $x_{90} = 520 + 75z_{90}$
      $= 520 + 75 \times 1.28$
      $= 616$

   b. $\Pr(-z \leq Z \leq z) = .90$
      $\Pr(Z \leq -z) = .05 \Rightarrow z_{05} \approx -1.65$
      $\dfrac{x - \mu}{\sigma} = \dfrac{x - 520}{75} = -1.65$
      $\Rightarrow x_{05} = 396.25 \approx 396$
      $\dfrac{x - \mu}{\sigma} = \dfrac{x - 520}{75} = 1.65$
      $\Rightarrow x_{95} = 643.75 \approx 644$
      Between 396 and 644

   c. $x_{98} = 520 + 75z_{98} = 520 + 75 \times 2.05 = 674$

41. $\mu = 30{,}000$, $\sigma = 5000$
    $\Pr(Z \leq z) = .02 \Rightarrow z_{02} \approx -2.05$
    $\dfrac{x - \mu}{\sigma} = \dfrac{x - 30{,}000}{5000} = -2.05 \Rightarrow x_{02} = 19{,}750$
    19,750 miles

43. True; As σ increases, the normal curve flattens out.

45. normalcdf(1,5,3,2)

    NORMDIST (5,3,2,TRUE) − NORMDIST (1,3,2,TRUE)

    $P(1 < x < 5)$ for x normal (3,2)

47. normalcdf(0,1000,1200,100)

    NORMDIST (1000,1200,100,TRUE)

    $P(x < 1000)$ for x normal (1200,100)

Exercises 7.7

1. $n = 25$, $p = \dfrac{1}{5}$

   $\mu = np = 25\left(\dfrac{1}{5}\right) = 5$,

   $\sigma = \sqrt{npq} = \sqrt{25\left(\dfrac{1}{5}\right)\left(\dfrac{4}{5}\right)} = 2$

   a. $\Pr(X = 5) \approx \Pr\left(\dfrac{4.5 - 5}{2} \leq Z \leq \dfrac{5.5 - 5}{2}\right)$
      $= \Pr(-.25 \leq Z \leq .25)$
      $= .5987 - .4013$
      $= .1974$

   b. $\Pr(3 \leq X \leq 7)$
      $\approx \Pr\left(\dfrac{2.5 - 5}{2} \leq Z \leq \dfrac{7.5 - 5}{2}\right)$
      $= \Pr(-1.25 \leq Z \leq 1.25)$
      $= .8944 - .1056$
      $= .7888$

   c. $\Pr(X < 10) \approx \Pr\left(Z \leq \dfrac{9.5 - 5}{2}\right)$
      $= \Pr(Z \leq 2.25)$
      $= .9878$

3. $n = 20$, $p = \dfrac{1}{6}$

   $\mu = 20\left(\dfrac{1}{6}\right) = \dfrac{10}{3}$, $\sigma = \sqrt{20\left(\dfrac{1}{6}\right)\left(\dfrac{5}{6}\right)} = \dfrac{5}{3}$

   $\Pr(X \geq 8) \approx \Pr\left(Z \geq \dfrac{7.5 - \frac{10}{3}}{\frac{5}{3}}\right)$
   $= \Pr(Z \geq 2.5)$
   $= 1 - .9938$
   $= .0062$

5. $n = 100$, $p = \dfrac{1}{2}$

$\mu = 100\left(\dfrac{1}{2}\right) = 50$, $\sigma = \sqrt{100\left(\dfrac{1}{2}\right)\left(\dfrac{1}{2}\right)} = 5$

$\Pr(X \geq 63) \approx \Pr\left(Z \geq \dfrac{62.5 - 50}{5}\right)$
$= \Pr(Z \geq 2.5)$
$= 1 - .9938$
$= .0062$

7. $n = 75$, $p = \dfrac{3}{4}$

$\mu = 75\left(\dfrac{3}{4}\right) = 56.25$, $\sigma = \sqrt{75\left(\dfrac{3}{4}\right)\left(\dfrac{1}{4}\right)} = 3.75$

$\Pr(X \geq 68) \approx \Pr\left(Z \geq \dfrac{67.5 - 56.25}{3.75}\right)$
$= \Pr(Z \geq 3)$
$= 1 - .9987$
$= .0013$

9. $n = 20$, $p = .310$
$\mu = 20(.310) = 6.2$,
$\sigma = \sqrt{20(.31)(.69)} \approx 2.068$

$\Pr(X \geq 6) \approx \Pr\left(Z \geq \dfrac{5.5 - 6.2}{2.068}\right)$
$\approx \Pr(Z \geq -.34)$
$\approx \Pr(Z \geq -.35)$
$= 1 - .3632$
$= .6368$

11. $n = 1000$, $p = .02$
$\mu = 1000(.02) = 20$,
$\sigma = \sqrt{1000(.02)(.98)} \approx 4.427$

$\Pr(X < 15) \approx \Pr\left(Z \leq \dfrac{14.5 - 20}{4.427}\right)$
$\approx \Pr(Z \leq -1.24)$
$\approx \Pr(Z \leq -1.25)$
$= .1056$

13. probability of failure $= (.01)(.02)(.01) = .000002$
$n = 1,000,000$,
$E(X) = \mu = 1,000,000(.000002) = 2$
$\sigma = \sqrt{1,000,000(.000002)(.999998)} \approx 1.414$

$\Pr(X > 3) \approx \Pr\left(Z \geq \dfrac{3.5 - 2}{1.414}\right)$
$\approx \Pr(Z \geq 1.06)$
$\approx \Pr(Z \geq 1.05)$
$= 1 - .8531$
$= .1469$

15. $n = 100$, $p = .35$
$\mu = 100(.35) = 35$, $\sigma = \sqrt{100(.35)(.65)} \approx 4.770$
$\Pr(30 \leq X \leq 40)$
$= \Pr\left(\dfrac{29.5 - 35}{4.77} \leq Z \leq \dfrac{40.5 - 35}{4.77}\right)$
$\approx \Pr(-1.15 \leq Z \leq 1.15)$
$= .8749 - .1251$
$= .7498$

17. $n = 1000$, $p = .03$
$\mu = 1000(.03) = 30$,
$\sigma = \sqrt{1000(.03)(.97)} \approx 5.394$

$\Pr(X \geq 29) \approx \Pr\left(Z \geq \dfrac{28.5 - 30}{5.394}\right)$
$\approx \Pr(Z \geq -0.278)$
$\approx \Pr(Z \geq -0.28)$
$= 1 - .3897$
$= .61$

19. $n = 100$, $p = \dfrac{1}{2}$
Exact: $\Pr(49 \leq X \leq 51) \approx .2356$
Normal Approximation:

$\mu = 100\left(\dfrac{1}{2}\right) = 50$, $\sigma = \sqrt{100\left(\dfrac{1}{2}\right)\left(\dfrac{1}{2}\right)} = 5$

$\Pr(48.5 \leq X \leq 51.5) \approx .2358$

21. $n = 150, p = .2$
    Exact: $\Pr(X = 30) \approx .0812$
    Normal Approximation: $\mu = 150(.2) = 30$,
    $\sigma = \sqrt{150(.2)(.8)} \approx 4.899$

    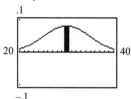

    $\Pr(29.5 \le X \le 30.5) \approx .0813$

23. .0410

    $1 - \text{binomcdf}(300, .02, 10)$

    $1 - \text{BINOMDIST}(10, 300, .02, 1)$

    binomial probabilities $n = 300, p = .02$,

    endpoint = 10

25. .0796

    $\text{binompdf}(100,.5,50)$

    $\text{BINOMDIST}(50, 100, .5, 0)$

    binomial probabilities $n = 100, p = .5$,

    endpoint = 50

Chapter 7 Fundamental Concept Check

1. *Bar Charts* and *pie charts* provide graphical ways of displaying qualitative data. A *histogram* is a graphical way of presenting a frequency distribution. A *box plot* is a graphical way of presenting a five–number summary of a collection of data.

2. The following definitions apply to a set of numbers. The *median* is the middle value when the numbers are ordered. The *first quartile* is a number for which roughly 25% of the numbers are less than that number; same as the 25th percentile. The *third quartile* is a number for which roughly 75% of the numbers are less than that that number, same as the 75th percentile. The *interquartile range* is the difference between the third quartile and the first quartile. The five–number summary consist of the minimum value, first quartile, median, third quartile, and maximum value.

3. A *frequency distribution* for a collection of numerical data is a table listing each number in the collection and the number of times it appears. A *relative frequency distribution* for a collection of numerical data is a table listing each number in the collection and the percentage of times it appears. A *probability distribution* is a table displaying the outcomes of an experiment and their probabilities.

4. Consider a table for the distribution. To construct a histogram for the table, draw a coordinate system, write the numbers from the left column of the table below the *x*–axis, and above each number draw a rectangle having height given in the second column of the table.

5. A *random variable* is a variable that assigns a number to each outcome of an experiment.

6. A *probability distribution* is a table whose first column lists the possible values of the random variable and whose second column gives the probabilities associated with each value.

7. A *binomial random variable* arises from observing the number of successes in an experiment consisting of a sequence of independent binomial trials.

8. $\binom{n}{k} p^k (1-p)^{n-k}$, where p is the probability of success on each binomial trial.

9. The three values give a general impression of the behavior of the random variable. The *expected value* gives the average you would expect when repeating the experiment many times. The *variance* gives you an idea of how closely concentrated the outcomes are likely to be around the expected value. The *standard deviation* serves the same purpose as the variance, but has the same unit of measure as the random variable.

10. The *Chebychev Inequality* gives a lower bound on the likelihood that the outcome of a random variable is within a specified distance from the expected value.

11. A *normal random variable* is a random variable whose probabilities are determined by calculating areas under a bell–shaped curve that is described by its mean μ and standard deviation σ.

12. The outcomes S for which $\Pr(X \leq S) = p\%$.

13. Probabilities associated with a *binomial random variable* having parameters n and p can be approximated with the use of a normal curve having $\mu = np$ and $\sigma = \sqrt{np(1-p)}$.
$\Pr(a \leq X \leq b)$ is approximately the area under the normal curve from $x = a - \frac{1}{2}$ to $x = b + \frac{1}{2}$.

Chapter 7 Review Exercises

1.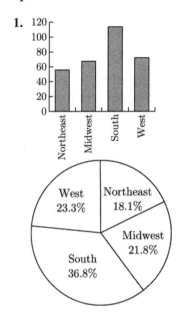

3.

Number Waiting in Line	Relative Frequency
0	.04
1	.10
2	.18
3	.26
4	.22
5	.14
6	.06

Pr(at most 3 customers in line)
$= .04 + .10 + .18 + .26$
$= .58$

4. a. Possible outcomes are HH, HT, TH, TT

Number of Heads, k	$\Pr(X = k)$
0	.25
1	.50
2	.25

b.

k	$\Pr(2X + 5 = k)$
5	.25
7	.50
9	.25

2. min = 1,
$Q_1 = 2.5, Q_2 = 4.5, Q_3 = 11.5, \max = 23;$
IQR $= Q_3 - Q_1 = 11.5 - 2.5 = 9;$

5. $n = 3$, $p = \dfrac{1}{3}$

   a.
k	$\Pr(X = k)$
0	$\binom{3}{0}\left(\dfrac{1}{3}\right)^0 \left(\dfrac{2}{3}\right)^3 = \dfrac{8}{27}$
1	$\binom{3}{1}\left(\dfrac{1}{3}\right)^1 \left(\dfrac{2}{3}\right)^2 = \dfrac{12}{27}$
2	$\binom{3}{2}\left(\dfrac{1}{3}\right)^2 \left(\dfrac{2}{3}\right)^1 = \dfrac{6}{27}$
3	$\binom{3}{3}\left(\dfrac{1}{3}\right)^3 \left(\dfrac{2}{3}\right)^0 = \dfrac{1}{27}$

   b. $\mu = 0\left(\dfrac{8}{27}\right) + 1\left(\dfrac{12}{27}\right) + 2\left(\dfrac{6}{27}\right) + 3\left(\dfrac{1}{27}\right) = 1$

   $\sigma^2 = (0-1)^2\left(\dfrac{8}{27}\right) + (1-1)^2\left(\dfrac{12}{27}\right) + (2-1)^2\left(\dfrac{6}{27}\right) + (3-1)^2\left(\dfrac{1}{27}\right)$

   $= \dfrac{2}{3}$

6. $n = 4$, $p = .3$

   $\Pr(X = 2) = \binom{4}{2}(.3)^2(.7)^2 = .2646$

7. The student has a .6 probability of guessing correctly on the six questions with answer *true* and a .4 probability of guessing correctly on the four questions with answer *false*. Therefore the student's expected score is $6(.6) + 4(.4) = 5.2$ correct answers which gives 52 points or 52%.
   A better strategy is to choose true for all the questions which guarantees a score of 60%.

8. a. $\Pr(\text{get 7 twice}) = \binom{12}{2}\left(\dfrac{1}{6}\right)^2\left(\dfrac{5}{6}\right)^{10} \approx .2961$

   b. $\Pr(\text{get 7 at least twice})$
   $= 1 - \Pr(\text{get 7 zero or one time})$
   $= 1 - \binom{12}{0}\left(\dfrac{1}{6}\right)^0\left(\dfrac{5}{6}\right)^{12} - \binom{12}{1}\left(\dfrac{1}{6}\right)^1\left(\dfrac{5}{6}\right)^{11}$
   $\approx .6187$

   c. The expected number of 7's is $12 \cdot \dfrac{1}{6} = 2$.

9. $\mu = 0(.2) + 1(.3) + 5(.1) + 10(.4) = 4.8$

   $\sigma^2 = (0-4.8)^2(.2) + (1-4.8)^2(.3) + (5-4.8)^2(.1) + (10-4.8)^2(.4)$
   $= 19.76$

10. Let X be the number of red balls.

k	$\Pr(X=k)$
0	$\dfrac{\binom{4}{0}\binom{4}{4}}{\binom{8}{4}} = \dfrac{1}{70}$
1	$\dfrac{\binom{4}{1}\binom{4}{3}}{\binom{8}{4}} = \dfrac{16}{70}$
2	$\dfrac{\binom{4}{2}\binom{4}{2}}{\binom{8}{4}} = \dfrac{36}{70}$
3	$\dfrac{\binom{4}{3}\binom{4}{1}}{\binom{8}{4}} = \dfrac{16}{70}$
4	$\dfrac{\binom{4}{4}\binom{4}{0}}{\binom{8}{4}} = \dfrac{1}{70}$

    $\mu = 0\left(\dfrac{1}{70}\right) + 1\left(\dfrac{16}{70}\right) + 2\left(\dfrac{36}{70}\right) + 3\left(\dfrac{16}{70}\right) + 4\left(\dfrac{1}{70}\right) = 2$

    $\sigma^2 = (0-2)^2\left(\dfrac{1}{70}\right) + (1-2)^2\left(\dfrac{16}{70}\right) + (2-2)^2\left(\dfrac{36}{70}\right) + (3-2)^2\left(\dfrac{16}{70}\right) + (4-2)^2\left(\dfrac{1}{70}\right)$
    $= \dfrac{4}{7}$

11. X has mean
    $\mu = (-2)(.3) + 0(.1) + 1(.4) + 3(.2) = .4$,
    variance
    $\sigma^2 = (-2-.4)^2(.3) + (0-.4)^2(.1) + (1-.4)^2(.4) + (3-.4)^2(.2)$
    $= 3.24$,
    and standard deviation
    $\sigma = \sqrt{3.24} = 1.8$.

12. When a pair of fair dice is rolled, the probabilities that the result is 7 or 11 are $\frac{1}{6}$ and $\frac{1}{18}$ respectively. Hence Lucy's expected winnings are $(-10)\frac{2}{9} + 3 \cdot \frac{7}{9} = \frac{1}{9} \approx .11$, or 11 cents per roll.

13. $\mu = 10, \sigma = \frac{1}{3}$
    $10 - c = 9$ and $10 + c = 11 \Rightarrow c = 1$
    Probability: $\geq 1 - \frac{\left(\frac{1}{3}\right)^2}{1^2} = \frac{8}{9}$

14. $\mu = 50, \sigma = 8$
    $50 - c = 38$ and $50 + c = 62 \Rightarrow c = 12$
    Probability $\geq 1 - \frac{8^2}{12^2} = \frac{5}{9}$

15. $\Pr(6.5 \leq X \leq 11) = \Pr\left(\frac{6.5-5}{3} \leq Z \leq \frac{11-5}{3}\right)$
    $= A(2) - A(.5)$
    $= .9772 - .6915$
    $= .2857$

16. $\Pr(Z \geq .75) = 1 - .7734 = .2266$

17. $\mu = 5.75, \sigma = .2$
    $\Pr(X \geq 6) = \Pr\left(Z \geq \frac{6-5.75}{.2}\right)$
    $= \Pr(Z \geq 1.25)$
    $= 1 - .8944$
    $= .1056$
    10.56%

18. $\Pr(Z \geq z) = .7734$
    $\Pr(Z < z) = 1 - .7734 = .2266$
    $z = -.75$

19. $\mu = 80, \sigma = 15$

$\Pr(80 - h \leq X \leq 80 + h) = .8664$

$\dfrac{1 - .8664}{2} = .0668 \Rightarrow$ (area left of $80 - h$)

$\Pr(Z \leq z) = .0668$ when $z = -1.5$

$\Pr(-1.5 \leq Z \leq 1.5) = .8664$

Therefore, $\dfrac{x - \mu}{\sigma} = -1.5$ and $\dfrac{x + \mu}{\sigma} = 1.5$.

$\dfrac{(80 - h) - 80}{15} = -1.5$ and $\dfrac{(80 + h) - 80}{15} = 1.5$

$h = 22.5$

20. a. $\Pr(133 \leq X) \approx \Pr\left(\dfrac{132.5 - 100}{15} \leq Z\right)$

$\approx \Pr(2.167 \leq Z)$

$\approx \Pr(2.20 \leq Z)$

$= 1 - .9861$

$= .0139$

$= 1.39\%$

b. $x_{95} = 100 + 15 z_{95}$

$= 100 + 15 \cdot 1.65$

$= 124.75$

21. $n = 54,\ p = \dfrac{2}{5}$

$\mu = 54\left(\dfrac{2}{5}\right) = 21.6$

$\sigma = \sqrt{54\left(\dfrac{2}{5}\right)\left(\dfrac{3}{5}\right)} = 3.6$

$\Pr(X \leq 13) \approx \Pr\left(Z \leq \dfrac{13.5 - 21.6}{3.6}\right) = \Pr(Z \leq -2.25) = .0122$

22. $n = 75,\ p = \dfrac{1}{4}$

$\mu = 75\left(\dfrac{1}{4}\right) = 18.75$

$\sigma = \sqrt{75\left(\dfrac{1}{4}\right)\left(\dfrac{3}{4}\right)} = 3.75$

$\Pr(8 \leq X \leq 22) \approx \Pr\left(\dfrac{7.5 - 18.75}{3.75} \leq Z \leq \dfrac{22.5 - 18.75}{3.75}\right) = \Pr(-3 \leq Z \leq 1) = .8413 - .0013 = .84$

Conceptual Exercises

23. a. scoring in the third quartile is not very good: 100, 40, 40, 40,

    b. scoring in the third quartile corresponds to a perfect grade: 100, 100, 90, 80, 70

24. a. The mean and median are equal: 1, 2, 3, 4, 5, 6, 7, 8, 9, 10 : The mean is 5.5; the median is 5.5

    b. the mean is less than the median: 1, 1, 1, 1, 4, 5, 6, 7, 8, 9 : The mean is 4.3; the median is 4.5

    c. the median is less than the mean 1, 2, 3, 4, 5, 6, 10, 12, 14, 100. The median is 5.5; the mean is 15.7

25. A population mean is the average of all the data in the entire population. When a sample is taken from a population, the sample mean is the average of all the data in that particular sample. Sample means vary whereas the population mean is fixed.

26. Expected value is a concept similar to the mean. It is the long range number you would expect to occur if the experiment was repeated a great many times. An example: the expected value might be the long range profit or loss of an insurance company due to the probability that an individual lives for an additional year.

27. Yes; in general, if we add a constant to each number in a set, then the mean will increase by that constant.

28. Yes; in general, if we multiply each number in a set by some constant, then the standard deviation will be multiplied by that constant.

29. The binomial probability distribution applies when there is a fixed number of independent trials when the probability of success is constant. The outcome of each trial is classified as either a "success" or a "failure".

30. Repeated trials that do not produce a binomial distribution: 1) tossing a coin until a head appears. 2) Having children until a girl is born.

Chapter 8

Exercises 8.1

1. Yes; the matrix is square, all entries are ≥ 0, and the sum of the entries in each column is 1.

3. No; the matrix is not square.

5. Yes; the matrix is square, all entries are ≥ 0, and the sum of the entries in each column is 1.

7. $\begin{array}{c} \,\,A\,\,\,B \\ \begin{array}{c}A\\B\end{array}\begin{bmatrix}.3 & .5\\.7 & .5\end{bmatrix}\end{array}$

9. $\begin{array}{c} \,\,A\,\,\,B\,\,\,C \\ \begin{array}{c}A\\B\\C\end{array}\begin{bmatrix}\frac{1}{3} & \frac{2}{9} & \frac{1}{3}\\ \frac{1}{3} & \frac{4}{9} & \frac{1}{6}\\ \frac{1}{3} & \frac{1}{3} & \frac{1}{2}\end{bmatrix}\end{array}$

11. $\begin{array}{c} \,\,A\,\,\,B\,\,\,C \\ \begin{array}{c}A\\B\\C\end{array}\begin{bmatrix}.4 & .2 & 0\\.5 & 0 & 0\\.1 & .8 & 1\end{bmatrix}\end{array}$

13.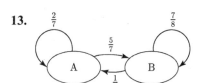

15.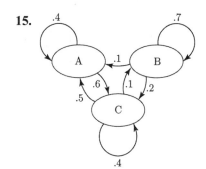

17.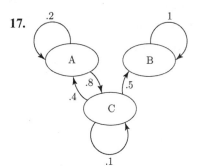

19. initial state matrix $=\begin{bmatrix}.47\\.53\end{bmatrix}_0$

next state $= A\begin{bmatrix}.47\\.53\end{bmatrix}_0 = \begin{bmatrix}.8 & .3\\.2 & .7\end{bmatrix}\begin{bmatrix}.47\\.53\end{bmatrix}_0$

$=\begin{bmatrix}.535\\.465\end{bmatrix}_1$

next state $= A^2\begin{bmatrix}.47\\.53\end{bmatrix}_0 = \begin{bmatrix}.70 & .45\\.30 & .55\end{bmatrix}\begin{bmatrix}.47\\.53\end{bmatrix}_0$

$=\begin{bmatrix}.5675\\.4325\end{bmatrix}_2$

After one generation, about 54% of French women will work. After two generations, about 57% of French women will work outside the home..

21. a. $A = \begin{array}{c}\,\,L\,\,\,H\\ \begin{array}{c}L\\H\end{array}\begin{bmatrix}.8 & .3\\.2 & .7\end{bmatrix}\end{array}$

b. initial state matrix $=\begin{bmatrix}.5\\.5\end{bmatrix}_0$

next state $= A\begin{bmatrix}.5\\.5\end{bmatrix}_0 = \begin{bmatrix}.8 & .3\\.2 & .7\end{bmatrix}\begin{bmatrix}.5\\.5\end{bmatrix}_0 = \begin{bmatrix}.55\\.45\end{bmatrix}$

Therefore, 55% of the customers would be Low users in February

next state $= A\begin{bmatrix}.55\\.45\end{bmatrix}_1 = \begin{bmatrix}.8 & .3\\.2 & .7\end{bmatrix}\begin{bmatrix}.55\\.45\end{bmatrix}_1 = \begin{bmatrix}.575\\.425\end{bmatrix}$

Therefore, 57.5% of the customers would be Low users in March.

8-1

23. a. $\begin{array}{c} \begin{array}{cc} S & O \end{array} \\ \begin{array}{c} S \\ O \end{array}\begin{bmatrix} .992 & .007 \\ .008 & .993 \end{bmatrix} \end{array}$

b. initial state $= \begin{bmatrix} .12 \\ .88 \end{bmatrix}_0$

next state $= \begin{bmatrix} .992 & .007 \\ .008 & .993 \end{bmatrix}\begin{bmatrix} .12 \\ .88 \end{bmatrix}_0 = \begin{bmatrix} .1252 \\ .8748 \end{bmatrix}_1$

next state $= \begin{bmatrix} .992 & .007 \\ .008 & .993 \end{bmatrix}\begin{bmatrix} .1252 \\ .8748 \end{bmatrix}_1$

$= \begin{bmatrix} .1303 \\ .8697 \end{bmatrix}_2$

Therefore, 12.52% and 13.03% will be the estimate of people living in the Southwest at the beginning of 2012 and 2013 respectively.

25. a. $\begin{array}{c} \begin{array}{cc} L & R \end{array} \\ \begin{array}{c} L \\ R \end{array}\begin{bmatrix} .9 & .7 \\ .1 & .3 \end{bmatrix} \end{array}$

b. $A^2 = \begin{bmatrix} .9 & .7 \\ .1 & .3 \end{bmatrix}\begin{bmatrix} .9 & .7 \\ .1 & .3 \end{bmatrix} = \begin{bmatrix} .88 & .84 \\ .12 & .16 \end{bmatrix}$

c. initial state $= \begin{bmatrix} .5 \\ .5 \end{bmatrix}_0$

$\begin{bmatrix} \end{bmatrix}_1 = \begin{bmatrix} .9 & .7 \\ .1 & .3 \end{bmatrix}\begin{bmatrix} .5 \\ .5 \end{bmatrix}_0$

$= \begin{bmatrix} .8 \\ .2 \end{bmatrix}_1$

$\begin{bmatrix} \end{bmatrix}_2 = A^2 \begin{bmatrix} .5 \\ .5 \end{bmatrix}_0$

$= \begin{bmatrix} .88 & .84 \\ .12 & .16 \end{bmatrix}\begin{bmatrix} .5 \\ .5 \end{bmatrix}_0$

$= \begin{bmatrix} .86 \\ .14 \end{bmatrix}_2$

d. Answers will vary. Correct answer is 87.5%.

27. initial state matrix $= \begin{bmatrix} .40 \\ .40 \\ .20 \end{bmatrix}_0$

next state $= \begin{bmatrix} .5 & .4 & .2 \\ .4 & .3 & .6 \\ .1 & .3 & .2 \end{bmatrix}\begin{bmatrix} .4 \\ .4 \\ .2 \end{bmatrix}$

$= \begin{bmatrix} .4 \\ .4 \\ .2 \end{bmatrix}_1$

40% will be Zone I, 40% will be in Zone II, 20% will be in Zone III.

29. a. 4%

b. 96% of the freshmen who held middle-of-the-road political views continued to hold these views as sophomores.

c.

d. initial state $= \begin{bmatrix} .30 \\ .48 \\ .22 \end{bmatrix}_0$

next state $= \begin{bmatrix} .94 & .02 & .01 \\ .05 & .96 & .04 \\ .01 & .02 & .95 \end{bmatrix}\begin{bmatrix} .30 \\ .48 \\ .22 \end{bmatrix}_0 = \begin{bmatrix} .2938 \\ .4846 \\ .2216 \end{bmatrix}_1$

next state $= \begin{bmatrix} .94 & .02 & .01 \\ .05 & .96 & .04 \\ .01 & .02 & .95 \end{bmatrix}\begin{bmatrix} .2938 \\ .4846 \\ .2216 \end{bmatrix}_1$

$\approx \begin{bmatrix} .2881 \\ .4888 \\ .2232 \end{bmatrix}_2$

48.46% of the students held middle-of-the-road political views as sophomores, 48.88% as juniors.

31. a.
$$\begin{array}{c c} & \begin{array}{c c c} U & S & R \end{array} \\ \begin{array}{c} U \\ S \\ R \end{array} & \begin{bmatrix} .86 & .05 & .03 \\ .08 & .86 & .05 \\ .06 & .09 & .92 \end{bmatrix} \end{array}$$

b. Since we are concerned with only the people who live in urban areas of 2013, we use

$$\begin{bmatrix} \end{bmatrix}_0 = \begin{bmatrix} 1 \\ 0 \\ 0 \end{bmatrix}_0.$$

$$\begin{bmatrix} \end{bmatrix}_2 = A^2 \begin{bmatrix} \end{bmatrix}_0$$

$$= \begin{bmatrix} .86 & .05 & .03 \\ .08 & .86 & .05 \\ .06 & .09 & .92 \end{bmatrix} \begin{bmatrix} .86 & .05 & .03 \\ .08 & .86 & .05 \\ .06 & .09 & .92 \end{bmatrix} \begin{bmatrix} 1 \\ 0 \\ 0 \end{bmatrix}_0$$

$$= \begin{bmatrix} .7454 \\ .1406 \\ .114 \end{bmatrix}_2$$

11.4% of people who live in urban areas in 2013 will live in rural areas in 2015.

33. a. initial state $= \begin{bmatrix} .4 \\ .4 \\ .2 \end{bmatrix}_0$

$$\begin{bmatrix} \end{bmatrix}_1 = \begin{bmatrix} .5 & .25 & .2 \\ .4 & .5 & .2 \\ .1 & .25 & .6 \end{bmatrix} \begin{bmatrix} .4 \\ .4 \\ .2 \end{bmatrix}_0$$

$$= \begin{bmatrix} .34 \\ .40 \\ .26 \end{bmatrix}_1$$

b. $\begin{bmatrix} \end{bmatrix}_2 = A \begin{bmatrix} \end{bmatrix}_1$

$$= \begin{bmatrix} .5 & .25 & .2 \\ .4 & .5 & .2 \\ .1 & .25 & .6 \end{bmatrix} \begin{bmatrix} .34 \\ .4 \\ .26 \end{bmatrix}_1$$

$$= \begin{bmatrix} .322 \\ .388 \\ .29 \end{bmatrix}_2$$

.322 is the probability that a woman that had high birth weight will have a granddaughter of high birth weight and .388 is the probability that her granddaughter will have average birth weights.

35. a.
$$\begin{array}{c c} & \begin{array}{c c c c c} 0 & 1 & 2 & 3 & 4 \end{array} \\ \begin{array}{c} 0 \\ 1 \\ 2 \\ 3 \\ 4 \end{array} & \begin{bmatrix} 0 & .25 & 0 & 0 & 0 \\ 1 & 0 & .5 & 0 & 0 \\ 0 & .75 & 0 & .75 & 0 \\ 0 & 0 & .5 & 0 & 1 \\ 0 & 0 & 0 & .25 & 0 \end{bmatrix} \end{array}$$

b. Using the initial matrix as follows:

$$\begin{bmatrix} 0 \\ 0 \\ 0 \\ 1 \\ 0 \end{bmatrix}_0$$

Calculating $A^2 \cdot (\text{initial})$ will yield:

$$\begin{bmatrix} 0 \\ .375 \\ 0 \\ .625 \\ 0 \end{bmatrix}_2$$

Therefore, there is a .625 probability that there will be 3 balls in Urn A after two time periods.

37.
$$A^2 = \begin{bmatrix} .45 & .44 \\ .55 & .56 \end{bmatrix}, A^3 = \begin{bmatrix} .445 & .444 \\ .555 & .556 \end{bmatrix},$$

$$A^4 = \begin{bmatrix} .4445 & .4444 \\ .5555 & .5556 \end{bmatrix}$$

$$\begin{bmatrix} \end{bmatrix}_3 = \begin{bmatrix} .445 & .444 \\ .555 & .556 \end{bmatrix} \begin{bmatrix} .3 \\ .7 \end{bmatrix}_0 = \begin{bmatrix} .4443 \\ .5557 \end{bmatrix}_3 \approx \begin{bmatrix} .44 \\ .56 \end{bmatrix}_3$$

$$\begin{bmatrix} \end{bmatrix}_4 = \begin{bmatrix} .4445 & .4444 \\ .5555 & .5556 \end{bmatrix} \begin{bmatrix} .3 \\ .7 \end{bmatrix}_0$$

$$= \begin{bmatrix} .44443 \\ .55557 \end{bmatrix}_4$$

$$\approx \begin{bmatrix} .44 \\ .56 \end{bmatrix}_4$$

39. $A^1 = A = \begin{bmatrix} \frac{1}{3} & \frac{1}{3} \\ \frac{2}{3} & \frac{2}{3} \end{bmatrix} \approx \begin{bmatrix} .33 & .33 \\ .67 & .67 \end{bmatrix}$

$A^2 = A \cdot A = \begin{bmatrix} \frac{1}{3} & \frac{1}{3} \\ \frac{2}{3} & \frac{2}{3} \end{bmatrix} \begin{bmatrix} \frac{1}{3} & \frac{1}{3} \\ \frac{2}{3} & \frac{2}{3} \end{bmatrix}$

$= \begin{bmatrix} \frac{1}{3} & \frac{1}{3} \\ \frac{2}{3} & \frac{2}{3} \end{bmatrix} \approx \begin{bmatrix} .33 & .33 \\ .67 & .67 \end{bmatrix}$

The pattern continues. All powers are $\begin{bmatrix} \frac{1}{3} & \frac{1}{3} \\ \frac{2}{3} & \frac{2}{3} \end{bmatrix}$.

41. $A^1 = A = \begin{bmatrix} .1 & .3 \\ .9 & .7 \end{bmatrix}$

$A^2 = A \cdot A = \begin{bmatrix} .1 & .3 \\ .9 & .7 \end{bmatrix} \begin{bmatrix} .1 & .3 \\ .9 & .7 \end{bmatrix} = \begin{bmatrix} .28 & .24 \\ .72 & .76 \end{bmatrix}$

$A^3 = A^2 \cdot A = \begin{bmatrix} .28 & .24 \\ .72 & .76 \end{bmatrix} \begin{bmatrix} .1 & .3 \\ .9 & .7 \end{bmatrix}$

$= \begin{bmatrix} .244 & .252 \\ .756 & .748 \end{bmatrix} \approx \begin{bmatrix} .24 & .25 \\ .76 & .75 \end{bmatrix}$

$A^4 = A^3 \cdot A = \begin{bmatrix} .244 & .252 \\ .756 & .748 \end{bmatrix} \begin{bmatrix} .1 & .3 \\ .9 & .7 \end{bmatrix}$

$= \begin{bmatrix} .2512 & .2496 \\ .7488 & .7504 \end{bmatrix} \approx \begin{bmatrix} .25 & .25 \\ .75 & .75 \end{bmatrix}$

$A^5 = A^4 \cdot A = \begin{bmatrix} .2512 & .2496 \\ .7488 & .7504 \end{bmatrix} \begin{bmatrix} .1 & .3 \\ .9 & .7 \end{bmatrix}$

$= \begin{bmatrix} .24976 & .25008 \\ .75024 & .74992 \end{bmatrix}$

$\approx \begin{bmatrix} .25 & .25 \\ .75 & .75 \end{bmatrix}$

43. $A^1 = A = \begin{bmatrix} .3 & .3 & .3 \\ .1 & .1 & .1 \\ .6 & .6 & .6 \end{bmatrix}$

$A^2 = A \cdot A = \begin{bmatrix} .3 & .3 & .3 \\ .1 & .1 & .1 \\ .6 & .6 & .6 \end{bmatrix} \begin{bmatrix} .3 & .3 & .3 \\ .1 & .1 & .1 \\ .6 & .6 & .6 \end{bmatrix}$

$= \begin{bmatrix} .3 & .3 & .3 \\ .1 & .1 & .1 \\ .6 & .6 & .6 \end{bmatrix}$

The pattern continues. All powers are
$\begin{bmatrix} .3 & .3 & .3 \\ .1 & .1 & .1 \\ .6 & .6 & .6 \end{bmatrix}$.

45. No; all powers have a zero entry in the upper right corner, so the matrix is not regular.

47. a. Use the method described in the text to generate the next four distribution matrices.
$\begin{bmatrix} .35 \\ .65 \end{bmatrix}, \begin{bmatrix} .425 \\ .575 \end{bmatrix}, \begin{bmatrix} .3875 \\ .6125 \end{bmatrix}, \begin{bmatrix} .40625 \\ .59375 \end{bmatrix}$

b. $A^4 B = \begin{bmatrix} .4375 & .375 \\ .5625 & .625 \end{bmatrix} \begin{bmatrix} .5 \\ .5 \end{bmatrix} = \begin{bmatrix} .40625 \\ .59375 \end{bmatrix}$

49. The distribution matrix gets closer and closer to $\begin{bmatrix} .4 \\ .6 \end{bmatrix}$.

51. The matrices get closer and closer to $\begin{bmatrix} .4 & .4 \\ .6 & .6 \end{bmatrix}$.

Each column of this matrix is the same as the 2 by 1 matrix found in Exercise 43.

Exercises 8.2

1. Yes; the matrix is regular, since all entries are positive.

3. Yes; the matrix is regular, since the second power is $\begin{bmatrix} .79 & .3 \\ .21 & .7 \end{bmatrix}$, which has all positive entries.

5. No; all powers have zero entries in the second columns.

7. $\begin{cases} x+y=1 \\ \begin{bmatrix} .5 & .1 \\ .5 & .9 \end{bmatrix}\begin{bmatrix} x \\ y \end{bmatrix} = \begin{bmatrix} x \\ y \end{bmatrix} \end{cases}$

$\begin{cases} x + y = 1 \\ .5x + .1y = x \\ .5x + .9y = y \end{cases}$

$\begin{cases} x + y = 1 \\ -.5x + .1y = 0 \\ .5x - .1y = 0 \end{cases}$

The second and third equations in this system are equivalent.

$\begin{bmatrix} 1 & 1 & | & 1 \\ -.5 & .1 & | & 0 \end{bmatrix} \xrightarrow{[2]+.5[1]} \begin{bmatrix} 1 & 1 & | & 1 \\ 0 & .6 & | & .5 \end{bmatrix}$

$\xrightarrow{\frac{5}{3}[2]} \begin{bmatrix} 1 & 1 & | & 1 \\ 0 & 1 & | & \frac{5}{6} \end{bmatrix}$

$\xrightarrow{[1]+(-1)[2]} \begin{bmatrix} 1 & 0 & | & \frac{1}{6} \\ 0 & 1 & | & \frac{5}{6} \end{bmatrix}$

$x = \frac{1}{6}, \ y = \frac{5}{6}$

The stable distribution is $\begin{bmatrix} x \\ y \end{bmatrix} = \begin{bmatrix} \frac{1}{6} \\ \frac{5}{6} \end{bmatrix}$.

9. $\begin{cases} x+y=1 \\ \begin{bmatrix} .8 & .3 \\ .2 & .7 \end{bmatrix}\begin{bmatrix} x \\ y \end{bmatrix} = \begin{bmatrix} x \\ y \end{bmatrix} \end{cases}$

$\begin{cases} x + y = 1 \\ .8x + .3y = x \\ .2x + .7y = y \end{cases}$

$\begin{cases} x + y = 1 \\ -.2x + .3y = 0 \\ .2x - .3y = 0 \end{cases}$

The second and third equations in this system are equivalent.

$\begin{bmatrix} 1 & 1 & | & 1 \\ -.2 & .3 & | & 0 \end{bmatrix} \xrightarrow{[2]+.2[1]} \begin{bmatrix} 1 & 1 & | & 1 \\ 0 & .5 & | & .2 \end{bmatrix}$

$\xrightarrow{2[2]} \begin{bmatrix} 1 & 1 & | & 1 \\ 0 & 1 & | & .4 \end{bmatrix}$

$\xrightarrow{[1]+(-1)[2]} \begin{bmatrix} 1 & 0 & | & .6 \\ 0 & 1 & | & .4 \end{bmatrix}$

$x = .6, \ y = .4$

The stable distribution is $\begin{bmatrix} x \\ y \end{bmatrix} = \begin{bmatrix} .6 \\ .4 \end{bmatrix}$.

11. $\begin{cases} x+y+z=1 \\ \begin{bmatrix} .1 & .4 & .7 \\ .6 & .4 & .2 \\ .3 & .2 & .1 \end{bmatrix}\begin{bmatrix} x \\ y \\ z \end{bmatrix} = \begin{bmatrix} x \\ y \\ z \end{bmatrix} \end{cases}$

$\begin{cases} x + y + z = 1 \\ .1x + .4y + .7z = x \\ .6x + .4y + .2z = y \\ .3x + .2y + .1z = z \end{cases}$

$\begin{cases} x + y + z = 1 \\ -.9x + .4y + .7z = 0 \\ .6x - .6y + .2z = 0 \\ .3x + .2y - .9z = 0 \end{cases}$

$\begin{bmatrix} 1 & 1 & 1 & | & 1 \\ -.9 & .4 & .7 & | & 0 \\ .6 & -.6 & .2 & | & 0 \\ .3 & .2 & -.9 & | & 0 \end{bmatrix}$

$\xrightarrow[10[4]]{\substack{10[2] \\ 10[3]}} \begin{bmatrix} 1 & 1 & 1 & | & 1 \\ -9 & 4 & 7 & | & 0 \\ 6 & -6 & 2 & | & 0 \\ 3 & 2 & -9 & | & 0 \end{bmatrix}$

$\xrightarrow[{[4]+(-3)[1]}]{\substack{[2]+9[1] \\ [3]+(-6)[1]}} \begin{bmatrix} 1 & 1 & 1 & | & 1 \\ 0 & 13 & 16 & | & 9 \\ 0 & -12 & -4 & | & -6 \\ 0 & -1 & -12 & | & -3 \end{bmatrix}$

$$\xrightarrow[\substack{[1]+(-1)[2] \\ [3]+12[2] \\ [4]+1[2]}]{\frac{1}{13}[2]} \begin{bmatrix} 1 & 0 & -\frac{3}{13} & \frac{4}{13} \\ 0 & 1 & \frac{16}{13} & \frac{9}{13} \\ 0 & 0 & \frac{140}{13} & \frac{30}{13} \\ 0 & 0 & -\frac{140}{13} & -\frac{30}{13} \end{bmatrix}$$

$$\xrightarrow[\substack{[1]+\frac{3}{13}[3] \\ [2]-\frac{16}{13}[3] \\ [4]+\frac{140}{13}[3]}]{\frac{13}{140}[3]} \begin{bmatrix} 1 & 0 & 0 & \frac{5}{14} \\ 0 & 1 & 0 & \frac{3}{7} \\ 0 & 0 & 1 & \frac{3}{14} \\ 0 & 0 & 0 & 0 \end{bmatrix}$$

$x = \frac{5}{14}, y = \frac{3}{7}, z = \frac{3}{14}$

The stable distribution is $\begin{bmatrix} x \\ y \\ z \end{bmatrix} = \begin{bmatrix} \frac{5}{14} \\ \frac{3}{7} \\ \frac{3}{14} \end{bmatrix}$.

13. The stochastic matrix is $\begin{array}{c} \\ L \\ H \end{array} \begin{bmatrix} L & H \\ .8 & .3 \\ .2 & .7 \end{bmatrix}$.

Find the stable distribution.

$\begin{cases} x + y = 1 \\ \begin{bmatrix} .8 & .3 \\ .2 & .7 \end{bmatrix} \begin{bmatrix} x \\ y \end{bmatrix} = \begin{bmatrix} x \\ y \end{bmatrix} \end{cases}$

$\begin{cases} x + y = 1 \\ .8x + .3y = x \\ .2x + .7y = y \end{cases}$

$\begin{cases} x + y = 1 \\ -.2x + .3y = 0 \\ .2x - .3y = 0 \end{cases}$

The second and third equations in this system are equivalent.

$\begin{bmatrix} 1 & 1 & | & 1 \\ -.2 & .3 & | & 0 \end{bmatrix} \xrightarrow{[2]+.2[1]} \begin{bmatrix} 1 & 1 & | & 1 \\ 0 & .5 & | & .2 \end{bmatrix}$

$\xrightarrow{2[2]} \begin{bmatrix} 1 & 1 & | & 1 \\ 0 & 1 & | & .4 \end{bmatrix}$

$\xrightarrow{[1]+(-1)[2]} \begin{bmatrix} 1 & 0 & | & .6 \\ 0 & 1 & | & .4 \end{bmatrix}$

$x = .6, y = .4$

The stable distribution is $\begin{bmatrix} x \\ y \end{bmatrix} = \begin{bmatrix} .6 \\ .4 \end{bmatrix}$.

In the long run, 40% of the customers will be High users.

15. The stochastic matrix is $\begin{array}{c} \\ L \\ R \end{array} \begin{bmatrix} L & R \\ .9 & .7 \\ .1 & .3 \end{bmatrix}$.

Find the stable distribution.

$\begin{cases} x + y = 1 \\ \begin{bmatrix} .9 & .7 \\ .1 & .3 \end{bmatrix} \begin{bmatrix} x \\ y \end{bmatrix} = \begin{bmatrix} x \\ y \end{bmatrix} \end{cases}$

$\begin{cases} x + y = 1 \\ .9x + .7y = x \\ .1x + .3y = y \end{cases}$

$\begin{cases} x + y = 1 \\ -.1x + .7y = 0 \\ .1x - .7y = 0 \end{cases}$

The second and third equations in this system are equivalent.

$\begin{bmatrix} 1 & 1 & | & 1 \\ -.1 & .7 & | & 0 \end{bmatrix} \xrightarrow{[2]+.1[1]} \begin{bmatrix} 1 & 1 & | & 1 \\ 0 & .8 & | & .1 \end{bmatrix}$

$\xrightarrow{1.25[2]} \begin{bmatrix} 1 & 1 & | & 1 \\ 0 & 1 & | & .125 \end{bmatrix}$

$\xrightarrow{[1]+(-1)[2]} \begin{bmatrix} 1 & 0 & | & .875 \\ 0 & 1 & | & .125 \end{bmatrix}$

$x = .785, y = .125$

The stable distribution is $\begin{bmatrix} x \\ y \end{bmatrix} = \begin{bmatrix} .875 \\ .125 \end{bmatrix}$.

After many days, 87.5% of the mice will be going to the left.

17. The stochastic matrix is $\begin{array}{c} \\ \text{GM} \\ \text{non-GM} \end{array} \begin{bmatrix} \text{GM} & \text{non-GM} \\ .6 & .1 \\ .4 & .9 \end{bmatrix}$.

Find the stable distribution.

$\begin{cases} x + y = 1 \\ \begin{bmatrix} .6 & .1 \\ .4 & .9 \end{bmatrix} \begin{bmatrix} x \\ y \end{bmatrix} = \begin{bmatrix} x \\ y \end{bmatrix} \end{cases}$

$\begin{cases} x + y = 1 \\ .6x + .1y = x \\ .4x + .9y = y \end{cases}$

$\begin{cases} x+y=1 \\ -.4x+.1y=0 \\ .4x-.1y=0 \end{cases}$

The second and third equations in the system are equivalent.

$\begin{bmatrix} 1 & 1 & | & 1 \\ -.4 & .1 & | & 0 \end{bmatrix} \xrightarrow{[2]+.4[1]} \begin{bmatrix} 1 & 1 & | & 1 \\ 0 & .5 & | & .4 \end{bmatrix}$

$\xrightarrow{2[2]} \begin{bmatrix} 1 & 1 & | & 1 \\ 0 & 1 & | & .8 \end{bmatrix} \xrightarrow{[1]+(-1)[2]} \begin{bmatrix} 1 & 0 & | & .2 \\ 0 & 1 & | & .8 \end{bmatrix}$

The stable distribution is $\begin{bmatrix} .2 \\ .8 \end{bmatrix}$.

In the long run General Motors market share is 20%.

19. The stochastic matrix is $\begin{array}{c} \\ R \\ S \end{array}\begin{bmatrix} R & S \\ .1 & .6 \\ .9 & .4 \end{bmatrix}$.

Find the stable distribution.

$\begin{cases} x+y=1 \\ \begin{bmatrix} .1 & .6 \\ .9 & .4 \end{bmatrix}\begin{bmatrix} x \\ y \end{bmatrix} = \begin{bmatrix} x \\ y \end{bmatrix} \end{cases}$

$\begin{cases} x + y = 1 \\ .1x + .6y = x \\ .9x + .4y = y \end{cases}$

$\begin{cases} x + y = 1 \\ -.9x + .6y = 0 \\ .9x - .6y = 0 \end{cases}$

The second and third equations in this system are equivalent.

$\begin{bmatrix} 1 & 1 & | & 1 \\ -.9 & .6 & | & 0 \end{bmatrix} \xrightarrow{[2]+.9[1]} \begin{bmatrix} 1 & 1 & | & 1 \\ 0 & 1.5 & | & .9 \end{bmatrix}$

$\xrightarrow{\frac{2}{3}[2]} \begin{bmatrix} 1 & 1 & | & 1 \\ 0 & 1 & | & .6 \end{bmatrix}$

$\xrightarrow{[1]+(-1)[2]} \begin{bmatrix} 1 & 0 & | & .4 \\ 0 & 1 & | & .6 \end{bmatrix}$

$x = .4, y = .6$

The stable distribution is $\begin{bmatrix} x \\ y \end{bmatrix} = \begin{bmatrix} .4 \\ .6 \end{bmatrix}$.

In the log run, the daily likelihood of rain is 40% or $\frac{2}{5}$.

21. a. $\begin{array}{c} \\ A \\ B \\ C \end{array}\begin{bmatrix} A & B & C \\ .7 & .1 & .1 \\ .2 & .8 & .3 \\ .1 & .1 & .6 \end{bmatrix}$

b. $\begin{bmatrix} \ \end{bmatrix}_1 = \begin{bmatrix} .7 & .1 & .1 \\ .2 & .8 & .3 \\ .1 & .1 & .6 \end{bmatrix}\begin{bmatrix} .4 \\ .3 \\ .3 \end{bmatrix}_0 = \begin{bmatrix} .34 \\ .41 \\ .25 \end{bmatrix}_1$

$\begin{bmatrix} \ \end{bmatrix}_2 = \begin{bmatrix} .7 & .1 & .1 \\ .2 & .8 & .3 \\ .1 & .1 & .6 \end{bmatrix}\begin{bmatrix} .34 \\ .41 \\ .25 \end{bmatrix}_1 = \begin{bmatrix} .304 \\ .471 \\ .225 \end{bmatrix}_2$

34% of the cars are at location A after one day, and 30.4% after two days.

c. Find the stable distribution.

$\begin{cases} x+y+z=1 \\ .7x+.1y+.1z=x \\ .2x+.8y+.3z=y \\ .1x+.1y+.6z=z \end{cases}$

$\begin{cases} x+y+z=1 \\ -.3x+.1y+.1z=0 \\ .2x-.2y+.3z=0 \\ .1x+.1y-.4z=0 \end{cases}$

The fourth equation is equivalent to the second plus the third and so is redundant.

$\begin{bmatrix} 1 & 1 & 1 & | & 1 \\ -.3 & .1 & .1 & | & 0 \\ .2 & -.2 & .3 & | & 0 \end{bmatrix} \xrightarrow{\substack{[2]+.3[1] \\ [3]-.2[1]}} \begin{bmatrix} 1 & 1 & 1 & | & 1 \\ 0 & .4 & .4 & | & .3 \\ 0 & -.4 & .1 & | & -.2 \end{bmatrix}$

$\xrightarrow{\substack{\frac{10}{4}[2] \\ -\frac{10}{4}[3]}} \begin{bmatrix} 1 & 1 & 1 & | & 1 \\ 0 & 1 & 1 & | & .75 \\ 0 & 1 & -.25 & | & .5 \end{bmatrix}$

$\xrightarrow{\substack{[1]-[2] \\ [3]-[2]}} \begin{bmatrix} 1 & 0 & 0 & | & .25 \\ 0 & 1 & 1 & | & .75 \\ 0 & 0 & -1.25 & | & -.25 \end{bmatrix}$

$\xrightarrow{\substack{[2]+\frac{8}{10}[3] \\ -\frac{8}{10}[3]}} \begin{bmatrix} 1 & 0 & 0 & | & .25 \\ 0 & 1 & 0 & | & .55 \\ 0 & 0 & 1 & | & .2 \end{bmatrix}$

In the long run there are $\frac{1}{4}$ at location A, $\frac{11}{20}$ at location B and $\frac{1}{5}$ at location C.

23. The stochastic matrix is $\begin{array}{c} \\ D \\ R \\ H \end{array}\begin{bmatrix} D & R & H \\ .5 & 0 & .25 \\ 0 & .5 & .25 \\ .5 & .5 & .5 \end{bmatrix}$.

Find the stable distribution.

$\begin{cases} x+y+z=1 \\ \begin{bmatrix} .5 & 0 & .25 \\ 0 & .5 & .25 \\ .5 & .5 & .5 \end{bmatrix}\begin{bmatrix} x \\ y \\ z \end{bmatrix} = \begin{bmatrix} x \\ y \\ z \end{bmatrix} \end{cases}$

$\begin{cases} x + y + z = 1 \\ .5x \quad\quad + .25z = x \\ \quad\quad .5y + .25z = y \\ .5x + .5y + .5z = z \end{cases}$

$\begin{cases} x + y + z = 1 \\ -.5x \quad\quad + .25z = 0 \\ \quad\quad -.5y + .25z = 0 \\ .5x + .5y - .5z = 0 \end{cases}$

$\begin{bmatrix} 1 & 1 & 1 & | & 1 \\ -.5 & 0 & .25 & | & 0 \\ 0 & -.5 & .25 & | & 0 \\ .5 & .5 & -.5 & | & 0 \end{bmatrix}$

$\xrightarrow[2[4]]{\substack{4[2] \\ 4[3]}} \begin{bmatrix} 1 & 1 & 1 & | & 1 \\ -2 & 0 & 1 & | & 0 \\ 0 & -2 & 1 & | & 0 \\ 1 & 1 & -1 & | & 0 \end{bmatrix}$

$\xrightarrow[{[4]+(-1)[1]}]{[2]+2[1]} \begin{bmatrix} 1 & 1 & 1 & | & 1 \\ 0 & 2 & 3 & | & 2 \\ 0 & -2 & 1 & | & 0 \\ 0 & 0 & -2 & | & -1 \end{bmatrix}$

$\xrightarrow[{[3]+2[2]}]{\substack{.5[2] \\ [1]+(-1)[2]}} \begin{bmatrix} 1 & 0 & -.5 & | & 0 \\ 0 & 1 & 1.5 & | & 1 \\ 0 & 0 & 4 & | & 2 \\ 0 & 0 & -2 & | & -1 \end{bmatrix}$

$\xrightarrow[{[4]+2[3]}]{\substack{.25[3] \\ [1]+(.5)[3] \\ [2]+(-1.5)[3]}} \begin{bmatrix} 1 & 0 & 0 & | & .25 \\ 0 & 1 & 0 & | & .25 \\ 0 & 0 & 1 & | & .5 \\ 0 & 0 & 0 & | & 0 \end{bmatrix}$

$x = .25, y = .25, z = .5$

The stable distribution is $\begin{bmatrix} x \\ y \\ z \end{bmatrix} = \begin{bmatrix} .25 \\ .25 \\ .5 \end{bmatrix}$.

In the long run, 25% will be dominant.

25. $\begin{bmatrix} .6 \\ 0 \\ .4 \end{bmatrix}$ and $\begin{bmatrix} .3 \\ .5 \\ .2 \end{bmatrix}$ are stable distributions for the

matrix $A = \begin{bmatrix} .4 & 0 & .9 \\ 0 & 1 & 0 \\ .6 & 0 & .1 \end{bmatrix}$ because their values add

to 1 and

$\begin{bmatrix} .4 & 0 & .9 \\ 0 & 1 & 0 \\ .6 & 0 & .1 \end{bmatrix}\begin{bmatrix} .6 \\ 0 \\ .4 \end{bmatrix} = \begin{bmatrix} .6 \\ 0 \\ .4 \end{bmatrix}$ and $\begin{bmatrix} .4 & 0 & .9 \\ 0 & 1 & 0 \\ .6 & 0 & .1 \end{bmatrix}\begin{bmatrix} .3 \\ .5 \\ .2 \end{bmatrix} = \begin{bmatrix} .3 \\ .5 \\ .2 \end{bmatrix}$.

However, given an arbitrary initial distribution

$\left[\begin{bmatrix} \\ \\ \\ \end{bmatrix} \neq \begin{bmatrix} .6 \\ 0 \\ .4 \end{bmatrix} \text{ or } \begin{bmatrix} .3 \\ .5 \\ .2 \end{bmatrix}, A^n\left[\begin{bmatrix} \\ \\ \\ \end{bmatrix}\right]\right.$ will not approach

$\begin{bmatrix} .6 \\ 0 \\ .4 \end{bmatrix}$ or $\begin{bmatrix} .3 \\ .5 \\ .2 \end{bmatrix}$ as n gets large, so the existence of a

stable distribution for A does not contradict the main premise of this section.

27. The stochastic matrix is

$\begin{array}{c} \\ L \\ A \\ H \end{array}\begin{bmatrix} H & A & L \\ .5 & .25 & .2 \\ .4 & .5 & .2 \\ .1 & .25 & .6 \end{bmatrix}$.

Find the stable distribution.

$\begin{cases} x + y + z = 1 \\ .5x + .25y + .2z = x \\ .4x + .5y + .2z = y \\ .1x + .25y + .6z = z \end{cases}$

$\begin{cases} x + y + z = 1 \\ -.5x + .25y + .2z = 0 \\ .4x - .5y + .2z = 0 \\ .1x + .25y - .4z = 0 \end{cases}$

The fourth equation is equivalent to the second plus the third and so is redundant.

$$\begin{bmatrix} 1 & 1 & 1 & | & 1 \\ -.5 & .25 & .2 & | & 0 \\ .4 & -.5 & .2 & | & 0 \end{bmatrix} \xrightarrow[{[3]-.4[1]}]{[2]+.5[1]} \begin{bmatrix} 1 & 1 & 1 & | & 1 \\ 0 & .75 & .7 & | & .5 \\ 0 & -.9 & -.2 & | & -.4 \end{bmatrix}$$

$$\xrightarrow{\frac{1}{.75}[2]} \begin{bmatrix} 1 & 1 & 1 & | & 1 \\ 0 & 1 & \frac{14}{15} & | & \frac{2}{3} \\ 0 & -.9 & -.2 & | & -.4 \end{bmatrix}$$

$$\xrightarrow[{[3]+0.9[2]}]{[1]-[2]} \begin{bmatrix} 1 & 0 & \frac{1}{15} & | & \frac{1}{3} \\ 0 & 1 & \frac{14}{15} & | & \frac{2}{3} \\ 0 & 0 & \frac{16}{25} & | & \frac{1}{5} \end{bmatrix}$$

$$\xrightarrow{\frac{25}{16}[3]} \begin{bmatrix} 1 & 0 & \frac{1}{15} & | & \frac{1}{3} \\ 0 & 1 & \frac{14}{15} & | & \frac{2}{3} \\ 0 & 0 & 1 & | & \frac{5}{16} \end{bmatrix}$$

$$\xrightarrow[{[2]-\frac{14}{15}[3]}]{[1]-\frac{1}{15}[3]} \begin{bmatrix} 1 & 0 & 0 & | & \frac{5}{16} \\ 0 & 1 & 0 & | & \frac{3}{8} \\ 0 & 0 & 1 & | & \frac{5}{16} \end{bmatrix}$$

In the long run, the $\frac{5}{16}$ of female babies will have High birth weight and $\frac{3}{8}$ will have Average.

29. Calculating [A] ^ 255 gives $\begin{bmatrix} .7 & .7 \\ .3 & .3 \end{bmatrix}$.

This suggests that the stable distribution is $\begin{bmatrix} .7 \\ .3 \end{bmatrix}$.

Check: $x = .7, y = .3$ is indeed a solution to the system $\begin{cases} x+y=1 \\ \begin{bmatrix} .85 & .35 \\ .15 & .65 \end{bmatrix}\begin{bmatrix} x \\ y \end{bmatrix} = \begin{bmatrix} x \\ y \end{bmatrix} \end{cases}$.

Also, $\begin{bmatrix} .85 & .35 \\ .15 & .65 \end{bmatrix}\begin{bmatrix} .7 \\ .3 \end{bmatrix} = \begin{bmatrix} .7 \\ .3 \end{bmatrix}$.

31. Calculating [A] ^ 255 → Frac gives

$\begin{bmatrix} \frac{8}{35} & \frac{8}{35} & \frac{8}{35} \\ \frac{3}{7} & \frac{3}{7} & \frac{3}{7} \\ \frac{12}{35} & \frac{12}{35} & \frac{12}{35} \end{bmatrix}$. This suggests that the stable

distribution is $\begin{bmatrix} \frac{8}{35} \\ \frac{3}{7} \\ \frac{12}{35} \end{bmatrix}$.

Check: $x = \frac{8}{35}, y = \frac{3}{7}, z = \frac{12}{35}$ is indeed a

solution to the system $\begin{cases} x+y+z=1 \\ \begin{bmatrix} .1 & .4 & .1 \\ .3 & .2 & .8 \\ .6 & .4 & .1 \end{bmatrix}\begin{bmatrix} x \\ y \\ z \end{bmatrix} = \begin{bmatrix} x \\ y \\ z \end{bmatrix} \end{cases}$.

Also, $\begin{bmatrix} .1 & .4 & .1 \\ .3 & .2 & .8 \\ .6 & .4 & .1 \end{bmatrix}\begin{bmatrix} \frac{8}{35} \\ \frac{3}{7} \\ \frac{12}{35} \end{bmatrix} = \begin{bmatrix} \frac{8}{35} \\ \frac{3}{7} \\ \frac{12}{35} \end{bmatrix}$.

Exercises 8.3

1. Yes; the states *A* and *B* are absorbing, and it's possible to get *A* and *B* from states *C* and *D*.

3. No; the state *A* is absorbing, but it's not possible to get to *A* from *C* or *D*.

5. No; states 1 and 2 are absorbing states, but states 3 and 4 do not lead to absorbing states.

7. Yes; state 1 is an absorbing state, and state 3 leads to state 1. Furthermore, it is possible to go from state 2 to state 1 through an intermediate step (state 2 to state 3 to state 1).

9. $\begin{array}{c} \begin{array}{ccc} B & A & C \end{array} \\ \begin{array}{c} B \\ A \\ C \end{array}\begin{bmatrix} 1 & .3 & .4 \\ 0 & .2 & .5 \\ 0 & .5 & .1 \end{bmatrix} \end{array}$

11. $\begin{array}{c} \begin{array}{cccc} D & A & B & C \end{array} \\ \begin{array}{c} D \\ A \\ B \\ C \end{array}\begin{bmatrix} 1 & .4 & 0 & .1 \\ 0 & .1 & 1 & .6 \\ 0 & .2 & 0 & .1 \\ 0 & .3 & 0 & .2 \end{bmatrix} \end{array}$

13. $\begin{bmatrix} 1 & 0 & | & .3 \\ 0 & 1 & | & .2 \\ \hline 0 & 0 & | & .5 \end{bmatrix} = \left[\begin{array}{c|c} I & S \\ \hline 0 & R \end{array}\right]$

$R = [.5]; S = \begin{bmatrix} .3 \\ .2 \end{bmatrix};$

$F = (I-R)^{-1} = [1-.5]^{-1} = [2]$

Find the stable matrix:

$S(I-R)^{-1} = \begin{bmatrix} .3 \\ .2 \end{bmatrix}[2] = \begin{bmatrix} .6 \\ .4 \end{bmatrix}$

$\left[\begin{array}{c|c} I & S(I-R)^{-1} \\ \hline 0 & 0 \end{array}\right] = \begin{bmatrix} 1 & 0 & | & .6 \\ 0 & 1 & | & .4 \\ \hline 0 & 0 & | & 0 \end{bmatrix}$

15. $\begin{bmatrix} 1 & 0 & | & \frac{1}{4} & \frac{1}{6} \\ 0 & 1 & | & \frac{1}{6} & 0 \\ \hline 0 & 0 & | & \frac{1}{4} & \frac{1}{2} \\ 0 & 0 & | & \frac{1}{3} & \frac{1}{3} \end{bmatrix} = \left[\begin{array}{c|c} I & S \\ \hline 0 & R \end{array}\right]$

$R = \begin{bmatrix} \frac{1}{4} & \frac{1}{2} \\ \frac{1}{3} & \frac{1}{3} \end{bmatrix}; S = \begin{bmatrix} \frac{1}{4} & \frac{1}{6} \\ \frac{1}{6} & 0 \end{bmatrix}$

Find the fundamental matrix:

$I - R = \begin{bmatrix} 1 & 0 \\ 0 & 1 \end{bmatrix} - \begin{bmatrix} \frac{1}{4} & \frac{1}{2} \\ \frac{1}{3} & \frac{1}{3} \end{bmatrix} = \begin{bmatrix} \frac{3}{4} & -\frac{1}{2} \\ -\frac{1}{3} & \frac{2}{3} \end{bmatrix}$

$= \begin{bmatrix} a & b \\ c & d \end{bmatrix}$

$\Delta = ad - bc = \left(\frac{3}{4}\right)\left(\frac{2}{3}\right) - \left(-\frac{1}{2}\right)\left(-\frac{1}{3}\right) = \frac{1}{3}$

$F = (I-R)^{-1} = \begin{bmatrix} \frac{d}{\Delta} & -\frac{b}{\Delta} \\ -\frac{c}{\Delta} & \frac{a}{\Delta} \end{bmatrix}$

$= \begin{bmatrix} \frac{2/3}{1/3} & -\frac{-1/2}{1/3} \\ -\frac{-1/3}{1/3} & \frac{3/4}{1/3} \end{bmatrix} = \begin{bmatrix} 2 & \frac{3}{2} \\ 1 & \frac{9}{4} \end{bmatrix}$

Find the stable matrix:

$S(I-R)^{-1} = \begin{bmatrix} \frac{1}{4} & \frac{1}{6} \\ \frac{1}{6} & 0 \end{bmatrix}\begin{bmatrix} 2 & \frac{3}{2} \\ 1 & \frac{9}{4} \end{bmatrix} = \begin{bmatrix} \frac{2}{3} & \frac{3}{4} \\ \frac{1}{3} & \frac{1}{4} \end{bmatrix}$

$\left[\begin{array}{c|c} I & S(I-R)^{-1} \\ \hline 0 & 0 \end{array}\right] = \begin{bmatrix} 1 & 0 & | & \frac{2}{3} & \frac{3}{4} \\ 0 & 1 & | & \frac{1}{3} & \frac{1}{4} \\ \hline 0 & 0 & | & 0 & 0 \\ 0 & 0 & | & 0 & 0 \end{bmatrix}$

17. $\begin{bmatrix} 1 & 0 & 0 & | & .1 & .2 \\ 0 & 1 & 0 & | & .3 & 0 \\ 0 & 0 & 1 & | & 0 & .2 \\ \hline 0 & 0 & 0 & | & .5 & 0 \\ 0 & 0 & 0 & | & .1 & .6 \end{bmatrix} = \left[\begin{array}{c|c} I & S \\ \hline 0 & R \end{array}\right]$

$R = \begin{bmatrix} .5 & 0 \\ .1 & .6 \end{bmatrix}; S = \begin{bmatrix} .1 & .2 \\ .3 & 0 \\ 0 & .2 \end{bmatrix}$

Find the fundamental matrix:

$I - R = \begin{bmatrix} 1 & 0 \\ 0 & 1 \end{bmatrix} - \begin{bmatrix} .5 & 0 \\ .1 & .6 \end{bmatrix} = \begin{bmatrix} .5 & 0 \\ -.1 & .4 \end{bmatrix}$

$= \begin{bmatrix} a & b \\ c & d \end{bmatrix}$

$\Delta = ad - bc = (.5)(.4) - (0)(-.1) = .2$

$F = (I-R)^{-1} = \begin{bmatrix} \frac{d}{\Delta} & -\frac{b}{\Delta} \\ -\frac{c}{\Delta} & \frac{a}{\Delta} \end{bmatrix} = \begin{bmatrix} \frac{.4}{.2} & -\frac{0}{.2} \\ -\frac{-.1}{.2} & \frac{.5}{.2} \end{bmatrix}$

$= \begin{bmatrix} 2 & 0 \\ .5 & 2.5 \end{bmatrix}$

Find the stable matrix:

$S(I-R)^{-1} = \begin{bmatrix} .1 & .2 \\ .3 & 0 \\ 0 & .2 \end{bmatrix}\begin{bmatrix} 2 & 0 \\ .5 & 2.5 \end{bmatrix}$

$= \begin{bmatrix} .3 & .5 \\ .6 & 0 \\ .1 & .5 \end{bmatrix}$

$\left[\begin{array}{c|c} I & S(I-R)^{-1} \\ \hline 0 & 0 \end{array}\right] = \begin{bmatrix} 1 & 0 & 0 & | & .3 & .5 \\ 0 & 1 & 0 & | & .6 & 0 \\ 0 & 0 & 1 & | & .1 & .5 \\ \hline 0 & 0 & 0 & | & 0 & 0 \\ 0 & 0 & 0 & | & 0 & 0 \end{bmatrix}$

19. If the gambler begins with $2, he should have $1 for an expected number of .79 plays.

21. a.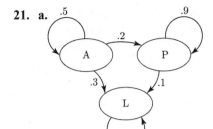

b. $\begin{array}{c} \\ L \\ A \\ P \end{array} \begin{array}{c} L\ \ \ A\ \ \ P \\ \left[\begin{array}{c|cc} 1 & .3 & .1 \\ \hline 0 & .5 & 0 \\ 0 & .2 & .9 \end{array}\right] \end{array}$

c. $R = \begin{bmatrix} .5 & 0 \\ .2 & .9 \end{bmatrix}; \ S = [.3 \ \ .1]$

$I - R = \begin{bmatrix} 1 & 0 \\ 0 & 1 \end{bmatrix} - \begin{bmatrix} .5 & 0 \\ .2 & .9 \end{bmatrix} = \begin{bmatrix} .5 & 0 \\ -.2 & .1 \end{bmatrix}$

$\Delta = (.5)(.1) - (0)(-.2) = .05$

$(I - R)^{-1} = \begin{bmatrix} \frac{.1}{.05} & 0 \\ \frac{.2}{.05} & \frac{.5}{.05} \end{bmatrix} = \begin{bmatrix} 2 & 0 \\ 4 & 10 \end{bmatrix}$

$S(I - R)^{-1} = [.3 \ \ .1] \begin{bmatrix} 2 & 0 \\ 4 & 10 \end{bmatrix} = [1 \ \ 1]$

The stable matrix is $\begin{array}{c} \\ L \\ A \\ P \end{array} \begin{array}{c} L\ \ A\ \ P \\ \left[\begin{array}{c|cc} 1 & 1 & 1 \\ \hline 0 & 0 & 0 \\ 0 & 0 & 0 \end{array}\right] \end{array}$.

d. Add the numbers in the A column of the fundamental matrix: 2 + 4 = 6 yrs.

e. In the long term, all the lawyers will be partners, therefore 0% will be associates.

23. a. (diagram)

b. $\begin{array}{c} \\ E \\ A \\ B \\ C \end{array} \begin{array}{c} E\ \ A\ \ B\ \ C \\ \left[\begin{array}{c|ccc} 1 & 0 & \frac{1}{4} & 0 \\ \hline 0 & 0 & \frac{1}{2} & \frac{1}{2} \\ 0 & \frac{2}{3} & 0 & \frac{1}{2} \\ 0 & \frac{1}{3} & \frac{1}{4} & 0 \end{array}\right] \end{array}$

c. $R = \begin{bmatrix} 0 & \frac{1}{2} & \frac{1}{2} \\ \frac{2}{3} & 0 & \frac{1}{2} \\ \frac{1}{3} & \frac{1}{4} & 0 \end{bmatrix}; \ S = \begin{bmatrix} 0 & \frac{1}{4} & 0 \end{bmatrix}$

$I - R = \begin{bmatrix} 1 & 0 & 0 \\ 0 & 1 & 0 \\ 0 & 0 & 1 \end{bmatrix} - \begin{bmatrix} 0 & \frac{1}{2} & \frac{1}{2} \\ \frac{2}{3} & 0 & \frac{1}{2} \\ \frac{1}{3} & \frac{1}{4} & 0 \end{bmatrix} = \begin{bmatrix} 1 & -\frac{1}{2} & -\frac{1}{2} \\ -\frac{2}{3} & 1 & -\frac{1}{2} \\ -\frac{1}{3} & -\frac{1}{4} & 1 \end{bmatrix}$

$(I - R)^{-1} = \begin{bmatrix} 4.2 & 3 & 3.6 \\ 4 & 4 & 4 \\ 2.4 & 2 & 3.2 \end{bmatrix}$

$S(I - R)^{-1} = \begin{bmatrix} 0 & \frac{1}{4} & 0 \end{bmatrix} \begin{bmatrix} 4.2 & 3 & 3.6 \\ 4 & 4 & 4 \\ 2.4 & 2 & 3.2 \end{bmatrix} = [1 \ \ 1 \ \ 1]$

The stable matrix is $\begin{array}{c} \\ E \\ A \\ B \\ C \end{array} \begin{array}{c} E\ \ A\ \ B\ \ C \\ \left[\begin{array}{c|ccc} 1 & 1 & 1 & 1 \\ \hline 0 & 0 & 0 & 0 \\ 0 & 0 & 0 & 0 \\ 0 & 0 & 0 & 0 \end{array}\right] \end{array}$.

d. Add the numbers in the A column of the fundamental matrix: 10.6 minutes.

25. **a.**
$$\begin{array}{c}\begin{array}{cccc}D & G & F & S\end{array}\\\begin{array}{c}D\\G\\F\\S\end{array}\left[\begin{array}{cc|cc}1 & 0 & .2 & .1\\0 & 1 & 0 & .9\\\hline 0 & 0 & 0 & 0\\0 & 0 & .8 & 0\end{array}\right]\end{array}$$

b. $R = \begin{bmatrix} 0 & 0 \\ .8 & 0 \end{bmatrix}; S = \begin{bmatrix} .2 & .1 \\ 0 & .9 \end{bmatrix}$

$I - R = \begin{bmatrix} 1 & 0 \\ 0 & 1 \end{bmatrix} - \begin{bmatrix} 0 & 0 \\ .8 & 0 \end{bmatrix} = \begin{bmatrix} 1 & 0 \\ -.8 & 1 \end{bmatrix}$

$= \begin{bmatrix} a & b \\ c & d \end{bmatrix}$

$\Delta = ad - bc = (1)(1) - (0)(-.8) = 1$

$(I-R)^{-1} = \begin{bmatrix} \frac{d}{\Delta} & -\frac{b}{\Delta} \\ -\frac{c}{\Delta} & \frac{a}{\Delta} \end{bmatrix} = \begin{bmatrix} \frac{1}{1} & -\frac{0}{1} \\ -\frac{-.8}{1} & \frac{1}{1} \end{bmatrix}$

$= \begin{bmatrix} 1 & 0 \\ .8 & 1 \end{bmatrix}$

$S(I-R)^{-1} = \begin{bmatrix} .2 & .1 \\ 0 & .9 \end{bmatrix}\begin{bmatrix} 1 & 0 \\ .8 & 1 \end{bmatrix}$

$= \begin{bmatrix} .28 & .1 \\ .72 & .9 \end{bmatrix}$

$\left[\begin{array}{c|c} I & S(I-R)^{-1} \\ \hline 0 & 0 \end{array}\right] = \begin{array}{c}\begin{array}{cccc}D & G & F & S\end{array}\\\begin{array}{c}D\\G\\F\\S\end{array}\left[\begin{array}{cc|cc}1 & 0 & .28 & .1\\0 & 1 & .72 & .9\\\hline 0 & 0 & 0 & 0\\0 & 0 & 0 & 0\end{array}\right]\end{array}$

c. From the F column of the stable matrix, the probability that a freshman will eventually graduate is .72.

d. The fundamental matrix is

$F = (I-R)^{-1} = \begin{array}{c}\begin{array}{cc}F & S\end{array}\\\begin{array}{c}F\\S\end{array}\begin{bmatrix} 1 & 0 \\ .8 & 1 \end{bmatrix}\end{array}$.

Add the numbers in the F column:
$1 + .8 = 1.8$.
The student will attend an expected number of 1.8 years.

27. First, arrange the matrix so that the absorbing states come first and calculate the fundamental and stable matrices.

$$\begin{array}{c}\begin{array}{cccc}\text{Paid} & \text{Bad} & \leq 30 & < 60\end{array}\\\begin{array}{c}\text{Paid}\\\text{Bad}\\\leq 30\\< 60\end{array}\left[\begin{array}{cc|cc}1 & 0 & .4 & .1\\0 & 1 & 0 & .1\\\hline 0 & 0 & .4 & .4\\0 & 0 & .2 & .4\end{array}\right] = \left[\begin{array}{c|c}I & S \\ \hline 0 & R\end{array}\right]\end{array}$$

$R = \begin{bmatrix} .4 & .4 \\ .2 & .4 \end{bmatrix}; S = \begin{bmatrix} .4 & .1 \\ 0 & .1 \end{bmatrix}$

$I - R = \begin{bmatrix} 1 & 0 \\ 0 & 1 \end{bmatrix} - \begin{bmatrix} .4 & .4 \\ .2 & .4 \end{bmatrix} = \begin{bmatrix} .6 & -.4 \\ -.2 & .6 \end{bmatrix}$

$= \begin{bmatrix} a & b \\ c & d \end{bmatrix}$

$\Delta = ad - bc = (.6)(.6) - (-.4)(-.2) = .28$

$F = (I-R)^{-1} = \begin{bmatrix} \frac{d}{\Delta} & -\frac{b}{\Delta} \\ -\frac{c}{\Delta} & \frac{a}{\Delta} \end{bmatrix}$

$= \begin{bmatrix} \frac{.6}{.28} & \frac{-(-.4)}{.28} \\ \frac{-(-.2)}{.28} & \frac{.6}{.28} \end{bmatrix} = \begin{bmatrix} \frac{15}{7} & \frac{10}{7} \\ \frac{5}{7} & \frac{15}{7} \end{bmatrix}$

$S(I-R)^{-1} = \begin{bmatrix} .4 & .1 \\ 0 & .1 \end{bmatrix}\begin{bmatrix} \frac{15}{7} & \frac{10}{7} \\ \frac{5}{7} & \frac{15}{7} \end{bmatrix}$

$= \begin{bmatrix} \frac{13}{14} & \frac{11}{14} \\ \frac{1}{14} & \frac{3}{14} \end{bmatrix}$

The stable matrix is

$$\begin{array}{c}\begin{array}{cccc}\text{Paid} & \text{Bad} & \leq 30 & < 60\end{array}\\\begin{array}{c}\text{Paid}\\\text{Bad}\\\leq 30\\< 60\end{array}\left[\begin{array}{cc|cc}1 & 0 & \frac{13}{14} & \frac{11}{14}\\0 & 1 & \frac{1}{14} & \frac{3}{14}\\\hline 0 & 0 & 0 & 0\\0 & 0 & 0 & 0\end{array}\right]\end{array}.$$

a. From the stable matrix, the probability of an account eventually being paid off is $\frac{13}{14}$ if it is currently at most 30 days overdue and $\frac{11}{14}$ if it is less than 60 days overdue (but more than 30 days overdue).

SSM: Finite Math   **Chapter 8:** Markov Processes

b. The fundamental matrix is

$\begin{array}{c} \leq 30 \\ < 60 \end{array} \begin{bmatrix} \frac{15}{7} & \frac{10}{7} \\ \frac{5}{7} & \frac{15}{7} \end{bmatrix}$. Add the numbers in the

first column: $\frac{15}{7} + \frac{5}{7} = \frac{20}{7}$. An account that is overdue at most 30 days is expected to reach an absorbing state (paid or bad) after $\frac{20}{7}$ months.

c. From the stable matrix, about $\frac{13}{14}$ of the "≤30 day" debt will be paid, and about $\frac{11}{14}$ of the "<60 day" debt will be paid.

$\frac{13}{14}(\$2000) + \frac{11}{14}(\$5000) \approx \$5786$. About $\$5786$ will be paid and about $\$1214$ will become bad debt.

29. First, find the absorbing stochastic matrix and calculate the fundamental and stable matrices.

$$\begin{array}{c} \\ \$0 \\ \$4 \\ \$1 \\ \$2 \\ \$3 \end{array} \begin{bmatrix} \$0 & \$4 & \$1 & \$2 & \$3 \\ 1 & 0 & \frac{1}{2} & 0 & 0 \\ 0 & 1 & 0 & 0 & \frac{1}{2} \\ \hline 0 & 0 & 0 & \frac{1}{2} & 0 \\ 0 & 0 & \frac{1}{2} & 0 & \frac{1}{2} \\ 0 & 0 & 0 & \frac{1}{2} & 0 \end{bmatrix} = \begin{bmatrix} I & S \\ \hline 0 & R \end{bmatrix}$$

$R = \begin{bmatrix} 0 & \frac{1}{2} & 0 \\ \frac{1}{2} & 0 & \frac{1}{2} \\ 0 & \frac{1}{2} & 0 \end{bmatrix}; S = \begin{bmatrix} \frac{1}{2} & 0 & 0 \\ 0 & 0 & \frac{1}{2} \end{bmatrix}$

$I - R = \begin{bmatrix} 1 & 0 & 0 \\ 0 & 1 & 0 \\ 0 & 0 & 1 \end{bmatrix} - \begin{bmatrix} 0 & \frac{1}{2} & 0 \\ \frac{1}{2} & 0 & \frac{1}{2} \\ 0 & \frac{1}{2} & 0 \end{bmatrix}$

$= \begin{bmatrix} 1 & -\frac{1}{2} & 0 \\ -\frac{1}{2} & 1 & -\frac{1}{2} \\ 0 & -\frac{1}{2} & 1 \end{bmatrix}$

$F = (I-R)^{-1} = \begin{bmatrix} 1 & -\frac{1}{2} & 0 \\ -\frac{1}{2} & 1 & -\frac{1}{2} \\ 0 & -\frac{1}{2} & 1 \end{bmatrix}^{-1}$

$= \begin{bmatrix} \frac{3}{2} & 1 & \frac{1}{2} \\ 1 & 2 & 1 \\ \frac{1}{2} & 1 & \frac{3}{2} \end{bmatrix}$

$S(I-R)^{-1} = \begin{bmatrix} \frac{1}{2} & 0 & 0 \\ 0 & 0 & \frac{1}{2} \end{bmatrix} \begin{bmatrix} \frac{3}{2} & 1 & \frac{1}{2} \\ 1 & 2 & 1 \\ \frac{1}{2} & 1 & \frac{3}{2} \end{bmatrix}$

$= \begin{bmatrix} \frac{3}{4} & \frac{1}{2} & \frac{1}{4} \\ \frac{1}{4} & \frac{1}{2} & \frac{3}{4} \end{bmatrix}$

The stable matrix is
$\begin{array}{c} \$0 \\ \$4 \\ \$1 \\ \$2 \\ \$3 \end{array} \begin{bmatrix} \$0 & \$4 & \$1 & \$2 & \$3 \\ 1 & 0 & \frac{3}{4} & \frac{1}{2} & \frac{1}{4} \\ 0 & 1 & \frac{1}{4} & \frac{1}{2} & \frac{3}{4} \\ 0 & 0 & 0 & 0 & 0 \\ 0 & 0 & 0 & 0 & 0 \\ 0 & 0 & 0 & 0 & 0 \end{bmatrix}$.

a. From the top row of the stable matrix, the probability of eventually going broke is $\frac{3}{4}$ if he starts with $\$1$, $\frac{1}{2}$ if he starts with $\$2$, and $\frac{1}{4}$ if he starts with $\$3$.

b. The fundamental matrix is
$\begin{array}{c} \$1 \\ \$2 \\ \$3 \end{array} \begin{bmatrix} \$1 & \$2 & \$3 \\ \frac{3}{2} & 1 & \frac{1}{2} \\ 1 & 2 & 1 \\ \frac{1}{2} & 1 & \frac{3}{2} \end{bmatrix}$.

Add the entries in the middle column:
$1 + 2 + 1 = 4$
Starting with $\$2$, he will play for an expected number of 4 times.

Chapter 8: Markov Processes

SSM: Finite Math

31. Consider the Markov process with states 1, 2 and 3 corresponding to the number of different quotations received so far. This process has absorbing stochastic matrix

$$A = \begin{array}{c} 3 \\ 2 \\ 1 \end{array} \left[\begin{array}{c|cc} 1 & \frac{1}{3} & 0 \\ \hline 0 & \frac{2}{3} & \frac{2}{3} \\ 0 & 0 & \frac{1}{3} \end{array} \right].$$

with columns labeled 3 2 1.

The distribution matrix after purchasing one bottle is $B = \begin{bmatrix} 0 \\ 0 \\ 1 \end{bmatrix}$, so the distribution matrix after purchasing four more bottles (for a total of five) is

$$A^4 B = \begin{pmatrix} 1 & \frac{65}{81} & \frac{50}{81} \\ 0 & \frac{16}{81} & \frac{10}{27} \\ 0 & 0 & \frac{1}{81} \end{pmatrix} \begin{pmatrix} 0 \\ 0 \\ 1 \end{pmatrix} = \begin{pmatrix} \frac{50}{81} \\ \frac{10}{27} \\ \frac{1}{81} \end{pmatrix}.$$

Therefore, the probability of receiving all three quotations after purchasing five bottles is $\frac{50}{81}$.

We have $R = \begin{bmatrix} \frac{2}{3} & \frac{2}{3} \\ 0 & \frac{1}{3} \end{bmatrix}$ so the fundamental matrix

$$(I - R)^{-1} = \begin{bmatrix} \frac{1}{3} & -\frac{2}{3} \\ 0 & \frac{2}{3} \end{bmatrix}^{-1} = \begin{bmatrix} 3 & 3 \\ 0 & \frac{3}{2} \end{bmatrix}.$$

The expected number of soft drinks you have to purchase to get all three quotations is the sum of the numbers in the "1" column of the fundamental matrix plus one for the first bottle, so it's $3 + \frac{3}{2} + 1 = 5\frac{1}{2}$.

33. Calculating $[A] \wedge 225 \rightarrow$ Frac gives

$$\begin{bmatrix} 1 & 0 & 0 & \frac{9}{37} & \frac{24}{37} \\ 0 & 1 & 0 & \frac{53}{74} & \frac{9}{37} \\ 0 & 0 & 1 & \frac{3}{74} & \frac{4}{37} \\ 0 & 0 & 0 & 0 & 0 \\ 0 & 0 & 0 & 0 & 0 \end{bmatrix}.$$

Let $[B] = S = \begin{bmatrix} 0 & .6 \\ .5 & .1 \\ 0 & .1 \end{bmatrix}$ and $[C] = R = \begin{bmatrix} .2 & .2 \\ .3 & 0 \end{bmatrix}$.

Then calculating
$[B] * (\text{identity}(2) - [C])^{-1} \rightarrow$ Frac

gives $\begin{bmatrix} \frac{9}{37} & \frac{24}{37} \\ \frac{53}{74} & \frac{9}{37} \\ \frac{3}{74} & \frac{4}{37} \end{bmatrix}$. Therefore, the matrix given

above is the exact stable matrix.

Chapter 8 Fundamental Concept Check

1. A process in which a system makes transitions from a set of states to another, with specified probabilities.

2. A transition matrix is a matrix whose columns and rows are labeled with the different states and whose ij^{th} entry is the probability of going from the state labeling the jth column to the state labeling the ith row. A stochastic matrix is a matrix whose entries are each greater than or equal to 0 and for which the sum of the entries in each column is 1. A distribution matrix is a column matrix giving the percent of items in each state at a specified time.

3. The matrix A multiplied by itself n times. Its entries give the probabilities of going from each state to each other state in n periods.

4. It is calculated by multiplying the initial distribution matrix by A^n.

5. A stochastic matrix for which some power has all positive entries.

6. The stable matrix of a regular stochastic matrix A is the matrix approached by successive powers of A. The stable distribution is the column matrix approached by successive distribution matrices.

7. Solve the system of linear equations
$\begin{cases} \text{sum of the entries of } X = 1 \\ AX = X \end{cases}$ for the column matrix X.

8. A state that does not change during successive transitions.

9. A stochastic matrix having one or more absorbing states and for which, from any state, it is possible to get to at least one absorbing state, either directly or through one or more intermediate states.

10. The stable matrix is $\begin{bmatrix} I & S(I-R)^{-1} \\ \hline 0 & 0 \end{bmatrix}$ where the absorbing matrix has the form $\begin{bmatrix} I & S \\ \hline 0 & R \end{bmatrix}$.

11. The matrix $(I-R)^{-1}$ in item 10; its ij^{th} entry is the expected number of times the process will be in non-absorbing state i, given that it started in non-absorbing state j.

Chapter 8 Review Exercises

1. Stochastic, neither; the matrix is stochastic because it is square, the entries are all ≥ 0, and the sum of the entries in each column is 1. It is not regular because all powers of the matrix include zero entries. The first state is an absorbing state, but the matrix is not an absorbing matrix because an object that begins in the third or fourth state will always remain within these two (nonabsorbing) states.

2. Stochastic, regular; the matrix is stochastic because it is square, the entries are all ≥ 0, and the sum of the entries in each column is 1. It is regular because it has no zero entries.

3. Stochastic, regular; the matrix is stochastic because it is square, the entries are all ≥ 0, and the sum of the entries in each column is 1. It is regular because the second power, $\begin{bmatrix} .3 & .21 \\ .7 & .79 \end{bmatrix}$, contains no zero entries.

4. Stochastic, absorbing; the matrix is stochastic because it is square, the entries are all ≥ 0, and the sum of the entries in each column is 1. It is absorbing because the first and second states are absorbing states, and it is possible for an object to get from the third state to an absorbing state.

5. Not stochastic because the sum of the entries in the middle column is not 1.

6. Stochastic, absorbing; the matrix is stochastic because it is square, the entries are all ≥ 0, and the sum of the entries in each column is 1. It is absorbing because the first and second states are absorbing states and it is possible to get from the third state (indirectly) or the fourth state (directly) to an absorbing state.

7. $\begin{cases} x+y=1 \\ \begin{bmatrix} .6 & .5 \\ .4 & .5 \end{bmatrix} \begin{bmatrix} x \\ y \end{bmatrix} = \begin{bmatrix} x \\ y \end{bmatrix} \end{cases}$

$\begin{cases} x + y = 1 \\ .6x + .5y = x \\ .4x + .5y = y \end{cases}$

$\begin{cases} x + y = 1 \\ -.4x + .5y = 0 \\ .4x - .5y = 0 \end{cases}$

The second and third equations in this system are equivalent.

$\begin{bmatrix} 1 & 1 & | & 1 \\ -.4 & .5 & | & 0 \end{bmatrix} \xrightarrow{[2]+.4[1]} \begin{bmatrix} 1 & 1 & | & 1 \\ 0 & .9 & | & .4 \end{bmatrix}$

$\xrightarrow{\frac{10}{9}[2]} \begin{bmatrix} 1 & 1 & | & 1 \\ 0 & 1 & | & \frac{4}{9} \end{bmatrix} \xrightarrow{[1]+(-1)[2]} \begin{bmatrix} 1 & 0 & | & \frac{5}{9} \\ 0 & 1 & | & \frac{4}{9} \end{bmatrix}$

$x = \frac{5}{9}, y = \frac{4}{9}$

The stable distribution is $\begin{bmatrix} \frac{5}{9} \\ \frac{4}{9} \end{bmatrix}$.

8. $\begin{bmatrix} 1 & 0 & 0 & | & \frac{1}{8} & \frac{1}{4} \\ 0 & 1 & 0 & | & \frac{1}{8} & 0 \\ 0 & 0 & 1 & | & 0 & \frac{1}{4} \\ \hline 0 & 0 & 0 & | & \frac{1}{4} & \frac{1}{2} \\ 0 & 0 & 0 & | & \frac{1}{2} & 0 \end{bmatrix} = \begin{bmatrix} I & | & S \\ \hline 0 & | & R \end{bmatrix}$

$R = \begin{bmatrix} \frac{1}{4} & \frac{1}{2} \\ \frac{1}{2} & 0 \end{bmatrix}; S = \begin{bmatrix} \frac{1}{8} & \frac{1}{4} \\ \frac{1}{8} & 0 \\ 0 & \frac{1}{4} \end{bmatrix}$

$I - R = \begin{bmatrix} 1 & 0 \\ 0 & 1 \end{bmatrix} - \begin{bmatrix} \frac{1}{4} & \frac{1}{2} \\ \frac{1}{2} & 0 \end{bmatrix}$

$= \begin{bmatrix} \frac{3}{4} & -\frac{1}{2} \\ -\frac{1}{2} & 1 \end{bmatrix}$

$= \begin{bmatrix} a & b \\ c & d \end{bmatrix}$

$\Delta = ad - bc = \left(\frac{3}{4}\right)(1) - \left(-\frac{1}{2}\right)\left(-\frac{1}{2}\right)$

$= \frac{1}{2}$

$(I - R)^{-1} = \begin{bmatrix} \frac{d}{\Delta} & -\frac{b}{\Delta} \\ -\frac{c}{\Delta} & \frac{a}{\Delta} \end{bmatrix}$

$= \begin{bmatrix} \frac{1}{1/2} & \frac{-1/2}{1/2} \\ -\frac{-1/2}{1/2} & \frac{3/4}{1/2} \end{bmatrix}$

$= \begin{bmatrix} 2 & 1 \\ 1 & \frac{3}{2} \end{bmatrix}$

$S(I - R)^{-1} = \begin{bmatrix} \frac{1}{8} & \frac{1}{4} \\ \frac{1}{8} & 0 \\ 0 & \frac{1}{4} \end{bmatrix} \begin{bmatrix} 2 & 1 \\ 1 & \frac{3}{2} \end{bmatrix}$

$= \begin{bmatrix} \frac{1}{2} & \frac{1}{2} \\ \frac{1}{4} & \frac{1}{8} \\ \frac{1}{4} & \frac{3}{8} \end{bmatrix}$

$\begin{bmatrix} I & | & S(I-R)^{-1} \\ \hline 0 & | & 0 \end{bmatrix} = \begin{bmatrix} 1 & 0 & 0 & | & \frac{1}{2} & \frac{1}{2} \\ 0 & 1 & 0 & | & \frac{1}{4} & \frac{1}{8} \\ 0 & 0 & 1 & | & \frac{1}{4} & \frac{3}{8} \\ \hline 0 & 0 & 0 & | & 0 & 0 \\ 0 & 0 & 0 & | & 0 & 0 \end{bmatrix}$

9. a.
$\begin{array}{c} \\ H \\ M \\ L \end{array} \begin{array}{ccc} H & M & L \end{array}$
$\begin{bmatrix} .5 & .4 & .3 \\ .4 & .3 & .5 \\ .1 & .3 & .2 \end{bmatrix}$

b. $\begin{bmatrix} .5 & .4 & .3 \\ .4 & .3 & .5 \\ .1 & .3 & .2 \end{bmatrix} \begin{bmatrix} .1 \\ .6 \\ .3 \end{bmatrix} = \begin{bmatrix} .38 \\ .37 \\ .25 \end{bmatrix}$

38% of the children of the current generation will have high incomes.

c. Find the stable distribution.

$\begin{cases} x + y + z = 1 \\ \begin{bmatrix} .5 & .4 & .3 \\ .4 & .3 & .5 \\ .1 & .3 & .2 \end{bmatrix} \begin{bmatrix} x \\ y \\ z \end{bmatrix} = \begin{bmatrix} x \\ y \\ z \end{bmatrix} \end{cases}$

$\begin{cases} x + y + z = 1 \\ .5x + .4y + .3z = x \\ .4x + .3y + .5z = y \\ .1x + .3y + .2z = z \end{cases}$

$\begin{cases} x + y + z = 1 \\ -.5x + .4y + .3z = 0 \\ .4x - .7y + .5z = 0 \\ .1x + .3y - .8z = 0 \end{cases}$

The fourth equation is equivalent to the second plus third and so is redundant.

$$\begin{bmatrix} 1 & 1 & 1 & | & 1 \\ -.5 & .4 & .3 & | & 0 \\ .4 & -.7 & .5 & | & 0 \end{bmatrix} \xrightarrow[10[3]]{10[2]} \begin{bmatrix} 1 & 1 & 1 & | & 1 \\ -5 & 4 & 3 & | & 0 \\ 4 & -7 & 5 & | & 0 \end{bmatrix}$$

$$\xrightarrow[[3]+(-4)[1]]{[2]+5[1]} \begin{bmatrix} 1 & 1 & 1 & | & 1 \\ 0 & 9 & 8 & | & 5 \\ 0 & -11 & 1 & | & -4 \end{bmatrix}$$

$$\xrightarrow[\substack{\frac{1}{9}[2] \\ [1]+(-1)[2] \\ [3]+(11)[2]}]{} \begin{bmatrix} 1 & 0 & \frac{1}{9} & | & \frac{4}{9} \\ 0 & 1 & \frac{8}{9} & | & \frac{5}{9} \\ 0 & 0 & \frac{97}{9} & | & \frac{19}{9} \end{bmatrix}$$

$$\xrightarrow[\substack{\frac{9}{97}[3] \\ [1]+\left(-\frac{1}{9}\right)[3] \\ [2]+\left(-\frac{8}{9}\right)[3]}]{} \begin{bmatrix} 1 & 0 & 0 & | & \frac{41}{97} \\ 0 & 1 & 0 & | & \frac{37}{97} \\ 0 & 0 & 1 & | & \frac{19}{97} \end{bmatrix}$$

$x = \dfrac{41}{97},\ y = \dfrac{37}{97},\ z = \dfrac{19}{97}.$

The stable distribution is $\begin{bmatrix} \frac{41}{97} \\ \frac{37}{97} \\ \frac{19}{97} \end{bmatrix}$. In the long run, $\dfrac{19}{97}$ of the population will have low incomes.

10. **a.** $\begin{array}{c} \ P\ \ \ N \\ \begin{array}{c}P \\ N\end{array}\!\begin{bmatrix} .8 & .3 \\ .2 & .7 \end{bmatrix} \end{array}$

  b. $A^2 \begin{bmatrix} 1 \\ 0 \end{bmatrix}_0 = \begin{bmatrix} .8 & .3 \\ .2 & .7 \end{bmatrix}\begin{bmatrix} .8 & .3 \\ .2 & .7 \end{bmatrix}\begin{bmatrix} 1 \\ 0 \end{bmatrix}_0 = \begin{bmatrix} .7 \\ .3 \end{bmatrix}_2$

  30% will need adjusting after 2 days.

  c. Find the stable distribution.
  $$\begin{cases} x + y = 1 \\ \begin{bmatrix} .8 & .3 \\ .2 & .7 \end{bmatrix}\begin{bmatrix} x \\ y \end{bmatrix} = \begin{bmatrix} x \\ y \end{bmatrix} \end{cases}$$

  $$\begin{cases} x + y = 1 \\ .8x + .3y = x \\ .2x + .7y = y \end{cases}$$

  $$\begin{cases} x + y = 1 \\ -.2x + .3y = 0 \\ .2x - .3y = 0 \end{cases}$$

  The second and third equations in this system are equivalent.

  $$\begin{bmatrix} 1 & 1 & | & 1 \\ -.2 & .3 & | & 0 \end{bmatrix} \xrightarrow{[2]+.2[1]} \begin{bmatrix} 1 & 1 & | & 1 \\ 0 & .5 & | & .2 \end{bmatrix}$$

  $$\xrightarrow{2[2]} \begin{bmatrix} 1 & 1 & | & 1 \\ 0 & 1 & | & .4 \end{bmatrix}$$

  $$\xrightarrow{[1]+(-1)[2]} \begin{bmatrix} 1 & 0 & | & .6 \\ 0 & 1 & | & .4 \end{bmatrix}$$

  $x = .6,\ y = .4$

  The stable distribution is $\begin{bmatrix} .6 \\ .4 \end{bmatrix}$. In the long run, 60% will be properly adjusted.

11. $\begin{bmatrix} 1 & 0 & \frac{1}{6} & \frac{1}{2} & \frac{2}{5} \\ 0 & 1 & 0 & 0 & \frac{2}{5} \\ 0 & 0 & 0 & 0 & 0 \\ 0 & 0 & \frac{2}{3} & \frac{1}{2} & 0 \\ 0 & 0 & \frac{1}{6} & 0 & \frac{1}{5} \end{bmatrix} = \begin{bmatrix} I & S \\ 0 & R \end{bmatrix}$

  $R = \begin{bmatrix} 0 & 0 & 0 \\ \frac{2}{3} & \frac{1}{2} & 0 \\ \frac{1}{6} & 0 & \frac{1}{5} \end{bmatrix};\ S = \begin{bmatrix} \frac{1}{6} & \frac{1}{2} & \frac{2}{5} \\ 0 & 0 & \frac{2}{5} \end{bmatrix}$

  $I - R = \begin{bmatrix} 1 & 0 & 0 \\ 0 & 1 & 0 \\ 0 & 0 & 1 \end{bmatrix} - \begin{bmatrix} 0 & 0 & 0 \\ \frac{2}{3} & \frac{1}{2} & 0 \\ \frac{1}{6} & 0 & \frac{1}{5} \end{bmatrix}$

  $= \begin{bmatrix} 1 & 0 & 0 \\ -\frac{2}{3} & \frac{1}{2} & 0 \\ -\frac{1}{6} & 0 & \frac{4}{5} \end{bmatrix}$

Use the Gauss-Jordan method to find $(I-R)^{-1}$.

$$\begin{bmatrix} 1 & 0 & 0 & | & 1 & 0 & 0 \\ -\frac{2}{3} & \frac{1}{2} & 0 & | & 0 & 1 & 0 \\ -\frac{1}{6} & 0 & \frac{4}{5} & | & 0 & 0 & 1 \end{bmatrix}$$

$$\xrightarrow[{[3]+\frac{1}{6}[1]}]{[2]+\frac{2}{3}[1]} \begin{bmatrix} 1 & 0 & 0 & | & 1 & 0 & 0 \\ 0 & \frac{1}{2} & 0 & | & \frac{2}{3} & 1 & 0 \\ 0 & 0 & \frac{4}{5} & | & \frac{1}{6} & 0 & 1 \end{bmatrix}$$

$$\xrightarrow[{\frac{5}{4}[3]}]{2[2]} \begin{bmatrix} 1 & 0 & 0 & | & 1 & 0 & 0 \\ 0 & 1 & 0 & | & \frac{4}{3} & 2 & 0 \\ 0 & 0 & 1 & | & \frac{5}{24} & 0 & \frac{5}{4} \end{bmatrix}$$

$$S(I-R)^{-1} = \begin{bmatrix} \frac{1}{6} & \frac{1}{2} & \frac{2}{5} \\ 0 & 0 & \frac{2}{5} \end{bmatrix} \begin{bmatrix} 1 & 0 & 0 \\ \frac{4}{3} & 2 & 0 \\ \frac{5}{24} & 0 & \frac{5}{4} \end{bmatrix}$$

$$= \begin{bmatrix} \frac{11}{12} & 1 & \frac{1}{2} \\ \frac{1}{12} & 0 & \frac{1}{2} \end{bmatrix}$$

$$\left[\begin{array}{c|c} I & S(I-R)^{-1} \\ \hline 0 & 0 \end{array}\right] = \begin{bmatrix} 1 & 0 & \frac{11}{12} & 1 & \frac{1}{2} \\ 0 & 1 & \frac{1}{12} & 0 & \frac{1}{2} \\ 0 & 0 & 0 & 0 & 0 \\ 0 & 0 & 0 & 0 & 0 \\ 0 & 0 & 0 & 0 & 0 \end{bmatrix}$$

12. a.
$$\begin{array}{c} \\ I \\ II \\ III \\ IV \end{array} \begin{array}{cccc} I & II & III & IV \end{array} \\ \begin{bmatrix} 1 & 0 & 0 & \frac{1}{4} \\ 0 & 1 & \frac{1}{3} & \frac{1}{4} \\ \hline 0 & 0 & 0 & \frac{1}{2} \\ 0 & 0 & \frac{2}{3} & 0 \end{bmatrix}$$

b. The mouse will find the cheese after two minutes if he either goes to room II after one minute, or goes to room III after one minute and then room II after two minutes. Therefore the probability that the mouse finds the cheese after two minutes is $\frac{1}{4} + \frac{1}{2} \cdot \frac{1}{3} = \frac{5}{12}$.

c. Find the stable matrix.

$$R = \begin{bmatrix} 0 & \frac{1}{2} \\ \frac{2}{3} & 0 \end{bmatrix}, S = \begin{bmatrix} 0 & \frac{1}{4} \\ \frac{1}{3} & \frac{1}{4} \end{bmatrix}$$

$$I - R = \begin{bmatrix} 1 & 0 \\ 0 & 1 \end{bmatrix} - \begin{bmatrix} 0 & \frac{1}{2} \\ \frac{2}{3} & 0 \end{bmatrix}$$

$$= \begin{bmatrix} 1 & -\frac{1}{2} \\ -\frac{2}{3} & 1 \end{bmatrix}$$

$$= \begin{bmatrix} a & b \\ c & d \end{bmatrix}$$

$$\Delta = ad - bc = (1)(1) - \left(-\frac{2}{3}\right)\left(-\frac{1}{2}\right) = \frac{2}{3}$$

$$F = (I-R)^{-1} = \begin{bmatrix} \frac{d}{\Delta} & -\frac{b}{\Delta} \\ -\frac{c}{\Delta} & \frac{a}{\Delta} \end{bmatrix}$$

$$= \begin{bmatrix} \frac{1}{2/3} & \frac{-1/2}{2/3} \\ \frac{-2/3}{2/3} & \frac{1}{2/3} \end{bmatrix}$$

$$= \begin{bmatrix} \frac{3}{2} & \frac{3}{4} \\ 1 & \frac{3}{2} \end{bmatrix}$$

$$S(I-R)^{-1} = \begin{bmatrix} 0 & \frac{1}{4} \\ \frac{1}{3} & \frac{1}{4} \end{bmatrix} \begin{bmatrix} \frac{3}{2} & \frac{3}{4} \\ 1 & \frac{3}{2} \end{bmatrix} = \begin{bmatrix} \frac{1}{4} & \frac{3}{8} \\ \frac{3}{4} & \frac{5}{8} \end{bmatrix}$$

The stable matrix is

$$\left[\begin{array}{c|c} I & S(I-R)^{-1} \\ \hline 0 & 0 \end{array}\right] = \begin{array}{c} I \\ II \\ III \\ IV \end{array} \begin{bmatrix} 1 & 0 & \frac{1}{4} & \frac{3}{8} \\ 0 & 1 & \frac{3}{4} & \frac{5}{8} \\ \hline 0 & 0 & 0 & 0 \\ 0 & 0 & 0 & 0 \end{bmatrix}$$

with column headers $I\ II\ III\ IV$.

From the fourth column, if he starts in room IV the probability of finding cheese in the long run is $\frac{5}{8} = .625$.

d. The fundamental matrix is
$\begin{array}{c}\text{III}\\\text{IV}\end{array}\begin{bmatrix}\text{III}&\text{IV}\\\frac{3}{2}&\frac{3}{4}\\1&\frac{3}{2}\end{bmatrix}$.

Add the entries in the column for room III:
$\frac{3}{2}+1=\frac{5}{2}$.

A mouse that starts in room III will spend an expected $2\frac{1}{2}$ minutes before finding the cheese or being trapped.

13. (c) is the correct choice because it satisfies the system $\begin{cases}x+y+z=1\\\begin{bmatrix}.4&.4&.2\\.1&.1&.3\\.5&.5&.5\end{bmatrix}\begin{bmatrix}x\\y\\z\end{bmatrix}=\begin{bmatrix}x\\y\\z\end{bmatrix}\end{cases}$

14. $A=\begin{array}{c}\text{A}\\\text{B}\end{array}\begin{bmatrix}\text{A}&\text{B}\\.9&.2\\.1&.8\end{bmatrix}$

$A^2\begin{bmatrix}\;\\\;\end{bmatrix}_0=\begin{bmatrix}.9&.2\\.1&.8\end{bmatrix}\begin{bmatrix}.9&.2\\.1&.8\end{bmatrix}\begin{bmatrix}.5\\.5\end{bmatrix}_0$

$=\begin{bmatrix}.585\\.415\end{bmatrix}_2$

58.5% of the regular listeners will listen to station A two days from now.

15. a. If the traffic is moderate on a particular day then for the next day the probability of light traffic is .2, the probability of moderate traffic is .75, and the probability of heavy traffic is .05.

    b. Find the stable distribution.
    $\begin{cases}x+y+z=1\\\begin{bmatrix}.70&.20&.10\\.20&.75&.30\\.10&.05&.60\end{bmatrix}\begin{bmatrix}x\\y\\z\end{bmatrix}=\begin{bmatrix}x\\y\\z\end{bmatrix}\end{cases}$

    $\begin{cases}x+y+z=1\\.7x+.2y+.1z=x\\.2x+.75y+.3z=y\\.1x+.05y+.6z=z\end{cases}$

$\begin{cases}x+y+z=1\\-.3x+.2y+.1z=0\\.2x-.25y+.3z=0\\.1x+.05y-.4z=0\end{cases}$

The fourth equation is equivalent to the second plus the third and so is redundant.

$\begin{bmatrix}1&1&1&|&1\\-.3&.2&.1&|&0\\.2&-.25&.3&|&0\end{bmatrix}\xrightarrow[20[3]]{10[2]}\begin{bmatrix}1&1&1&|&1\\-3&2&1&|&0\\4&-5&6&|&0\end{bmatrix}$

$\xrightarrow[\substack{[2]+3[1]\\[3]+(-4)[1]}]{}\begin{bmatrix}1&1&1&|&1\\0&5&4&|&3\\0&-9&2&|&-4\end{bmatrix}$

$\xrightarrow[\substack{\frac{1}{5}[2]\\[1]+(-1)[2]\\[3]+9[2]}]{}\begin{bmatrix}1&0&\frac{1}{5}&|&\frac{2}{5}\\0&1&\frac{4}{5}&|&\frac{3}{5}\\0&0&\frac{46}{5}&|&\frac{7}{5}\end{bmatrix}$

$\xrightarrow[\substack{\frac{5}{46}[3]\\[1]+(-\frac{1}{5})[3]\\[2]+(-\frac{4}{5})[3]}]{}\begin{bmatrix}1&0&0&|&\frac{17}{46}\\0&1&0&|&\frac{11}{23}\\0&0&1&|&\frac{7}{46}\end{bmatrix}$

The stable distribution is $\begin{bmatrix}\frac{17}{46}\\\frac{11}{23}\\\frac{7}{46}\end{bmatrix}$ or about $\begin{bmatrix}.370\\.478\\.152\end{bmatrix}$.

About 37.0% of workdays will have light traffic, 47.8% will have moderate traffic, and 15.2% will have heavy traffic.

c. $\frac{7}{46}\cdot 20\approx 3.04$

About 3 workdays will have heavy traffic.

16.
$$\begin{array}{c} & \begin{array}{cccc} C & L & G & S \end{array} \\ \begin{array}{c} C \\ L \\ G \\ S \end{array} & \left[\begin{array}{cc|cc} 1 & 0 & .60 & .05 \\ 0 & 1 & .10 & .40 \\ 0 & 0 & .20 & .50 \\ 0 & 0 & .10 & .05 \end{array}\right] \end{array}$$

$$R = \begin{bmatrix} .20 & .50 \\ .10 & .05 \end{bmatrix}$$

$$I - R = \begin{bmatrix} 1 & 0 \\ 0 & 1 \end{bmatrix} - \begin{bmatrix} .20 & .50 \\ .10 & .05 \end{bmatrix}$$

$$= \begin{bmatrix} .80 & -.50 \\ -.10 & .95 \end{bmatrix}$$

$$= \begin{bmatrix} a & b \\ c & d \end{bmatrix}$$

$\Delta = ad - bc = (.80)(.95) - (-.50)(-.10) = .71$

$F = (I - R)^{-1}$

$$= \begin{bmatrix} \frac{d}{\Delta} & -\frac{b}{\Delta} \\ -\frac{c}{\Delta} & \frac{a}{\Delta} \end{bmatrix}$$

$$= \begin{bmatrix} \frac{.95}{.71} & \frac{-.50}{.71} \\ -\frac{-.10}{.71} & \frac{.80}{.71} \end{bmatrix}$$

$$= \begin{array}{c} & \begin{array}{cc} G & S \end{array} \\ \begin{array}{c} G \\ S \end{array} & \left[\begin{array}{cc} \frac{95}{71} & \frac{50}{71} \\ \frac{10}{71} & \frac{80}{71} \end{array}\right] \end{array}$$

Add the entries in each column:

$\frac{95}{71} + \frac{10}{71} = \frac{105}{71} = 1\frac{34}{71} \approx 1.48$

$\frac{50}{71} + \frac{80}{71} = \frac{130}{71} = 1\frac{59}{71} \approx 1.83$

A patient who begins in state G has an expected number of approximately 1.48 months; a patient who begins in state S has an expected number of approximately 1.83 months.

17. **a.** $\begin{cases} x + y + z + w = 1 \\ \begin{bmatrix} .20 & .10 & .05 & .05 \\ .30 & .20 & .20 & .30 \\ .40 & .40 & .50 & .40 \\ .10 & .30 & .25 & .25 \end{bmatrix} \begin{bmatrix} x \\ y \\ z \\ w \end{bmatrix} = \begin{bmatrix} x \\ y \\ z \\ w \end{bmatrix} \end{cases}$

$\begin{cases} x + y + z + w = 1 \\ .2x + .1y + .05z + .05w = x \\ .3x + .2y + .2z + .3w = y \\ .4x + .4y + .5z + .4w = z \\ .1x + .3y + .25z + .25w = w \end{cases}$

$\begin{cases} x + y + z + w = 1 \\ -.8x + .1y + .05z + .05w = 0 \\ .3x - .8y + .2z + .3w = 0 \\ .4x + .4y - .5z + .4w = 0 \\ .1x + .3y + .25z - .75w = 0 \end{cases}$

The fifth equation is equivalent to the sum of the second, third and fourth and so is redundant.

SSM: Finite Math   Chapter 8: Markov Processes

$$\begin{bmatrix} 1 & 1 & 1 & 1 & | & 1 \\ -.8 & .1 & .05 & .05 & | & 0 \\ .3 & -.8 & .2 & .3 & | & 0 \\ .4 & .4 & -.5 & .4 & | & 0 \end{bmatrix} \begin{matrix} 20[2] \\ 10[3] \\ 10[4] \end{matrix} \rightarrow \begin{bmatrix} 1 & 1 & 1 & 1 & | & 1 \\ -16 & 2 & 1 & 1 & | & 0 \\ 3 & -8 & 2 & 3 & | & 0 \\ 4 & 4 & -5 & 4 & | & 0 \end{bmatrix} \begin{matrix} [2]+16[1] \\ [3]+(-3)[1] \\ [4]+(-4)[1] \end{matrix} \rightarrow \begin{bmatrix} 1 & 1 & 1 & 1 & | & 1 \\ 0 & 18 & 17 & 17 & | & 16 \\ 0 & -11 & -1 & 0 & | & -3 \\ 0 & 0 & -9 & 0 & | & -4 \end{bmatrix}$$

$$\begin{matrix} \frac{1}{18}[2] \\ [1]+(-1)[2] \\ [3]+11[2] \end{matrix} \rightarrow \begin{bmatrix} 1 & 0 & \frac{1}{18} & \frac{1}{18} & | & \frac{1}{9} \\ 0 & 1 & \frac{17}{18} & \frac{17}{18} & | & \frac{8}{9} \\ 0 & 0 & \frac{169}{18} & \frac{187}{18} & | & \frac{61}{9} \\ 0 & 0 & -9 & 0 & | & -4 \end{bmatrix} \begin{matrix} \frac{18}{169}[3] \\ [1]+\left(-\frac{1}{18}\right)[3] \\ [2]+\left(-\frac{17}{18}\right)[3] \\ [4]+9[3] \end{matrix} \rightarrow \begin{bmatrix} 1 & 0 & 0 & -\frac{1}{169} & | & \frac{12}{169} \\ 0 & 1 & 0 & -\frac{17}{169} & | & \frac{35}{169} \\ 0 & 0 & 1 & \frac{187}{169} & | & \frac{122}{169} \\ 0 & 0 & 0 & \frac{1683}{169} & | & \frac{422}{169} \end{bmatrix}$$

$$\begin{matrix} \frac{169}{1683}[4] \\ [1]+\frac{1}{169}[4] \\ [2]+\frac{17}{169}[4] \\ [3]+\left(-\frac{187}{169}\right)[4] \end{matrix} \rightarrow \begin{bmatrix} 1 & 0 & 0 & 0 & | & \frac{122}{1683} \\ 0 & 1 & 0 & 0 & | & \frac{23}{99} \\ 0 & 0 & 1 & 0 & | & \frac{4}{9} \\ 0 & 0 & 0 & 1 & | & \frac{422}{1683} \end{bmatrix}$$

The stable distribution is $\begin{bmatrix} \frac{122}{1683} \\ \frac{23}{99} \\ \frac{4}{9} \\ \frac{422}{1683} \end{bmatrix}$.

In the long run, the probability of having 1, 2, 3, or 4 units of water in the reservoir at any given time will be $\frac{122}{1683}, \frac{23}{99}, \frac{4}{9}$, or $\frac{422}{1683}$, respectively.

  b. $\frac{122}{1683}(\$4000) + \frac{23}{99}(\$6000) + \frac{4}{9}(\$10,000) + \frac{422}{1683}(\$3000) \approx \$6881$
The average weekly benefits will be about $6881.

Conceptual Exercises

18. The entries in each column are the probabilities of the transitions to the various states from the state associated to the column. These must add up to 1, just as the probabilities of all branches emanating from any node in a tree diagram must add up to 1.

19. The probability of going from state A to state B after four time periods.

20. The stochastic matrix A represents a Markov process; A^2 represents this Markov process after two times periods, and so is also a stochastic matrix.

21. True

22. False

Chapter 9

Exercises 9.1

1. $R: \begin{bmatrix} -1 & -2 \\ \underline{0} & \underline{3} \end{bmatrix}$, row 2;

   $C: \begin{bmatrix} -1 & -2 \\ \underline{0} & 3 \end{bmatrix}$, column 1

3. $R: \begin{bmatrix} -2 & 4 & 1 \\ \underline{-1} & \underline{3} & \underline{5} \\ -3 & 5 & 2 \end{bmatrix}$, row 2;

   $C: \begin{bmatrix} -2 & 4 & 1 \\ \underline{-1} & 3 & \underline{5} \\ -3 & \underline{5} & 2 \end{bmatrix}$, column 1

5. $R: \begin{bmatrix} \underline{0} & \underline{3} \\ -1 & 1 \\ -2 & -4 \end{bmatrix}$, row 1;

   $C: \begin{bmatrix} \underline{0} & 3 \\ -1 & 1 \\ -2 & -4 \end{bmatrix}$, column 1

7. Row minima: $\begin{bmatrix} 1 & 0 \\ 0 & \underline{-1} \end{bmatrix}$,

   column maxima: $\begin{bmatrix} \underline{1} & 0 \\ 0 & -1 \end{bmatrix}$

   a. Row 1, column 2

   b. 0

9. $\begin{array}{cc} & H \quad T \\ \begin{matrix} H \\ T \end{matrix} & \begin{bmatrix} 2 & -1 \\ -1 & -4 \end{bmatrix} \end{array}$

   Row minima: $\begin{bmatrix} 2 & \underline{-1} \\ -1 & \underline{-4} \end{bmatrix}$,

   column maxima: $\begin{bmatrix} \underline{2} & -1 \\ -1 & -4 \end{bmatrix}$

   Row 1, column 2 is a saddle point, so the game is strictly determined. R should show heads, C should show tails.

11. $\begin{array}{c} \\ F \\ A \\ N \end{array} \begin{array}{c} F \quad\quad A \quad\quad N \\ \begin{bmatrix} 8000 & -1000 & 1000 \\ -7000 & 4000 & -2000 \\ 3000 & 3000 & 2000 \end{bmatrix} \end{array}$

   Row minima: $\begin{bmatrix} 8000 & -1000 & 1000 \\ \underline{-7000} & 4000 & -2000 \\ 3000 & 3000 & \underline{2000} \end{bmatrix}$

   Column maxima: $\begin{bmatrix} \underline{8000} & -1000 & 1000 \\ -7000 & \underline{4000} & -2000 \\ 3000 & 3000 & \underline{2000} \end{bmatrix}$

   Row 3, column 3 is a saddle point, so the game is strictly determined. Both candidates should be neutral.

13. $\begin{array}{c} \\ 5 \\ 10 \end{array} \begin{array}{c} 6 \quad 7 \quad 8 \\ \begin{bmatrix} 1 & -5 & -5 \\ -6 & 3 & 2 \end{bmatrix} \end{array}$

   Row minima: $\begin{bmatrix} 1 & -5 & \underline{-5} \\ \underline{-6} & 3 & 2 \end{bmatrix}$

   Column maxima: $\begin{bmatrix} \underline{1} & -5 & -5 \\ -6 & \underline{3} & \underline{2} \end{bmatrix}$

   No saddle point, so the game is not strictly determined.

Exercises 9.2

1. **a.** $[.5 \; .5]\begin{bmatrix} 3 & -1 \\ -7 & 5 \end{bmatrix}\begin{bmatrix} .5 \\ .5 \end{bmatrix} = [0]$

   b. $[1 \; 0]\begin{bmatrix} 3 & -1 \\ -7 & 5 \end{bmatrix}\begin{bmatrix} .5 \\ .5 \end{bmatrix} = [1]$

   c. $[.3 \; .7]\begin{bmatrix} 3 & -1 \\ -7 & 5 \end{bmatrix}\begin{bmatrix} .6 \\ .4 \end{bmatrix} = [-1.12]$

   d. $[.75 \; .25]\begin{bmatrix} 3 & -1 \\ -7 & 5 \end{bmatrix}\begin{bmatrix} .2 \\ .8 \end{bmatrix} = [.5]$

   (b) is most advantageous to R.

3. $[.3 \; .7]\begin{bmatrix} -20,000 & 0 \\ 0 & -50,000 \end{bmatrix}\begin{bmatrix} .2 \\ .8 \end{bmatrix} = [-29,200]$

   $29,200

5. The payoff matrix is $\begin{array}{c} \\ V \\ C \end{array}\begin{array}{cc} V & C \\ \begin{bmatrix} 0 & 2 \\ -1 & 0 \end{bmatrix} \end{array}$.

   $[.25 \; .75]\begin{bmatrix} 0 & 2 \\ -1 & 0 \end{bmatrix}\begin{bmatrix} .4 \\ .6 \end{bmatrix} = [0]$

   Zero

7. $\begin{array}{c} \\ R \end{array}\begin{array}{c} C \\ \begin{bmatrix} 3 & -1 \\ -2 & 2 \end{bmatrix} \end{array}$ max min of rows = -1

   $\begin{array}{c} \\ R \end{array}\begin{array}{c} C \\ \begin{bmatrix} 3 & -1 \\ -2 & 2 \end{bmatrix} \end{array}$ min max of columns = 2

   $[.3 \; .7]\begin{bmatrix} 3 & -1 \\ -2 & 2 \end{bmatrix}\begin{bmatrix} c_1 \\ c_2 \end{bmatrix} \to [-.5 \; 1.1]\begin{bmatrix} c_1 \\ c_2 \end{bmatrix} \to -.5c_1 + 1.1c_2 = 0 \to c_1 = \frac{1.1c_2}{.5}$ (to be fair)

   $c_1 + c_2 = 1$

   Further, $\frac{11}{5}c_2 + c_2 = 1 \to \frac{16}{5}c_2 = 1 \to c_2 = \frac{5}{16} \to c_1 = \frac{11}{16}$

   For fair game, Carol's strategy is $\begin{bmatrix} \frac{11}{16} \\ \frac{5}{16} \end{bmatrix}$

9.

$$R \begin{matrix} & C \\ & \begin{bmatrix} 5 & -1 \\ -2 & 2 \end{bmatrix} \end{matrix} \quad \text{max min of rows} = -1$$

$$R \begin{matrix} & C \\ & \begin{bmatrix} 5 & -1 \\ -2 & 2 \end{bmatrix} \end{matrix} \quad \text{min max of columns} = 2 \qquad \text{Not strictly determined.}$$

$$[.7 \quad .3]\begin{bmatrix} 5 & -1 \\ -2 & 2 \end{bmatrix} = [2.9 \quad -.1]$$

$$[2.9 \quad -.1]\begin{bmatrix} c_1 \\ c_2 \end{bmatrix} \to 2.9c_1 - 0.1c_2 = 0 \to c_2 = 29c_1$$

Carol's strategy: $\begin{bmatrix} \dfrac{1}{30} \\ \dfrac{29}{30} \end{bmatrix}$

11.

$$R \begin{matrix} & C \\ & \begin{bmatrix} 1 & 2 & 4 \\ 1 & 0 & 5 \\ 0 & 1 & -1 \end{bmatrix} \end{matrix} \quad \text{max min of rows} = 1$$

$$R \begin{matrix} & C \\ & \begin{bmatrix} 1 & 2 & 4 \\ 1 & 0 & 5 \\ 0 & 1 & -1 \end{bmatrix} \end{matrix} \quad \text{min max of rows} = 1$$

Game strictly determined. Optimal strategy- play row 1, column 1.

$R \to [1 \quad 0 \quad 0]$

$C \begin{bmatrix} 1 \\ 0 \\ 0 \end{bmatrix}$

Value of game = $1.00 in favor of Renee'. The game is not fair since the value of the game is different from zero.

$$[1 \ 0 \ 0]\begin{bmatrix} -3 & -2 & 6 \\ 2 & 0 & 2 \\ 5 & -2 & -4 \end{bmatrix} = [-3 \ -2 \ 6]$$

$$[-3 \ -2 \ 6]\begin{bmatrix} \frac{1}{3} \\ c_2 \\ c_3 \end{bmatrix} = 0 \rightarrow$$

$-1 - 2c_2 + 6c_3 = 0$

$-2c_2 = -6c_3 + 1$

$c_2 = 3c_3 - \frac{1}{2}$

$c_2 + c_3 = \frac{2}{3}$

$3c_3 - \frac{1}{2} + c_3 = \frac{2}{3}$

$4c_3 = \frac{7}{6}$

$c_3 = \frac{7}{24} \rightarrow c_2 = \frac{9}{24}$

Strategy: $\begin{bmatrix} \frac{1}{3} & \frac{3}{8} & \frac{7}{24} \end{bmatrix}$

Exercises 9.3

1. Objective function : Minimize $y_1 + y_2$

   Constraints: $\begin{cases} y_1 + 4y_2 \geq 1 \\ 6y_1 + 3y_2 \geq 1 \\ y_1 \geq 0 \\ y_2 \geq 0 \end{cases}$

3. Maximize $M = z_1 + z_2$ subject to

   $\begin{cases} 2z_1 + 4z_2 \leq 1 \\ 5z_1 + 3z_2 \leq 1 \\ z_1 \geq 0, z_2 \geq 0 \end{cases}$

   $\begin{array}{c|ccccc|c} & z_1 & z_2 & t & u & M & \\ \hline t & 2 & 4 & 1 & 0 & 0 & 1 \\ u & 5 & 3 & 0 & 1 & 0 & 1 \\ M & -1 & -1 & 0 & 0 & 1 & 0 \end{array}$

   $\begin{array}{c|ccccc|c} & z_1 & z_2 & t & u & M & \\ \hline t & 0 & \frac{14}{5} & 1 & -\frac{2}{5} & 0 & \frac{3}{5} \\ z_1 & 1 & \frac{3}{5} & 0 & \frac{1}{5} & 0 & \frac{1}{5} \\ M & 0 & -\frac{2}{5} & 0 & \frac{1}{5} & 1 & \frac{1}{5} \end{array}$

   $\begin{array}{c|ccccc|c} & z_1 & z_2 & t & u & M & \\ \hline z_2 & 0 & 1 & \frac{5}{14} & -\frac{1}{7} & 0 & \frac{3}{14} \\ z_1 & 1 & 0 & -\frac{3}{14} & \frac{2}{7} & 0 & \frac{1}{14} \\ M & 0 & 0 & \frac{1}{7} & \frac{1}{7} & 1 & \frac{2}{7} \end{array}$

   $z_1 = \frac{1}{14}, z_2 = \frac{3}{14}, M = \frac{2}{7}, v = \frac{1}{M} = \frac{7}{2}$,

   and the optimal strategy for C is

   $\begin{bmatrix} vz_1 \\ vz_2 \end{bmatrix} = \begin{bmatrix} \frac{1}{4} \\ \frac{3}{4} \end{bmatrix}$. The optimal strategy for

   R is given by the bottom entries under t and u:

   $[vt \ vu] = \begin{bmatrix} \frac{7}{2} \cdot \frac{1}{7} & \frac{7}{2} \cdot \frac{1}{7} \end{bmatrix} = \begin{bmatrix} \frac{1}{2} & \frac{1}{2} \end{bmatrix}$

5. Add 7 to each entry to make all the entries positive. We get $\begin{bmatrix} 10 & 1 \\ 2 & 11 \end{bmatrix}$. Then maximize
$M = z_1 + z_2$ subject to $\begin{cases} 10z_1 + z_2 \leq 1 \\ 2z_1 + 11z_2 \leq 1. \\ z_1 \geq 0, z_2 \geq 0 \end{cases}$

$\begin{array}{c} \\ t \\ u \\ M \end{array} \begin{array}{cccccc} z_1 & z_2 & t & u & M & \\ \left[\begin{array}{ccccc|c} 10 & 1 & 1 & 0 & 0 & 1 \\ 2 & 11 & 0 & 1 & 0 & 1 \\ \hline -1 & -1 & 0 & 0 & 1 & 0 \end{array}\right] \end{array}$

$\begin{array}{c} \\ z_1 \\ u \\ M \end{array} \begin{array}{cccccc} z_1 & z_2 & t & u & M & \\ \left[\begin{array}{ccccc|c} 1 & \frac{1}{10} & \frac{1}{10} & 0 & 0 & \frac{1}{10} \\ 0 & \frac{54}{5} & -\frac{1}{5} & 1 & 0 & \frac{4}{5} \\ \hline 0 & -\frac{9}{10} & \frac{1}{10} & 0 & 1 & \frac{1}{10} \end{array}\right] \end{array}$

$\begin{array}{c} \\ z_1 \\ z_2 \\ M \end{array} \begin{array}{cccccc} z_1 & z_2 & t & u & M & \\ \left[\begin{array}{ccccc|c} 1 & 0 & \frac{11}{108} & -\frac{1}{108} & 0 & \frac{5}{54} \\ 0 & 1 & -\frac{1}{54} & \frac{5}{54} & 0 & \frac{2}{27} \\ \hline 0 & 0 & \frac{1}{12} & \frac{1}{12} & 1 & \frac{1}{6} \end{array}\right] \end{array}$

$z_1 = \frac{5}{54}$, $z_2 = \frac{2}{27}$, $M = \frac{1}{6}$, $v = \frac{1}{M} = 6$,

and the optimal strategy for C is

$\begin{bmatrix} vz_1 \\ vz_2 \end{bmatrix} = \begin{bmatrix} \frac{5}{9} \\ \frac{4}{9} \end{bmatrix}$. The optimal strategy for R is given by the bottom entries under t and u:

$[vt \ vu] = [6 \cdot \frac{1}{12} \ \ 6 \cdot \frac{1}{12}] = [\frac{1}{2} \ \frac{1}{2}]$

7. Maximize $M = z_1 + z_2$ subject to
$\begin{cases} 4z_1 + z_2 \leq 1 \\ 2z_1 + 4z_2 \leq 1 \\ z_1 \geq 0, z_2 \geq 0 \end{cases}$

$\begin{array}{c} \\ t \\ u \\ M \end{array} \begin{array}{cccccc} z_1 & z_2 & t & u & M & \\ \left[\begin{array}{ccccc|c} 4 & 1 & 1 & 0 & 0 & 1 \\ 2 & 4 & 0 & 1 & 0 & 1 \\ \hline -1 & -1 & 0 & 0 & 1 & 0 \end{array}\right] \end{array}$

$\begin{array}{c} \\ t \\ z_2 \\ M \end{array} \begin{array}{cccccc} z_1 & z_2 & t & u & M & \\ \left[\begin{array}{ccccc|c} \frac{7}{2} & 0 & 1 & -\frac{1}{4} & 0 & \frac{3}{4} \\ \frac{1}{2} & 1 & 0 & \frac{1}{4} & 0 & \frac{1}{4} \\ \hline -\frac{1}{2} & 0 & 0 & \frac{1}{4} & 1 & \frac{1}{4} \end{array}\right] \end{array}$

$\begin{array}{c} \\ z_1 \\ z_2 \\ M \end{array} \begin{array}{cccccc} z_1 & z_2 & t & u & M & \\ \left[\begin{array}{ccccc|c} 1 & 0 & \frac{2}{7} & -\frac{1}{14} & 0 & \frac{3}{14} \\ 0 & 1 & -\frac{1}{7} & \frac{2}{7} & 0 & \frac{1}{7} \\ \hline 0 & 0 & \frac{1}{7} & \frac{3}{14} & 1 & \frac{5}{14} \end{array}\right] \end{array}$

$z_1 = \frac{3}{14}$, $z_2 = \frac{1}{7}$, $M = \frac{5}{14}$, $v = \frac{1}{M} = \frac{14}{5}$,

and the optimal strategy for C is

$\begin{bmatrix} vz_1 \\ vz_2 \end{bmatrix} = \begin{bmatrix} \frac{3}{5} \\ \frac{2}{5} \end{bmatrix}$.

The optimal strategy for R is given by the bottom entries under t and u:

$[vt \ vu] = [\frac{14}{5} \cdot \frac{1}{7} \ \ \frac{14}{5} \cdot \frac{3}{14}] = [\frac{2}{5} \ \frac{3}{5}]$

9. Add 2 to each entry to make all entries of the payoff matrix positive.
$\begin{bmatrix} 5 & 7 & 1 \\ 6 & 1 & 8 \end{bmatrix}$

Set the tableaux up to find C's optimal strategy, then read the dual's solution off the bottom row:

$\begin{array}{c} \\ t \\ u \\ M \end{array} \begin{array}{ccccccc} z_1 & z_2 & z_3 & t & u & M & \\ \left[\begin{array}{cccccc|c} 5 & 7 & 1 & 1 & 0 & 0 & 1 \\ 6 & 1 & 8 & 0 & 1 & 0 & 1 \\ \hline -1 & -1 & -1 & 0 & 0 & 1 & 0 \end{array}\right] \end{array}$

$\begin{array}{c} \\ z_2 \\ u \\ M \end{array} \begin{array}{ccccccc} z_1 & z_2 & z_3 & t & u & M & \\ \left[\begin{array}{cccccc|c} \frac{5}{7} & 1 & \frac{1}{7} & \frac{1}{7} & 0 & 0 & \frac{1}{7} \\ \frac{37}{7} & 0 & \frac{55}{7} & -\frac{1}{7} & 1 & 0 & \frac{6}{7} \\ \hline -\frac{2}{7} & 0 & -\frac{6}{7} & \frac{1}{7} & 0 & 1 & \frac{1}{7} \end{array}\right] \end{array}$

$\begin{array}{c} \\ z_2 \\ z_3 \\ M \end{array} \begin{array}{ccccccc} z_1 & z_2 & z_3 & t & u & M & \\ \left[\begin{array}{cccccc|c} \frac{34}{55} & 1 & 0 & \frac{8}{55} & -\frac{1}{55} & 0 & \frac{7}{55} \\ \frac{37}{55} & 0 & 1 & -\frac{1}{55} & \frac{7}{55} & 0 & \frac{6}{55} \\ \hline \frac{16}{55} & 0 & 0 & \frac{7}{55} & \frac{6}{55} & 1 & \frac{13}{55} \end{array}\right] \end{array}$

$t = \frac{7}{55}$, $u = \frac{6}{55}$, $M = \frac{13}{55}$, $v = \frac{1}{M} = \frac{55}{13}$,

and R's optimal strategy is
$[vt \ vu] = \begin{bmatrix} \frac{7}{13} & \frac{6}{13} \end{bmatrix}$.

11. Add 4 to each entry to make them all positive. Maximize $M = z_1 + z_2$ subject

to $\begin{cases} z_1 + 5z_2 \leq 1 \\ 8z_1 + 3z_2 \leq 1 \\ 5z_1 + 4z_2 \leq 1 \\ z_1 \geq 0, z_2 \geq 0 \end{cases}$.

$$\begin{array}{c|ccccc|c} & z_1 & z_2 & s & t & u & M \\ \hline s & 1 & 5 & 1 & 0 & 0 & 0 & 1 \\ t & 8 & 3 & 0 & 1 & 0 & 0 & 1 \\ u & 5 & 4 & 0 & 0 & 1 & 0 & 1 \\ \hline M & -1 & -1 & 0 & 0 & 0 & 1 & 0 \end{array}$$

$$\begin{array}{c|ccccc|c} & z_1 & z_2 & s & t & u & M \\ \hline z_2 & \frac{1}{5} & 1 & \frac{1}{5} & 0 & 0 & 0 & \frac{1}{5} \\ t & \frac{37}{5} & 0 & -\frac{3}{5} & 1 & 0 & 0 & \frac{2}{5} \\ u & \frac{21}{5} & 0 & -\frac{4}{5} & 0 & 1 & 0 & \frac{1}{5} \\ \hline M & -\frac{4}{5} & 0 & \frac{1}{5} & 0 & 0 & 1 & \frac{1}{5} \end{array}$$

$$\begin{array}{c|ccccc|c} & z_1 & z_2 & s & t & u & M \\ \hline z_2 & 0 & 1 & \frac{8}{37} & -\frac{1}{37} & 0 & 0 & \frac{7}{37} \\ z_1 & 1 & 0 & -\frac{3}{37} & \frac{5}{37} & 0 & 0 & \frac{2}{37} \\ u & 0 & 0 & -\frac{17}{37} & -\frac{21}{37} & 1 & 0 & -\frac{1}{37} \\ \hline M & 0 & 0 & \frac{5}{37} & \frac{4}{37} & 0 & 1 & \frac{9}{37} \end{array}$$

$z_1 = \frac{2}{37}, z_2 = \frac{7}{37}, M = \frac{9}{37}, v = \frac{1}{M} = \frac{37}{9}$,

and C's optimal strategy is $\begin{bmatrix} vz_1 \\ vz_2 \end{bmatrix} = \begin{bmatrix} \frac{2}{9} \\ \frac{7}{9} \end{bmatrix}$.

13. The original matrix is not strictly determined.

The original matrix is $\begin{bmatrix} 5 & -3 \\ -3 & 1 \end{bmatrix}$. Add 4 to each entry to make all the entries positive. We get $\begin{bmatrix} 9 & 1 \\ 1 & 5 \end{bmatrix}$. Then maximize

$M = z_1 + z_2$ subject to $\begin{cases} 9z_1 + z_2 \leq 1 \\ z_1 + 5z_2 \leq 1 \\ z_1 \geq 0, z_2 \geq 0 \end{cases}$.

$$\begin{array}{c|cccc|c} & z_1 & z_2 & t & u & M \\ \hline t & 9 & 1 & 1 & 0 & 0 & 1 \\ u & 1 & 5 & 0 & 1 & 0 & 1 \\ \hline M & -1 & -1 & 0 & 0 & 1 & 0 \end{array}$$

$$\begin{array}{c|cccc|c} & z_1 & z_2 & t & u & M \\ \hline z_2 & 9 & 1 & 1 & 0 & 0 & 1 \\ u & -44 & 0 & -5 & 1 & 0 & -4 \\ \hline M & 8 & 0 & 1 & 0 & 1 & 1 \end{array}$$

$$\begin{array}{c|cccc|c} & z_1 & z_2 & t & u & M \\ \hline z_2 & 0 & 1 & -\frac{1}{44} & \frac{9}{44} & 0 & \frac{2}{11} \\ z_1 & 1 & 0 & \frac{5}{44} & -\frac{1}{44} & 0 & \frac{1}{11} \\ \hline M & 0 & 0 & \frac{1}{11} & \frac{2}{11} & 1 & \frac{3}{11} \end{array}$$

$v = \frac{1}{M} = \frac{11}{3}$

Renee's optimal strategy is given by the bottom entries under t and u:

$[vt \ vu] = \begin{bmatrix} \frac{1}{3} & \frac{2}{3} \end{bmatrix}$

Carlos' optimal strategy is $\begin{bmatrix} vz_1 \\ vz_2 \end{bmatrix} = \begin{bmatrix} \frac{1}{3} \\ \frac{2}{3} \end{bmatrix}$.

The value of the game is

$v - 4 = \frac{1}{M} - 4 = \frac{11}{3} - 4 = -\frac{1}{3}$.

15. The payoff matrix, with entries in thousands of dollars, is $\begin{matrix} & 1 & 2 \\ 1 & \\ 2 & \end{matrix}\begin{bmatrix} -2 & 7 \\ 7 & -1 \end{bmatrix}$. Add 3 to each entry to make all the entries positive. We get $\begin{bmatrix} 1 & 10 \\ 10 & 2 \end{bmatrix}$. Then maximize $M = z_1 + z_2$ subject to
$$\begin{cases} z_1 + 10z_2 \leq 1 \\ 10z_1 + 2z_2 \leq 1. \\ z_1 \geq 0, z_2 \geq 0 \end{cases}$$

$$\begin{array}{c|ccccc|c} & z_1 & z_2 & t & u & M & \\ \hline t & 1 & 10 & 1 & 0 & 0 & 1 \\ u & 10 & 2 & 0 & 1 & 0 & 1 \\ \hline M & -1 & -1 & 0 & 0 & 1 & 0 \end{array}$$

$$\begin{array}{c|ccccc|c} & z_1 & z_2 & t & u & M & \\ \hline z_2 & \frac{1}{10} & 1 & \frac{1}{10} & 0 & 0 & \frac{1}{10} \\ u & \frac{49}{5} & 0 & -\frac{1}{5} & 1 & 0 & \frac{4}{5} \\ \hline M & -\frac{9}{10} & 0 & \frac{1}{10} & 0 & 1 & \frac{1}{10} \end{array}$$

$$\begin{array}{c|ccccc|c} & z_1 & z_2 & t & u & M & \\ \hline z_2 & 0 & 1 & \frac{5}{49} & -\frac{1}{98} & 0 & \frac{9}{98} \\ z_1 & 1 & 0 & -\frac{1}{49} & \frac{5}{49} & 0 & \frac{4}{49} \\ \hline M & 0 & 0 & \frac{4}{49} & \frac{9}{98} & 1 & \frac{17}{98} \end{array}$$

$$v = \frac{1}{M} = \frac{98}{17}$$

a. R's optimal strategy is given by the bottom entries under t and u:
$$[vt \ vu] = \begin{bmatrix} \frac{8}{17} & \frac{9}{17} \end{bmatrix}$$

b. C's optimal strategy is $\begin{bmatrix} vz_1 \\ vz_2 \end{bmatrix} = \begin{bmatrix} \frac{8}{17} \\ \frac{9}{17} \end{bmatrix}$.

c. $v - 3 = \frac{1}{M} - 3 = \frac{98}{17} - 3 \approx 2.765$

The value of the game is about $2765.

17. Add 4 to each entry to make them all positive. Maximize $M = z_1 + z_2$
$$\text{subject to } \begin{cases} 2z_1 + 5z_2 \leq 1 \\ 6z_1 + z_2 \leq 1 \\ 5z_1 + 2z_2 \leq 1 \\ z_1 \geq 0, z_2 \geq 0 \end{cases}.$$

$$\begin{array}{c|ccccc|c} & z_1 & z_2 & s & t & u & M & \\ \hline s & 2 & 5 & 1 & 0 & 0 & 0 & 1 \\ t & 6 & 1 & 0 & 1 & 0 & 0 & 1 \\ u & 5 & 2 & 0 & 0 & 1 & 0 & 1 \\ \hline M & -1 & -1 & 0 & 0 & 0 & 1 & 0 \end{array}$$

$$\begin{array}{c|ccccc|c} & z_1 & z_2 & s & t & u & M & \\ \hline z_2 & \frac{2}{5} & 1 & \frac{1}{5} & 0 & 0 & 0 & \frac{1}{5} \\ t & \frac{28}{5} & 0 & -\frac{1}{5} & 1 & 0 & 0 & \frac{4}{5} \\ u & \frac{21}{5} & 0 & -\frac{2}{5} & 0 & 1 & 0 & \frac{3}{5} \\ \hline M & -\frac{3}{5} & 0 & \frac{1}{5} & 0 & 0 & 1 & \frac{1}{5} \end{array}$$

$$\begin{array}{c|ccccc|c} & z_1 & z_2 & s & t & u & M & \\ \hline z_2 & 0 & 1 & \frac{3}{14} & -\frac{1}{14} & 0 & 0 & \frac{1}{7} \\ z_1 & 1 & 0 & -\frac{1}{28} & \frac{5}{28} & 0 & 0 & \frac{1}{7} \\ u & 0 & 0 & -\frac{1}{4} & -\frac{3}{4} & 1 & 0 & 0 \\ \hline M & 0 & 0 & \frac{5}{28} & \frac{3}{28} & 0 & 1 & \frac{2}{7} \end{array}$$

$$v = \frac{1}{M} = \frac{7}{2}$$

Rosedale's optimal strategy is given by the bottom entries under s, t and u:
$$[vs \ vt \ vu] = \begin{bmatrix} \frac{5}{8} & \frac{3}{8} & 0 \end{bmatrix}$$

Carter's optimal strategy is
$$\begin{bmatrix} vz_1 \\ vz_2 \end{bmatrix} = \begin{bmatrix} \frac{1}{2} \\ \frac{1}{2} \end{bmatrix}.$$

Chapter 9 Fundamental Concept Check

1. The payoffs from player C to player R corresponding to the various moves.

2. Pure Strategy: each player repeatedly makes the same move.

   Mixed strategy: each move is selected with a specified probability.

3. A game in which a payoff to one player results in a loss of the same amount to the other player.

4. Row player: choose the row whose least element is maximal.

   Column player: choose the column whose greatest element is minimal.

5. When the entry is simultaneously the least element of its row and the greatest element of its column.

6. A game in which, if both players use optimal pure strategies, then the saddle point gives the payoff to the row player; the value of the game is the value of the saddle point.

7. The average payoff per game to the row player when the pair of mixed strategies are used; the expected value of the pair R, C of mixed strategies is found by computing the product RAC, where A is the payoff matrix for the game.

8. Row player: the strategy for which the column player's best counterstrategy results in the greatest possible expected value.

   Column player: the strategy for which the row player's best counterstrategy results in the least possible expected value.

   Optimal mixed strategies for each player are found by solving a pair of dual linear programming problems, as described on pages 419 – 420.

Chapter 9 Review Exercises

1. Row minima: $\begin{bmatrix} 5 & \underline{-1} & 1 \\ \underline{-3} & 5 & 1 \\ 4 & 3 & \underline{2} \end{bmatrix}$,

   column maxima: $\begin{bmatrix} \underline{5} & -1 & 1 \\ -3 & \underline{5} & 1 \\ 4 & 3 & \underline{2} \end{bmatrix}$

   The game is strictly determined, with a saddle point at row 3, column 3 and a value of 2.

2. Row minima: $\begin{bmatrix} \underline{1} & 2 & 3 \\ 3 & 2 & \underline{1} \end{bmatrix}$,

   column maxima: $\begin{bmatrix} 1 & 2 & \underline{3} \\ \underline{3} & \underline{2} & 1 \end{bmatrix}$

   The game has no saddle point and so is not strictly determined.

3. Row minima: $\begin{bmatrix} \underline{0} & 1 \\ 1 & \underline{0} \\ 2 & \underline{-1} \end{bmatrix}$,

   column maxima: $\begin{bmatrix} 0 & \underline{1} \\ 1 & 0 \\ \underline{2} & -1 \end{bmatrix}$

   The game has no saddle point and so is not strictly determined.

4. Row minima: $\begin{bmatrix} 2 & \underline{1} & 2 \\ \underline{-1} & 0 & 3 \\ 4 & 1 & \underline{-4} \end{bmatrix}$,

   column maxima: $\begin{bmatrix} 2 & \underline{1} & 2 \\ -1 & 0 & \underline{3} \\ \underline{4} & \underline{1} & -4 \end{bmatrix}$

   The game is strictly determined, with a saddle point at row 1, column 2 and a value of 1.

5. $\begin{bmatrix} \frac{3}{4} & \frac{1}{4} \end{bmatrix} \begin{bmatrix} 0 & 24 \\ 12 & -36 \end{bmatrix} \begin{bmatrix} \frac{1}{3} \\ \frac{2}{3} \end{bmatrix} = [7]$

   7

6. $\begin{bmatrix} \frac{1}{2} & \frac{1}{2} \end{bmatrix} \begin{bmatrix} -6 & 6 & 0 \\ 0 & -12 & 24 \end{bmatrix} \begin{bmatrix} \frac{1}{3} \\ \frac{1}{3} \\ \frac{1}{3} \end{bmatrix} = [2]$

2

7. $[.2 \ .3 \ .5] \begin{bmatrix} 1 & 0 \\ -3 & 1 \\ 0 & 5 \end{bmatrix} \begin{bmatrix} .4 \\ .6 \end{bmatrix} = [1.4]$

1.4

8. $[.1 \ .1 \ .8] \begin{bmatrix} 0 & 1 & 3 \\ -1 & 0 & 2 \\ -3 & -2 & 0 \end{bmatrix} \begin{bmatrix} .4 \\ .3 \\ .3 \end{bmatrix} = [-1.3]$

−1.3

9. Add 4 to each entry to get $\begin{bmatrix} 1 & 8 \\ 6 & 2 \end{bmatrix}$. Then use the simplex method.

$\begin{array}{c} \\ t \\ u \\ M \end{array} \begin{bmatrix} z_1 & z_2 & t & u & M & \\ 1 & 8 & 1 & 0 & 0 & 1 \\ 6 & 2 & 0 & 1 & 0 & 1 \\ \hline -1 & -1 & 0 & 0 & 1 & 0 \end{bmatrix}$

$\begin{array}{c} \\ t \\ z_1 \\ M \end{array} \begin{bmatrix} z_1 & z_2 & t & u & M & \\ 0 & \frac{23}{3} & 1 & -\frac{1}{6} & 0 & \frac{5}{6} \\ 1 & \frac{1}{3} & 0 & \frac{1}{6} & 0 & \frac{1}{6} \\ \hline 0 & -\frac{2}{3} & 0 & \frac{1}{6} & 1 & \frac{1}{6} \end{bmatrix}$

$\begin{array}{c} \\ z_2 \\ z_1 \\ M \end{array} \begin{bmatrix} z_1 & z_2 & t & u & M & \\ 0 & 1 & \frac{3}{23} & -\frac{1}{46} & 0 & \frac{5}{46} \\ 1 & 0 & -\frac{1}{23} & \frac{4}{23} & 0 & \frac{3}{23} \\ \hline 0 & 0 & \frac{2}{23} & \frac{7}{46} & 1 & \frac{11}{46} \end{bmatrix}$

$v = \frac{1}{M} = \frac{46}{11}$

R's optimal strategy:
$[vt \ vu] = \begin{bmatrix} \frac{46}{11} \cdot \frac{2}{23} & \frac{46}{11} \cdot \frac{7}{46} \end{bmatrix} = \begin{bmatrix} \frac{4}{11} & \frac{7}{11} \end{bmatrix}$

C's optimal strategy:
$\begin{bmatrix} vz_1 \\ vz_2 \end{bmatrix} = \begin{bmatrix} \frac{46}{11} \cdot \frac{3}{23} \\ \frac{46}{11} \cdot \frac{5}{46} \end{bmatrix} = \begin{bmatrix} \frac{6}{11} \\ \frac{5}{11} \end{bmatrix}$

10. Add 7 to each entry to get $\begin{bmatrix} 10 & 1 \\ 3 & 11 \end{bmatrix}$. Then apply the simplex method.

$\begin{array}{c} \\ t \\ u \\ M \end{array} \begin{bmatrix} z_1 & z_2 & t & u & M & \\ 10 & 1 & 1 & 0 & 0 & 1 \\ 3 & 11 & 0 & 1 & 0 & 1 \\ \hline -1 & -1 & 0 & 0 & 1 & 0 \end{bmatrix}$

$\begin{array}{c} \\ z_1 \\ u \\ M \end{array} \begin{bmatrix} z_1 & z_2 & t & u & M & \\ 1 & \frac{1}{10} & \frac{1}{10} & 0 & 0 & \frac{1}{10} \\ 0 & \frac{107}{10} & -\frac{3}{10} & 1 & 0 & \frac{7}{10} \\ \hline 0 & -\frac{9}{10} & \frac{1}{10} & 0 & 1 & \frac{1}{10} \end{bmatrix}$

$\begin{array}{c} \\ z_1 \\ z_2 \\ M \end{array} \begin{bmatrix} z_1 & z_2 & t & u & M & \\ 1 & 0 & \frac{11}{107} & -\frac{1}{107} & 0 & \frac{10}{107} \\ 0 & 1 & -\frac{3}{107} & \frac{10}{107} & 0 & \frac{7}{107} \\ \hline 0 & 0 & \frac{8}{107} & \frac{9}{107} & 1 & \frac{17}{107} \end{bmatrix}$

$v = \frac{1}{M} = \frac{107}{17}$

R's optimal strategy:
$[vt \ vu] = \begin{bmatrix} \frac{107}{17} \cdot \frac{8}{107} & \frac{107}{17} \cdot \frac{9}{107} \end{bmatrix} = \begin{bmatrix} \frac{8}{17} & \frac{9}{17} \end{bmatrix}$

C's optimal strategy:
$\begin{bmatrix} vz_1 \\ vz_2 \end{bmatrix} = \begin{bmatrix} \frac{107}{17} \cdot \frac{10}{107} \\ \frac{107}{17} \cdot \frac{7}{107} \end{bmatrix} = \begin{bmatrix} \frac{10}{17} \\ \frac{7}{17} \end{bmatrix}$

11. Row 2, column 3 is a saddle point: $[0 \ 1]$

12. $\begin{array}{c} \\ s \\ t \\ u \\ M \end{array} \begin{bmatrix} z_1 & z_2 & s & t & u & M & \\ 1 & 3 & 1 & 0 & 0 & 0 & 1 \\ 3 & 1 & 0 & 1 & 0 & 0 & 1 \\ 4 & 2 & 0 & 0 & 1 & 0 & 1 \\ \hline -1 & -1 & 0 & 0 & 0 & 1 & 0 \end{bmatrix}$

$\begin{array}{c} \\ s \\ t \\ z_1 \\ M \end{array} \begin{bmatrix} z_1 & z_2 & s & t & u & M & \\ 0 & \frac{5}{2} & 1 & 0 & -\frac{1}{4} & 0 & \frac{3}{4} \\ 0 & -\frac{1}{2} & 0 & 1 & -\frac{3}{4} & 0 & \frac{1}{4} \\ 1 & \frac{1}{2} & 0 & 0 & \frac{1}{4} & 0 & \frac{1}{4} \\ \hline 0 & -\frac{1}{2} & 0 & 0 & \frac{1}{4} & 1 & \frac{1}{4} \end{bmatrix}$

$$\begin{array}{c|cccccc|c}
 & z_1 & z_2 & s & t & u & M & \\
\hline
z_2 & 0 & 1 & \frac{2}{5} & 0 & -\frac{1}{10} & 0 & \frac{3}{10} \\
t & 0 & 0 & \frac{1}{5} & 1 & -\frac{4}{5} & 0 & \frac{2}{5} \\
z_1 & 1 & 0 & -\frac{1}{5} & 0 & \frac{3}{10} & 0 & \frac{1}{10} \\
\hline
M & 0 & 0 & \frac{1}{5} & 0 & \frac{1}{5} & 1 & \frac{2}{5}
\end{array}$$

$$v = \frac{1}{M} = \frac{5}{2}$$

$$\begin{bmatrix} vz_1 \\ vz_2 \end{bmatrix} = \begin{bmatrix} \frac{5}{2} \cdot \frac{1}{10} \\ \frac{5}{2} \cdot \frac{3}{10} \end{bmatrix} = \begin{bmatrix} \frac{1}{4} \\ \frac{3}{4} \end{bmatrix}$$

13. The payoff matrix is $\begin{array}{c} 2 \\ 6 \end{array}\begin{bmatrix} -3 & 2 \\ 6 & -3 \end{bmatrix}$. Add 4 to each entry to get $\begin{bmatrix} 1 & 6 \\ 10 & 1 \end{bmatrix}$. Then apply the simplex method.

$$\begin{array}{c|ccccc|c}
 & z_1 & z_2 & t & u & M & \\
\hline
t & 1 & 6 & 1 & 0 & 0 & 1 \\
u & 10 & 1 & 0 & 1 & 0 & 1 \\
\hline
M & -1 & -1 & 0 & 0 & 1 & 0
\end{array}$$

$$\begin{array}{c|ccccc|c}
 & z_1 & z_2 & t & u & M & \\
\hline
z_2 & \frac{1}{6} & 1 & \frac{1}{6} & 0 & 0 & \frac{1}{6} \\
u & \frac{59}{6} & 0 & -\frac{1}{6} & 1 & 0 & \frac{5}{6} \\
\hline
M & -\frac{5}{6} & 0 & \frac{1}{6} & 0 & 1 & \frac{1}{6}
\end{array}$$

$$\begin{array}{c|ccccc|c}
 & z_1 & z_2 & t & u & M & \\
\hline
z_2 & 0 & 1 & \frac{10}{59} & -\frac{1}{59} & 0 & \frac{9}{59} \\
z_1 & 1 & 0 & -\frac{1}{59} & \frac{6}{59} & 0 & \frac{5}{59} \\
\hline
M & 0 & 0 & \frac{9}{59} & \frac{5}{59} & 1 & \frac{14}{59}
\end{array}$$

$$v = \frac{1}{M} = \frac{59}{14}$$

18.
$$a_{11}r + a_{21}(1-r) = a_{12}r + a_{22}(1-r)$$
$$a_{11}r + a_{21} - a_{21}r = a_{12}r + a_{22} - a_{22}r$$
$$a_{11}r - a_{21}r - a_{12}r + a_{22}r = a_{22} - a_{21}$$
$$r(a_{11} - a_{21} - a_{12} + a_{22}) = a_{22} - a_{21}$$
$$r = \frac{a_{22} - a_{21}}{a_{11} - a_{21} - a_{12} + a_{22}}$$

a. Carol's optimal strategy is
$$\begin{bmatrix} \frac{59}{14} \cdot \frac{5}{59} \\ \frac{59}{14} \cdot \frac{9}{59} \end{bmatrix} = \begin{bmatrix} \frac{5}{14} \\ \frac{9}{14} \end{bmatrix}.$$
Ruth's optimal strategy is
$$\begin{bmatrix} \frac{59}{14} \cdot \frac{9}{59} & \frac{59}{14} \cdot \frac{5}{59} \end{bmatrix} = \begin{bmatrix} \frac{9}{14} & \frac{5}{14} \end{bmatrix}.$$

b. Since the value is positive, the game favors Ruth (the row player). The value of the game is $\frac{59}{14} - 4 = \frac{3}{14}$.

14. a.
$$\begin{array}{c} \\ A \\ B \\ C \end{array}\begin{array}{ccc} \text{Strong} & \text{Avg.} & \text{Weak} \end{array}\\ \begin{bmatrix} 3000 & 2000 & 1000 \\ 6000 & 2000 & -3000 \\ 15{,}000 & 1000 & -10{,}000 \end{bmatrix}$$

b. Row 1, column 3 is a saddle point. The investors optimal strategy is to buy stock A.

15. For the given conditions, the matrices will not have a saddle point and therefore, will not be strictly determined.

16. The optimal strategy will remain the same as long as the row 1, column 1 value is less than the smallest value in the matrix or $2 + h < 6$ or $h < 4$.

17. a. Since both lines are straight lines, the intersection of the two lines would be the only solution to the system of equations. Moving from that intersection point would increase the value of one line while decreasing the value of the other.

b. You would use the equations
$y = a_{11}r + a_{12}(1-r)$ and
$y = a_{21}r + a_{22}(1-r)$ and find the point of intersection as in part a.

Chapter 10

Exercises 10.1

1. $i = \dfrac{.03}{12} = .0025$

   $n = (12)(2) = 24$

3. $i = \dfrac{.022}{2} = .011$

   $n = (2)(20) = 40$

5. $i = \dfrac{.045}{12} = .00375$

   $n = (12)(3.5) = 42$

7. $i = \dfrac{.028}{1} = .028$

   $n = (1)(4) = 4$

   $P = \$500$

   $F = \$558.40$

9. $i = \dfrac{.024}{2} = .012$

   $n = (2)(9.5) = 19$

   $P = \$7174.85$

   $F = \$9000$

11. $i = \dfrac{.06}{12} = .005$

    $n = (12)(30) = 360$

    $P = \$3000$

    $F = \$18,067.73$

13. $\left(1 + \dfrac{.021}{12}\right)^{12 \times 2} (\$1000) = \$1042.86$

15. $F = \left(1 + \dfrac{.027}{12}\right)^{12 \times 3} (\$6000) = \$6505.63$

    $int. = F - P = \$6505.63 - \$6000 = \$505.63$

17. $\left(1 + \dfrac{.04}{4}\right)^{4 \times 6.25} (\$10,000) = \$12,824.32$

19. $\left[\dfrac{1}{\left(1 + \dfrac{.024}{12}\right)^{12 \times 25}}\right] (\$100,000) = \$54,914.06$

21. $\left[\dfrac{1}{\left(1 + \dfrac{.034}{4}\right)^{4 \times 3}}\right] (\$10,000) = \$9034.19$

23. For $P = \$1400$,

    $F = \left(1 + \dfrac{.025}{1}\right)^9 (\$1400) = \$1748.41$.

    $\$1400$ now is more profitable.

25. $r_{\text{eff}} = \left(1 + \dfrac{.622}{52}\right)^{52} - 1 \approx .8558$

    This interest rate is better than 85%.

27. $F = \left(1 + \dfrac{.016}{4}\right)^{0.25 \times 4} (\$1000) = \$1004$

    $int. = F - P = \$1004 - \$1000 = \$4.00$

    $F = \left(1 + \dfrac{.016}{4}\right)^{0.25 \times 4} (\$1004) = \$1008.02$

    $int. = F - P = \$1008.02 - \$1004 = \$4.02$

    $F = \left(1 + \dfrac{.016}{4}\right)^{0.25 \times 4} (\$1008.02) = \$1012.05$

    $int. = F - P = \$1012.05 - \$1008.02 = \$4.03$

29. $F = \left(1 + \dfrac{.03}{4}\right)^{2 \times 4} (\$2000) = \$2123.20$

    $int. = F - P = \$2123.20 - \$2000 = \$123.20$

    $F = \left(1 + \dfrac{.03}{4}\right)^{3 \times 4} (\$2000) = \$2187.61$

    $int. = F - P = \$2187.61 - \$2123.20 = \$64.41$

31. $B_5 = \left(1 + \dfrac{.026}{4}\right)^{4 \times 5} (\$1000) = \$1138.35$

$B_4 = \left(1 + \dfrac{.026}{4}\right)^{4 \times 4} (\$1000) = \$1109.23$

$\text{int.} = B_5 - B_4 = \$1138.35 - \$1109.23 = \$29.12$

33. $P = \left[\dfrac{1}{\left(1 + \frac{.04}{4}\right)^{4 \times 1} - 1}\right] (\$406.04) = \$10{,}000$

35. $1500(1+r)^7 = 2100$

$(1+r)^7 = 1.4$

$(1+r) = \sqrt[7]{1.4}$

$r \approx 4.92\%$

37. a. $\dfrac{r}{12}(\$10{,}000.00) = \$20.00$, so $r = .024 = 2.4\%$ compounded monthly

b. $(1.002)^3 (\$10{,}000.00) = \$10{,}060.12$
$\$10{,}060.12 - \$10{,}040.04 = \$20.08$

c. $(1.002)^{24}(\$10{,}000.00) = \$10{,}491.20$
$[(1.002)^{24} - (1.002)^{23}](\$10{,}000) = \$20.94$

39. a. $r = .015$
$n = \dfrac{6}{12} = \dfrac{1}{2}$
$P = \$500$
$F = \$503.75$

b. $r = .025$
$n = 2$
$P = \$500$
$F = \$525$

41. $F = (1 + 3 \cdot .012)(\$1000) = \$1036$

43. $P = \left[\dfrac{1}{(1 + 10 \cdot .02)}\right](\$3000) = \$2500$

45. $\left(1 + \dfrac{6}{12}r\right)(\$980) = \$1000$
$r \approx .0408 = 4.08\%$

47. $(1 + n \cdot .015)(\$500) = \$800$, $n = 40$ years

49. $(1 + n \cdot .02)P = 2P$, $n = 50$ years

51. $F = (1 + nr)P$; $P = \dfrac{F}{1 + nr}$

53. $\left(1 + \dfrac{.04}{4}\right)^{4 \times 1}(\$100) = \$104.06$

$\dfrac{\$4.06}{\$100} = .0406 = 4.06\%$

55. $r_{\text{eff}} = \left(1 + \dfrac{.04}{2}\right)^2 - 1 = .0404$; 4.04%

57. $r_{\text{eff}} = \left(1 + \dfrac{.044}{12}\right)^{12} - 1 \approx .0449$; 4.49%

59. $\left(1 + \dfrac{r}{4}\right)^4 - 1 = .0406$; $r \approx .04$; 4%

61. $\left(1 + \dfrac{r}{52}\right)^{52} - 1 = .03$; $r \approx .0296$; 2.96%

63. $\dfrac{r}{4}(\$10{,}000) = \$100$, so $r = .04$ and $i = .01$

$\dfrac{1}{(1.01)^{12}}(\$10{,}000) = \$8874.49$

65. (a)

67. Since we start with $1000 and this amount doubles every six years, we have $2000 at the end of six years, $4000 at the end of twelve years and $8000 at the end of eighteen years. So, it will take 18 years for the investment to grow to $8000.

69. Assume an initial investment of $100.00. Then over a 10 year period, the amount of growth would be $100(1 + .04)^{10} = 148.02$ about a 48% increase; so answer d) is correct.

SSM: Finite Math Chapter 10: The Mathematics of Finance

71. Assume $100 invested initially. Then 100 would increase to 102.5 after the first year. The total amount of the investment after three years would then be $100(1.025)(1.03)(1.084) = \$114.44$. If the same amount was invested at an interest rate of r % compounded annually, the amount would be the same, so;

$$100(1+r)^3 = 114.44$$
$$(1+r)^3 = 1.1444$$
$$(1+r) = \sqrt[3]{1.1444}$$
$$r \approx 4.6\%$$

73. $N = 24, I\% = 2.1, PV = -1000, PMT = 0,$
 $P/Y = 12,$ find $FV.$
 $= FV(0.175\%, 24, 0, -1000)$
 compound interest $PV = 1000, i = .175\%,$
 $n = 24,$ annual

75. $N = 300, I\% = 2.4, PMT = 0, FV = -100000,$
 $P/Y = 12,$ find $PV.$
 $= PV(0.2\%, 300, 0, -100000)$
 compound interest $FV = 100000, i = .2\%,$
 $n = 300,$ annual

77. $N = 7, PMT = 0, PV = 1500, FV = -2100,$
 $P/Y = 1,$ find $I\%.$
 $= RATE(7, 0, 1500, -2100)$
 compound interest $PV = 1500, FV = 2100,$
 $n = 7,$ annual

79. After 25 years: $(1.027)^{25}(\$1000) = \$1946.53$
 After 35 years: $(1.027)^{35}(\$1000) = \$2540.77$
 25; 35

81. $(1.019)^n(\$500,000)$ passes $\$1,000,000$ when $n = 37$ years.

83. After year 20, option A would have a value of $F = (1+20(.04))(\$1000) = \$1800$ and option B would have a value of $F = (1+.03)^{20}(\$1000) = \$1806.11$.

Exercises 10.2

1. $i = \dfrac{.021}{12} = .00175$

 $n = (12)(10) = 120$

 $R = \$100$

 $F = \$13,340.09$

3. $i = \dfrac{.034}{4} = .0085$

 $n = (4)(5) = 20$

 $R = \$2000$

 $P = \$36,642.08$

5. $\left[\dfrac{\left(1+\frac{.026}{2}\right)^{2\times 5}-1}{\frac{.026}{2}}\right](\$1500) = \$15,908.62$

7. $\left[\dfrac{\frac{.018}{12}}{\left(1+\frac{.018}{12}\right)^{12\times 1}-1}\right](\$1681.83) = \$139.00$

9. $\left[\dfrac{1-(1.00125)^{-24}}{.00125}\right](\$3000) = \$70,887.09$

11. $\left[\dfrac{\frac{.016}{4}}{1-\left(1+\frac{.016}{4}\right)^{-4\times 3}}\right](\$47,336.25) = \$4048.00$

13. a. $\left[\dfrac{\left(1+\frac{.03}{12}\right)^{12\times 4}-1}{\frac{.03}{12}}\right]($500) = $25,465.60$

b. $\$25,465.60 - 48(\$500) = \$1465.60$

c.
Month	Interest	Balance
1		$500.00
2	$.0025 \times 500 = \$1.25$	$(1.0025)(500) + 500 = \$1001.25$
3	$.0025 \times 1001.25 = \$2.50$	$(1.0025)(1.001.25) + 500 = \$1503.75$

15.
Quarter	Interest	Balance
1		$10000.00
2	$.0055 \times 10000 = \$55.00$	$10000 + 55 - 1000 = 9055$
3	$.0055 \times 9055 = \$49.80$	$9055 + 49.80 - 1000 = 8104.80$
4	$.0055 \times 8104.80 = \$44.58$	

17. $13,340.09 - (100)(120) = \$1340.09$

19. $(2000)(20) - 36,642.08 = \$3357.92$

21. $\left[\dfrac{\left(1+\frac{.03}{12}\right)^{12\times 10}-1}{\frac{.03}{12}}\right]($1000) = \$139,741.42$

Lump sum is better.

23. $\left[\dfrac{\left(1+\frac{.033}{12}\right)^{12\times 1}-1}{\frac{.033}{12}}\right]($200) = \$2436.63$

$200 a month is better.

25. Jack withdraws for $12(.75) = 9$ months.
$\left[\dfrac{1-\left(1+\frac{.018}{12}\right)^{-12\times 0.75}}{\frac{.018}{12}}\right]($200) = \$1786.57$

27. $\left(1+\frac{.08}{4}\right)^{4\times 11}($1000) + \left[\dfrac{\left(1+\frac{.08}{4}\right)^{4\times 11}-1}{\frac{.08}{4}}\right]($100) = \$9340.32$

29. $\left[\dfrac{\left(1+\frac{.021}{12}\right)^{12\times 10}-1}{\frac{.021}{12}}\right]($100) + \left(1+\frac{.021}{12}\right)^{12\times 3}($1000) = \$14,405.06$

31. The future value of the annuity will be $10R + \$3400$, so:

$$10R + \$3400 = \left[\frac{(1+.02)^{10} - 1}{.02}\right]R$$

$$10R + \$3400 = 10.949721R$$

$$\$3400 = .949721R$$

$$\$3580 = R$$

33. (a)

35. Present value: $\left[\frac{1-(1.015)^{-30}}{.015}\right]\left(\frac{.04}{2}\right)(\$5000) + \frac{\$5000}{(1.015)^{30}}$

$= \$5600.40$

37. a. $\left[\dfrac{\frac{.09}{12}}{1-\left(1+\frac{.09}{12}\right)^{-12\times 5}}\right](\$200,000)$

$= \$4151.67$

b. Treat as two annuities. The first has a present value of

$\left[\dfrac{1-(1.005)^{-60}}{.005}\right](.0075)(\$200,000)$

$= \$77,588.34$.

The second has a present value of

$\left[\dfrac{1-(1.005)^{-60}}{.005}\right](\$4151.67) \div (1.005)^{60}$

$= \$159,207.80$.

The total present value $= \$236,796.14$

39. $\left[\dfrac{\frac{.04}{2}}{\left(1+\frac{.04}{2}\right)^{2\times 15}-1}\right](\$1,000,000) = \$24,649.92$

41. $\left[\dfrac{\left(1+\frac{.048}{12}\right)^{12\times 15}-1}{\frac{.048}{12}}\right](\$30) = \$7886.136075$

Since the problem was in millions, the face value will be $7886.136075 million or $7,886,130,075.

43. $\left[\dfrac{\frac{.026}{4}}{\left(1+\frac{.026}{4}\right)^{4\times 15}-1}\right](\$5,000,000) = \$68,404.06$

45. $.05P = \$12,000$, so $P = \$240,000$.

47. $P = (1.03)^7(\$20,000) = \$24,597.48$

$R = \left[\dfrac{.03}{1-(1.03)^{-4}}\right](\$24,597.48) = \$6617.39$

49. $P = \dfrac{R}{i} = \dfrac{\$90,000}{.03} = \$3,000,000$

$\dfrac{\$3,000,000}{(1.03)^9} = \$2,299,250.20$

51. $N = 10, I\% = 2.6, PV = 0, PMT = -1500, P/Y = 2,$ find FV.

    $= FV(1.3\%, 10, -1500, 0)$

    annuity $FV, n = 10, i = 1.3\%, pmt = \$1500$

53. $N = 12, I\% = 1.8, PV = 0, FV = -1681.83, P/Y = 12,$ find PMT.

    $= PMT(0.15\%, 12, 0, -1681.83)$

    annuity $pmt, n = 12, i = .15\%, FV = \$1681.83$

55. $N = 24, I\% = 1.5, PMT = -3000, FV = 0, P/Y = 12,$ find PV.

    $= PV(0.125\%, 24, -3000, 0)$

    annuity $PV, i = .125\%, n = 24, pmt = \$3000$

57. $N = 12, I\% = 1.6, PV = -47336.25, FV = 0, P/Y = 4,$ find PMT.

    $= PMT(0.4\%, 12, -47336.25, 0)$

    annuity $pmt, n = 12, i = .4\%, PV = \$47336.25$

59. Set up a table with $Y_1 = (1.05 \wedge X - 1)/.05 * 1000$. After 2, 3, and 4 years, Y_1 equals $2050, $3152.50 and $4310.13. Y_1 equals $30,539 after 19 years, and exceeds $50,000 after 26 years.

61. Use $Y_1 = (1.001 \wedge X - 1)/.001 * 15$. $Y_1 = 503$ after 33 weeks.

63. The original $5000 will go through 10 years of interest and 10 years of payments. Therefore, the original money will be worth $5000(1.06)^{10}(0.997)^{10} = \$8689.21$. The $5000 deposited in the second year will go through 9 years of interest and 9 years of payment, therefore it will be worth $5000(1.06)^9(0.997)^9 = \$8222.03$. Continue this pattern for the 10 years and adding all the values together will give you a balance of $68,617.21 at the 0.3% payment. Using the same method for the 1.5% payment will yield a balance of $63,882.62 after the 10 years. The difference would then be $4734.59.

Exercises 10.3

1. $\left[\dfrac{\tfrac{.064}{4}}{1 - \left(1 + \tfrac{.064}{4}\right)^{-4 \times 5}} \right] (\$10,000) = \$588.22$

3. $\left[\dfrac{\dfrac{.052}{2}}{1-\left(1+\dfrac{.052}{2}\right)^{-2\times 3}}\right](\$4000) = \$728.63$

5. $\left[\dfrac{1-\left(1+\dfrac{.078}{52}\right)^{-52\times 2}}{\dfrac{.078}{52}}\right](\$23.59) = \$2270.00$

7. $\left[\dfrac{\dfrac{.054}{12}}{1-\left(1+\dfrac{.054}{12}\right)^{-12\times 25}}\right](\$100,000) = \$608.13$

9. $\left[\dfrac{1-\left(1+\dfrac{.045}{12}\right)^{-12\times 30}}{\dfrac{.045}{12}}\right](\$724.56) = \$143,000.00$

11. $\left(1+\dfrac{.042}{12}\right)(\$10,000) - \$1019.35 = \$9015.65$

13. $\left(1+\dfrac{r}{52}\right)(\$2000) - \$100 = \$1903$

$1+\dfrac{r}{52} = 1.0015$

$\dfrac{r}{52} = .0015$

$r = .078 = 7.8\%$

15. $\left[\dfrac{\dfrac{.06}{12}}{1-\left(1+\dfrac{.06}{12}\right)^{-4}}\right]830 = 210.10$

Payment	Amount	Interest	Applied to Principal	Unpaid balance
1	$210.10	$4.15	$205.95	$624.05
2	210.10	3.12	206.98	417.07
3	210.10	2.06	208.01	209.06
4	210.11	1.05	209.06	0.00

17. a. $\dfrac{.048}{12}(\$204{,}700) = \$818.80$

b. $\$1073.99 - \$818.80 = \$255.19$

c. $\$204{,}700 - \$255.19 = \$204{,}444.81$

d. $\left[\dfrac{1-\left(1+\frac{.048}{12}\right)^{-12(5)}}{\frac{.048}{12}}\right](\$1073.99) = \$57{,}188.75$

e. $\left[\dfrac{1-\left(1+\frac{.048}{12}\right)^{-12(4)}}{\frac{.048}{12}}\right](\$1073.99) = \$46{,}819.79$

$\$57{,}188.75 - \$46{,}819.79 = \$10{,}368.96$

f. Use result of part (d): .004($57,188.75) = $228.76

19. a. $\left[\dfrac{\frac{.06}{12}}{1-\left(1+\frac{.06}{12}\right)^{-12\times 3}}\right](\$9480) = \$288.40$

b. ($288.40)(3)(12) = $10,382.40

c. $10,382.40 − $9480 = $902.40

d. $\left[\dfrac{1-\left(1+\frac{.06}{12}\right)^{-12(2)}}{\frac{.06}{12}}\right](\$288.40) = \$6507.13$

e. $\left[\dfrac{1-\left(1+\frac{.06}{12}\right)^{-12(1)}}{\frac{.06}{12}}\right](\$288.40) = \$3350.90$

f. 12($288.40) − ($6507.13 − $3350.90) = $304.57

g.

Payment	Amount	Interest	Applied to Principal	Unpaid balance
1	$288.40	$47.40	$241.00	$9239.00
2	288.40	46.20	242.20	8996.80
3	288.40	44.98	243.42	8753.38
4	288.40	43.77	244.63	8508.75

21. a. Monthly payment $\left[\dfrac{\frac{.009}{12}}{1-\left(1+\frac{.009}{12}\right)^{-60}}\right]($18,000) = 306.91

    b. amount of loan = $18,000 − $500 = $17,500

       Monthly payment $\left[\dfrac{\frac{.06}{12}}{1-\left(1+\frac{.06}{12}\right)^{-12\times 5}}\right]($17,500) = 338.32

    c. Option a is more favorable.

23. Option 1: Pay 500 for first five months. Loan amount is 5000

    Monthly payment $R = \left[\dfrac{\frac{.09}{12}}{1-\left(1+\frac{.09}{12}\right)^{-13}}\right]($5000) = 405.11

    Total amount of payments: $($405.11)(13) + $500 = 5766.43

    Option 2: r = 0.06. n = 18 and P = 5500

    Monthly payment $R = \left[\dfrac{\frac{.06}{12}}{1-\left(1+\frac{.06}{12}\right)^{-18}}\right]($5500) = 320.27

    Total amount of payments: $($320.27)(18) = 5764.86, so Option 2 costs slightly less.

25. $\left[\dfrac{1-\left(1+\frac{.064}{2}\right)^{-2\times 8}}{\frac{.064}{2}}\right]($905.33) + \left[\dfrac{1}{\left(1+\frac{.064}{2}\right)^{2\times 8}}\right]($5,000) = $14,220.61$

27. $\left[\dfrac{1-\left(1+\frac{.06}{12}\right)^{-12\times 3}}{\frac{.06}{12}}\right]($100) = 3287.10

    $6287.10 − ($2000 + $3287.10) = 1000.00

    $\left(1+\dfrac{.06}{12}\right)^{12\times 3}($1000) = 1196.68

29. Assume you have a $100,000 mortgage, find your payments, and your balance after 15 years. The payment will be

    $\left[\dfrac{\frac{.068}{12}}{1-\left(1+\frac{.068}{12}\right)^{-12\times 30}}\right]($100,000) = $651.93.$

    The balance after 15 years will be $\left[\dfrac{1-\left(1+\frac{.068}{12}\right)^{-12\times 15}}{\frac{.068}{12}}\right]($651.93) = $73,441.68.$

    Therefore, the percent paid is $\dfrac{100,000 - 73,441.68}{100,000} = .2656 = 26.56\%.$

31. $\left[\dfrac{\frac{.06}{12}}{1-\left(1+\frac{.06}{12}\right)^{-12\times 25}}\right]($50,000) = 322.15 per month

$\left[\dfrac{1-\left(1+\frac{.06}{12}\right)^{-12(25-10)}}{\frac{.06}{12}}\right]($322.15) = $38,175.91$

$150,000 - $38,175.91 = $111,824.09$

33. The loan amount (F) is 36 times the payment amount (R) minus the interest. The loan amount is also

$F = \left[\dfrac{1-\left(1+\frac{.06}{12}\right)^{-12\times 3}}{\frac{.06}{12}}\right] R$

$F = 32.8710 R$
Therefore,
$36R - $1085.16 = 32.8710 R$
$3.1290 R = 1085.16
$R = 346.81

35. (a)

37. a. $I_n + Q_n = R$ and $I_n = iB_{n-1}$
$iB_{n-1} + Q_n = R$
$B_{n-1} = \dfrac{R - Q_n}{i}$

b. $(1+i)\dfrac{R-Q_n}{i} - R = \dfrac{R-Q_{n+1}}{i}$
$(1+i)(R-Q_n) - iR = R - Q_{n+1}$
$(1+i)R - (1+i)Q_n - iR = R - Q_{n+1}$
$R + iR - (1+i)Q_n - iR = R - Q_{n+1}$
$-(1+i)Q_n = -Q_{n+1}$
$(1+i)Q_n = Q_{n+1}$

c. The amount of the portion applied to the principal in the next month is equal to the amount of the portion applied to the principal in the previous month multiplied by 1 plus the interest rate.

d. $Q_{11} = (1+i)Q_{10}$ and $Q_{12} = (1+i)Q_{11}$
$= (1+0.01)($100)$ $\quad\quad = (1+0.01)($101)$
$= (1.01)($100)$ $\quad\quad\quad\, = (1.01)($101)$
$Q_{11} = 101.00 $\quad\quad\quad\quad\, Q_{11} = 102.01

39.
$N = 20, I\% = 6.4, PV = -10000,$
$FV = 0, \; P/Y = 4,$ find PMT.
$= PMT(1.6\%, 20, -10000, 0)$

annuity pmt, $PV = 10000, $i = 1.6\%$,
$n = 20$

41. $N = 104, I\% = 7.8, PMT = -23.59,$
$FV = 0, \; P/Y = 52,$ find PV.
$= PV(.15\%, 104, -23.59, 0)$

annuity PV, $pmt = 23.59, $i = .15\%$,
$n = 104$

43. $N = 300, I\% = 5.4, PV = -100000,$
    $FV = 0, P/Y = 12,$ find PMT.

    $= PMT(.45\%, 300, -100000, 0)$

    annuity pmt, $PV = \$100000, i = .45\%,$
    $n = 300$

45. $N = 360, I\% = 4.5, PMT = -724.56,$
    $FV = 0, P/Y = 12,$ find PV.

    $= PV(.375\%, 360, -724.56, 0)$

    annuity PV, $pmt = \$724.56, i = .375\%,$
    $n = 360$

47. After 1 month:
    $\left(1 + \frac{.048}{12}\right)(\$4193.97) - \$100 = \$4110.75$
    After 2 months:
    $\left(1 + \frac{.048}{12}\right)(\$4110.75) - \$100 = \$4027.19$
    After 3 months:
    $\left(1 + \frac{.048}{12}\right)(\$4027.19) - \$100 = \$3943.30$
    The loan will be paid off after 46 months.

49. Enter 10000, then run
    1.0075 * Ans − 166.68 repeatedly. The balance drops below $7500 after 25 months, below $5000 after 46 months, and below $2500 after 65 months.

51. The balance B must drop to where $\left(\frac{.085}{12}\right)B < \$250$, or $B < \$35,294.12$. Let $Y_1 = Y_6(300-X)*1000$ where $Y_6 = ((1+I)^{\wedge}X - 1)/(I(1+I)^{\wedge}X)$. Make a table. Then $B \leq \$35,294.12$ after 260 payments which means that the next payment, after 261 months, is the first one where at least 75% goes toward debt reduction.

Exercises 10.4

1. deferred

3. [amount after taxes] $= (1 - .45)(300,000)$
   $= \$165,000$

5. $\left[\dfrac{\left(1 + \frac{.06}{1}\right)^{1 \times 52} - 1}{\frac{.06}{1}}\right](\$5000) = \$1,641,407.11$

7. If we assume a marginal tax bracket of 20%,
   $\left[\dfrac{\left(1 + \frac{.06}{1}\right)^{1 \times 52} - 1}{\frac{.06}{1}}\right](.8)(\$5000) = \$1,313,125.69$

9. **a.** For Earl:
   [earnings after income tax]
   $= [1 - \text{tax bracket}] \cdot [\text{amount}]$
   $= [.60] \cdot [5000]$
   $= 3000$
   $$\left[\frac{\left(1 + \frac{.06}{1}\right)^{1 \times 12} - 1}{\frac{.06}{1}}\right](\$3000) = \$50,609.82$$
   This money then earns interest compounded annually for 36 years and grows to
   $\$50,609.82 \cdot (1.06)^{36} = \$50,609.82(8.147252)$
   $= \$412,330.96$

   b. For Larry:
   $$\left[\frac{\left(1 + \frac{.06}{1}\right)^{1 \times 36} - 1}{\frac{.06}{1}}\right](\$3000) = \$357,362.60$$

   c. Earl paid in $12 \times \$3000 = \$36,000$ while Larry paid in $36 \times \$3000 = \$108,000$. Larry paid in more.

   d. Earl has $54,968.36 more than Larry.

11. $R = \dfrac{P(1+rt)}{12t} = \dfrac{4000(1 + .10 \cdot 1)}{12(1)}$
    $= \$366.67$

13. $R = \dfrac{P(1+rt)}{12t} = \dfrac{3000(1 + .09 \cdot 3)}{12(3)}$
    $= \$105.83$

15. $r = \dfrac{12Rt - P}{Pt} = \dfrac{12(171.21)(1) - 2000}{2000(1)}$
    $\approx .0273$ or about 2.73%

17. $r = \dfrac{12Rt - P}{Pt} = \dfrac{12(608.44)(3) - 20,000}{20,000(3)}$
    $\approx .0317$ or about 3.17%

19. **a.** $[\text{total repayment}] = \dfrac{[\text{loan amount}]}{1 - rt} = \dfrac{880}{1 - .06(2)}$
    $= 1000$
    $[\text{monthly payment}] = \dfrac{[\text{total payment}]}{12t} = \dfrac{1000}{12(2)}$
    $= \$41.67$

    b. $R = \dfrac{P(1+rt)}{12t} = \dfrac{880(1 + .06 \cdot 2)}{12(2)}$
    $= \$41.07$
    The monthly payment is less.

21. False. The effective rate differs from the APR only when discount points are involved.

23. False. The longer the mortgage will be held, the lower the effective rate.

25. False. The up-front fees must change proportionally for there to be no effect on the APR.

27. $\text{Payment} = \left[\dfrac{\frac{.09}{12}}{1-\left(1+\frac{.09}{12}\right)^{-12\times 25}}\right]($250{,}000) = \$2097.99$

New $P = 250{,}000 - 5000 = 245{,}000$
Using the Excel function 12*RATE(300, −2097.99, 245000, 0) gives .09249 or 9.25%; (d).

29. $\text{Payment} = \left[\dfrac{\frac{.055}{12}}{1-\left(1+\frac{.055}{12}\right)^{-12\times 20}}\right]($250{,}000) = \$1719.72$

New $P = 250{,}000 - 10{,}000 = 240{,}000$
Using the Excel function 12*RATE(240, −1719.72, 240000, 0) gives .06002 or about 6%; (a).

31. [monthly payment] = \$581.03
Using the Excel function 12*RATE(48, −581.03, 99000, −94341.50) gives .05999 or about 6%; (a).

33. $\text{Payment} = \left[\dfrac{\frac{.06}{12}}{1-\left(1+\frac{.06}{12}\right)^{-12\times 30}}\right]($100{,}000) = \$599.55$

New $P = 100{,}000 - 3000 = 97{,}000$
The mortgage has $360 - 84 = 276$ months to go. Therefore,

$\text{balance} = \left[\dfrac{1-\left(1+\frac{.06}{12}\right)^{-276}}{\frac{.06}{12}}\right]($599.55) = \$89{,}639.31$

Using the Excel function 12*RATE(84, −599.55, 97000, −89639.31) gives .06560 or about 6.56%; (c).

35. APR: $n = 20 \cdot 12 = 240$

$R = \dfrac{.005(1.005)^{240}}{1.005^{240}-1} \cdot 80{,}000 = 573.14$

$P = 80{,}000 - .03(80{,}000) = 77{,}600$
12*RATE(n, R, P, 0) = 12*RATE(240, −573.14, 77600, 0) = 6.38%
effective rate: $m = 10 \cdot 12 = 120$
$R = 573.14$, $P = 77{,}600$

$B = \dfrac{1.005^{120}-1}{.005(1.005)^{120}} \cdot 573.14 = 51{,}624.70$

12*RATE(m, R, P, B) = 12*RATE(120, −573.14, 77600, −51624.70) = 6.47%

37. APR: $n = 15 \cdot 12 = 180$

$R = \dfrac{.0075(1.0075)^{180}}{1.0075^{180}-1} \cdot 120{,}000 = 1217.12$

$P = 120{,}000 - .01(120{,}000) = 118{,}800$

$\dfrac{(1+i)^{180}-1}{i(1+i)^{180}} \cdot 1217.12 = 118{,}800$ yields

9.17%.
effective rate: $m = 5 \cdot 12 = 60$
$R = 1217.12$, $P = 118{,}800$

$B = \dfrac{1.0075^{120}-1}{.0075(1.0075)^{120}} \cdot 1217.12 = 96{,}081.51$

$1217.12 - 96{,}081.51i$
$= (1217.12 - 118{,}8000i)\cdot (1+i)^{60}$ yields
9.27%.

39. The salesman is comparing the future value of the savings account to the sum of the present values of the loan payments at time of payment. A proper comparison would be the future value of the savings account, $1083.14, to the future value of the series of payments (assuming 4% interest)

$$\left[\frac{\left(1+\frac{.04}{12}\right)^{24}-1}{\frac{.04}{12}}\right](\$43.87) = \$1094.24$$

41. a. $P = \$200,000$ and $i = \frac{.069}{12} = .00575$

   Payment $= iP = (.00575)(\$200,000) = \$1150$

   b. Payment $= \left[\dfrac{\frac{.069}{12}}{1-\left(1+\frac{.069}{12}\right)^{-12\times 10}}\right](\$200,000) = \$2311.87$

43. a. For the first 5 years; Payment $= \left[\dfrac{\frac{.06}{12}}{1-\left(1+\frac{.06}{12}\right)^{-12\times 25}}\right](\$250,000) = \$1610.75$

   b. The balance after 5 years; balance $= \left[\dfrac{1-\left(1+\frac{.06}{12}\right)^{-12\times 20}}{\frac{.06}{12}}\right](\$1610.75) = \$224,829.73$

   c. For the sixth year, $P = \$224,829.73$, $n = 240$, and $i = .044 + .025 = .069$.

   Payment $= \left[\dfrac{\frac{.069}{12}}{1-\left(1+\frac{.069}{12}\right)^{-240}}\right](\$224,829.73) = \$1729.63$

45. a. The balance at the beginning of the 7th year;

   balance $= \left[\dfrac{1-\left(1+\frac{.069}{12}\right)^{-228}}{\frac{.069}{12}}\right](\$1729.63) = \$219,418.04$

   b. Without the cap, $P = \$219,418.04$, $n = 228$, and $i = .077 + .025 = .102$.

   Payment $= \left[\dfrac{\frac{.102}{12}}{1-\left(1+\frac{.102}{12}\right)^{-228}}\right](\$219,418.04) = \$2181.80$

   c. Without the cap, the percentage increase from the sixth year to the seventh year would be
   $\dfrac{2181.80 - 1729.63}{1729.63} = .2614 = 26.14\%$

   Since this percentage is greater than the 7% cap, the monthly payment will be $(1.07)(\$1729.63) = \$1850.70$.

d. The interest due in the 73rd month would be $(.0085)(\$219,418.04) = \$1865.05$.

e. Since the interest owed is more than the payment made, the balance will increase by $\$1865.05 - \$1850.70 = \$14.35$. Therefore the new balance will be $\$219,418.04 + \$14.35 = \$219,432.39$.

47.
```
N=36
I%=11.08292218
PV=10000
PMT=-327.78
FV=0
P/Y=12
C/Y=12
PMT:END BEGIN
```

The APR is about 11.08%.

49.

	A	B	C	D
1	n =	300		
2	i =	0.00542		
3	loan amount =	300,000.00		
4	points:	3.00		
5				
6	1/a$_n$ =	0.0067520716		
7	monthly payment =	2,025.62		
8	APR =	0.06831867		
9				
10	formula in B6:	=(B2*(1+B2)^B1)/(-1+(1+B2)^B1)		
11	formula in B7:	=B6*B3		
12	formula in B8:	=12*RATE(B1,-B7,B3-0.01*B4*B3,0)		

The APR is about 6.83.%.

51.

	A	B	C	D
1	n =	300		
2	i =	0.00542		
3	loan amount =	140,000.00		
4	points:	3		
5				
6	1/a$_n$ =	0.0067520716		
7	monthly payment =	945.29		
8	m =	84		
9	a$_m$=	127.1356748173		
10	unpaid balance =	120180.0853		
11	APR =	0.070787924		
12				
13				
14	formula in B6:	=(B2*(1+B2)^B1)/(-1+(1+B2)^B1)		
15	formula in B7:	=B6*B3		
16	formula in B9:	=(-1+(1+B2)^(B1-B8))/(B2*(1+B2)^(B1-B8))		
17	formula in B10:	=B9*B7		
18	formula in B11:	=12*RATE(B8,-B7,B3-0.01*B4*B3,-B10)		

The APR is about 7.08%.

53. Mortgage A costs an extra $1750 up front.
 [difference in monthly payments] = $1181.61 − $1159.84 = $21.77
 Using the Excel function NPER(.002, 21.77, −1750) gives 87.72 months.

55. Mortgage A costs an extra $2000 up front.
 [difference in monthly payments] = $1303.85 − $1283.93 = $19.92
 Using the Excel function NPER(.004, 19.92, −2000) gives 128.63 months.

Chapter 10 Fundamental Concept Check

1. When you deposit money into a savings account, the *principal* is the amount deposited.

2. *Simple Interest*: Interest is paid only on the principal

 compound interest: Interest is paid not only on the principal, but also on the interest previously earned.

3. At any time, the balance in a savings account is the amount of money in the account at that time. It includes the principal plus the interest earned. The future value of an amount of money is the quantity the money will grow to at a specified time in the future, given a specified rate.

4. Divide the annual interest rate by the number of times interest is compounded per year.

5. At the end of each interest period, the balance is increased by i % of the current balance, where i is the interest rate per period.

6. At the end of each year, the balance is increased by r % of the principal, where r is the annual interest rate.

7. The amount of money that will grow to that amount in a specified time period and at a specified interest rate.

8. nominal rate (aka APR): stated annual interest rate

 effective rate (aka APY): the simple interest rate that produces the same yield after one year as the compound amount produced using the nominal rate; the calculation of the APY takes into account the number of interest periods.

9. a stream of fixed payments over a specified period of time that is used to accumulate savings (increasing annuity) or to generate periodic income (decreasing annuity)

10. future value: the balance at a specified time in the future and at a specified interest rate.

 present value: the amount of money needed to generate the stream of payments at a specified interest rate.

 rent: the amount of the periodic payment.

11. increasing annuity: money is deposited periodically into a bank to accumulate future funds; the present value is 0 and the future value is the amount of money accumulated at the end of a specified time period.

 decreasing annuity: money is withdrawn periodically from a bank account; the present value is the amount initially deposited and the future value is usually 0.

12. increasing annuity: $B_{new} = (1+i)B_{previous} + R$

    decreasing annuity: $B_{new} = (1+i)B_{previous} - R$

13. $F = \dfrac{(1+i)^n - 1}{i} \cdot R$

14. $P = \dfrac{1 - (1+i)^{-n}}{i} \cdot R$

15. payment number: ranges from 0 to the total number of payments

    unpaid balance: amount still owed on the loan

    interest: amount of interest paid during the current interest period

    amount of payment applied to principal: amount of the loan that is paid off during the current interest period

16. $B_{new} = (1+i)B_{previous} - R$

17. certain types of mortgages do not fully amortize over the term of the loan, thus leaving a balance due at maturity. The final payment is called a balloon payment

18. An IRA is an increasing annuity into which money is deposited annually. After the owner reaches a certain age money must be withdrawn from the account. With a traditional IRA, the deposits are tax deductible, but the withdrawals are taxed. With a Roth IRA, the deposits are not tax deductible but the withdrawals are tax free.

19. the total interest paid is $P \cdot r \cdot t$ where P is the amount of the loan, r is the annual interest rate, and t is the number of years.

20. a fee equal to 1 percent, for each discount point, of the loan amount that is paid to the lender.

21. the effective mortgage rate takes into account the length of time the loan will be held and the unpaid balance at that time, while the APR assumes the loan will be held for the full term

22. the monthly payment for a certain number of years consists only of interest payments.

23. the interest rate changes periodically as determined by a measure of current interest rates

Chapter 10 Review Exercises

1. (d)

2. $\left[\dfrac{\frac{.027}{12}}{\left(1 + \frac{.027}{12}\right)^{12 \times 10} - 1} \right] (\$240{,}000) = \$1744.37$

SSM: Finite Math   Chapter 10: The Mathematics of Finance

3. The monthly mortgage payment should not exceed $\left(\dfrac{39{,}216}{12}\right)(.25) = \$817.00$.

$$\left[\dfrac{1-\left(1+\dfrac{.042}{12}\right)^{-12\times 30}}{\dfrac{.042}{12}}\right](\$817) = \$167{,}069.80$$

4. $50\left(1+\dfrac{.0219}{365}\right)^{365} = \$51.11$

5. 2.92% compounded daily yields $\left(1+\dfrac{.0292}{365}\right)^{365} - 1 = .0296 = 2.96\%$ annually. 3% compounded annually is better.

6. $\left[\dfrac{\left(1+\dfrac{.03}{12}\right)^{12\times 5} - 1}{\dfrac{.03}{12}}\right](\$200) = \$12{,}929.34$

7. a. $\left[\dfrac{\dfrac{.045}{12}}{1-\left(1+\dfrac{.045}{12}\right)^{-12\times 15}}\right](\$200{,}000) = \$1529.99$

    b. $\left[\dfrac{1-\left(1+\dfrac{.045}{12}\right)^{-12(15-5)}}{\dfrac{.045}{12}}\right](\$1529.99) = \$147{,}627.70$

8. $(\$35{,}000)(1.005)^{120} = \$63{,}678.89$

9. $\dfrac{\$50{,}000}{\left(1+\dfrac{.03}{12}\right)^{12\times 10}} = \$37{,}054.78$

10. $\dfrac{\$10{,}000}{\left(1+\dfrac{.027}{12}\right)^{12\times 2}} + \dfrac{\$5000}{\left(1+\dfrac{.027}{12}\right)^{12\times 3}} = \$14{,}086.28$

11. $\left[\dfrac{\dfrac{.06}{12}}{1-\left(1+\dfrac{.06}{12}\right)^{-12\times 4}}\right](\$12{,}000 - \$3{,}000) = \$211.37$

12. $\left[\dfrac{\dfrac{.04}{2}}{1-\left(1+\dfrac{.04}{2}\right)^{-2\times 5}}\right]\left(1+\dfrac{.04}{2}\right)^{2\times 2}(\$100{,}000) = \$12{,}050.34$

SSM: Finite Math   **Chapter 10:** The Mathematics of Finance

13. $\dfrac{\$30{,}000}{\left(1+\frac{.024}{12}\right)^{12\times 15}} = \$20{,}937.82$

    $\$105{,}003.50 - \$20{,}937.82 = \$84{,}065.68$

    $\left[\dfrac{\frac{.024}{12}}{1-\left(1+\frac{.024}{12}\right)^{-12\times 15}}\right](\$84{,}065.68) = \$556.59$

14. $\dfrac{\$100{,}000}{\left(1+\frac{.06}{12}\right)^{12\times 10}} = \$54{,}963.27$

    $\left[\dfrac{\frac{.06}{12}}{1-\left(1+\frac{.06}{12}\right)^{-12\times 10}}\right](\$500{,}000 - \$54{,}963.27) = \$4940.82$

15. $\left[\dfrac{\left(1+\frac{.06}{12}\right)^{12\times 30}-1}{\frac{.06}{12}}\right](\$100) = \$100{,}451.50$

16. $\left[\dfrac{1-\left(1+\frac{.045}{12}\right)^{-12\times 10}}{\frac{.045}{12}}\right](\$2000) = \$192{,}978.65$

17. Investment A: $\left[\dfrac{(1+.025)^{10}-1}{.025}\right]1000 = \$11{,}203.38$

    Investment B: $5000(1+.025)^5 + 5000 = \$10{,}657.04$

    Thus Investment A is the better investment.

18. Present value of annuity is $\left[\dfrac{1-\left(1+\frac{.048}{12}\right)^{-12\times 5}}{\frac{.048}{12}}\right](\$5) = \$266.24$

    The present value of $1000 is $\dfrac{\$1000}{\left(1+\frac{.048}{12}\right)^{12\times 5}} = \$787.00$

    Yes, it is a bargain, since the present value is $266.24 + 787.00 = \$1053.24$.

19. $\left(1+\dfrac{.022}{2}\right)^2 - 1 = .022121 = 2.2121\%$

20. $\left(1+\dfrac{.027}{12}\right)^{12} - 1 = .027337 = 2.7337\%$

21. $\left(1+\dfrac{.022}{4}\right)^{4\times 15}(\$10{,}000) + \left[\dfrac{\left(1+\frac{.022}{4}\right)^{4\times 15}-1}{\frac{.022}{4}}\right](\$1000) = \$84{,}753.66$

10-19
Copyright © 2014 Pearson Education, Inc.

22. $R = \left[\dfrac{\frac{.06}{12}}{1-\left(1+\frac{.06}{12}\right)^{-36}}\right]($10,000) = 304.22

Paym	Amount	Interest	Applied to Principal	Unpaid balance
1	$304.22	$50.00	$254.22	$9745.78
2	304.22	48.73	255.49	9490.29
3	304.22	47.45	256.77	9233.52
4	304.22	46.17	258.05	8975.47
5	304.22	44.88	259.34	8716.13
6	304.22	43.58	260.64	8455.49

23. $\left[\dfrac{(1.0015)^{120}-1}{.0015}\right]($200)(1.0015)^{120} = $31,451.59$

24. $\left[\dfrac{\frac{.018}{12}}{1-\left(1+\frac{.018}{12}\right)^{-12\times 5}}\right]($300,000) = 5232.12

25. $\left[\dfrac{\frac{.048}{12}}{1-\left(1+\frac{.048}{12}\right)^{-12\times 30}}\right]($150,000) = 787.00

26. a. [amount after taxes] $= (1-.30)(30,000)$
    $= $21,000$

    b. $30,000 \cdot (1.06)^5 = 40,146.77$
    [amount after taxes] $= (1-.35)(40,146.77) = $26,095.40$

27. a. [amount after taxes] $= (1-0)(30,000) = $30,000$

    b. $30,000 \cdot (1.06)^5 = $40,146.77$

28. $r = \dfrac{12Rt - P}{Pt} = \dfrac{12(228.42)(2) - 5000}{5000(2)} = .048208 = 4.82\%$

29. Loan A is better, because the monthly payments for Loan B will be $\dfrac{3000(1.06)}{12} = 265.

30. $P = $90,000 - .02($90,000) = $88,200$
    $n = 15 \cdot 12 = 180$
    $R = \dfrac{.005}{1-(1.005)^{-80}} \cdot $90,000 = 759.47

31. $m = 6 \cdot 12 = 72$
    $R = 716.43
    $P = $100,000 - 0.3($100,000) = $97,000$
    $B = $81,298.32$

32. Mortgage A costs an extra $2000 up front.
    [difference in monthly payments] = $1413.56 − $1350.41 = $63.15
    $i = .00291667$, $R = \$63.15$, $P = \$2000$
    Using the Excel function NPER(.00291667, 63.15, −2000) gives 33.3 months.

33. Through technology, you can find $x = .06$ on a graphing calculator by finding the intersection of
    $Y_1 = 245000$ and $Y_2 = ((1+X/12)^{\wedge}240 - 1)/(X/12(1+X/12)^{\wedge}240) * 1755.21$.

34. Solve $567.79 - i(86{,}837.98) = [567.79 - i(97{,}000)](1+i)^{96}$, using technology, to find $i = .005$; the interest rate is $12i = .06$. On a graphing calculator, find the intersection of
    $Y_1 = 567.79 - X(86837.98)$ and $Y_2 = (567.79 - X*97000)(1+X)^{\wedge}96$.

35. a. $P = \$380{,}000$ and $i = \dfrac{.069}{12} = .00575$
    Payment $= iP = (0.00575)(\$380{,}000) = \$2185$

    b. Payment $= \left[\dfrac{\frac{.069}{12}}{1-\left(1+\frac{.069}{12}\right)^{-12\times 15}}\right](\$380{,}000) = \$3394.34$

36. a. For the first 5 years; Payment $= \left[\dfrac{\frac{.063}{12}}{1-\left(1+\frac{.063}{12}\right)^{-12\times 25}}\right](\$220{,}000) = \$1458.08$

    b. The balance after 5 years; balance $= \left[\dfrac{1-\left(1+\frac{.063}{12}\right)^{-12\times 20}}{\frac{.063}{12}}\right](\$1458.08) = \$198{,}690.34$

    c. For the sixth year, $P = \$198{,}690.34$, $n = 240$, and $i = .0455 + .028 = .0735$.
    Payment $= \left[\dfrac{\frac{.0735}{12}}{1-\left(1+\frac{.0735}{12}\right)^{-240}}\right](\$198{,}690.34) = \$1582.46$

Conceptual Exercises

37. The effective rate will be slightly higher than the nominal rate.
38. No, not necessarily. It depends upon the number of compounding periods and the time.
39. No. Much more. For example, a house payment is a decreasing annuity. If you pay an additional 5% on the loan each month, the duration of the loan will decrease significantly more than just 5%.
40. The payment will decrease because the amount being applied to the principle will decrease. The total amount of interest paid will increase due to the length of time being added to the loan.
41. When you have successive payments, the interest on the loan is re-calculated more frequently. The interest on the loan is always the interest on the unpaid balance. If more frequent payments are made, the interest will decrease faster.
42. $B_{new} = (1+i)B_{previous} - R$

Chapter 11

Exercises 11.1

1. $a = 4, b = -6, \dfrac{b}{1-a} = \dfrac{-6}{1-4} = 2$

3. $a = -\dfrac{1}{2}, b = 0, \dfrac{b}{1-a} = \dfrac{0}{1+\frac{1}{2}} = 0$

5. $a = -\dfrac{2}{3}, b = 15, \dfrac{b}{1-a} = \dfrac{15}{1+\frac{2}{3}} = 9$

7. a. $y_0 = 10, y_1 = \dfrac{1}{2}(10) - 1 = 4,$

   $y_2 = \dfrac{1}{2}(4) - 1 = 1, y_3 = \dfrac{1}{2}(1) - 1 = -\dfrac{1}{2},$

   $y_4 = \dfrac{1}{2}\left(-\dfrac{1}{2}\right) - 1 = -\dfrac{5}{4}$

   b.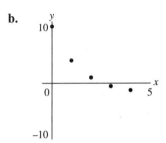

   c. $y_n = \dfrac{-1}{1-\frac{1}{2}} + \left(10 - \dfrac{-1}{1-\frac{1}{2}}\right)\left(\dfrac{1}{2}\right)^n$

   $= -2 + 12\left(\dfrac{1}{2}\right)^n$

9. a. $y_0 = 3.5, y_1 = 2(3.5) - 3 = 4,$
   $y_2 = 2(4) - 3 = 5, y_3 = 2(5) - 3 = 7,$
   $y_4 = 2(7) - 3 = 11$

   b.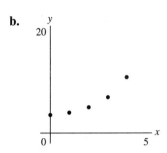

   c. $y_n = \dfrac{-3}{1-2} + \left(3.5 - \dfrac{-3}{1-2}\right)(2)^n$

   $= 3 + (.5)2^n$

11. a. $y_0 = 17.5, y_1 = -.4(17.5) + 7 = 0,$
    $y_2 = -.4(0) + 7 = 7,$
    $y_3 = -.4(7) + 7 = 4.2,$
    $y_4 = -.4(4.2) + 7 = 5.32$

    b.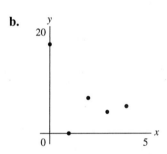

    c. $y_n = \dfrac{7}{1+.4} + \left(17.5 - \dfrac{7}{1+.4}\right)(-.4)^n$

    $= 5 + 12.5(-.4)^n$

13. a. $y_0 = 15$, $y_1 = 2(15) - 16 = 14$,
    $y_2 = 2(14) - 16 = 12$,
    $y_3 = 2(12) - 16 = 8$,
    $y_4 = 2(8) - 16 = 0$

   b.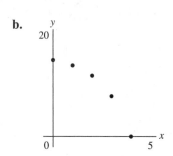

   c. $y_n = \dfrac{-16}{1-2} + \left(15 - \dfrac{-16}{1-2}\right)(2)^n$
      $= 16 - 2^n$

15. $y_0 = 6 - 5(.2)^0 = 1$
    $y_1 = 6 - 5(.2)^1 = 5$
    $y_2 = 6 - 5(.2)^2 = 5.8$
    $y_3 = 6 - 5(.2)^3 = 5.96$
    $y_4 = 6 - 5(.2)^4 = 5.992$

17. $y_n = 1.03 y_{n-1}$, $y_0 = 1000$

19. $y_n = .99 y_{n-1} - 1{,}000{,}000$, $y_0 = 70{,}000{,}000$

21. a. $y_0 = 1$, $y_1 = 1+2 = 3$, $y_2 = 3+2 = 5$,
$y_3 = 5+2 = 7$, $y_4 = 7+2 = 9$

b.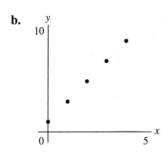

The points lie on a straight line.

c. $a = 1$, so the denominator of $\dfrac{b}{1-a}$ is zero.

23. $1.07(631) - 349 = \$326.17$

25. a. $y_n = 1.04 y_{n-1} + 250$, $y_0 = 800$

b. $y_n = \dfrac{250}{1-1.04} + \left(800 - \dfrac{250}{1-1.04}\right)(1.04)^n$

$y_n = -6250 + 7050(1.04)^n$

c. $y_7 = -6250 + 7050(1.04)^7 \approx 3027.32$
about $3027.32

27. $y_n = .978 y_{n-1} - 10$, $y_0 = 1000$

29. a.–b.

n	y_n
0	102.67
1	100.0035
2	97.203675
3	94.263859
4	91.177052
5	87.935904
6	84.532699
7	80.959334
8	77.207301
9	73.267666
10	69.13105
11	64.787602
12	60.226982
13	55.438331
14	50.410248
15	45.13076
16	39.587298
17	33.766663
18	27.654996
19	21.237746
20	14.499633
21	7.4246151
22	-0.004154
23	-7.804362
24	-15.99458

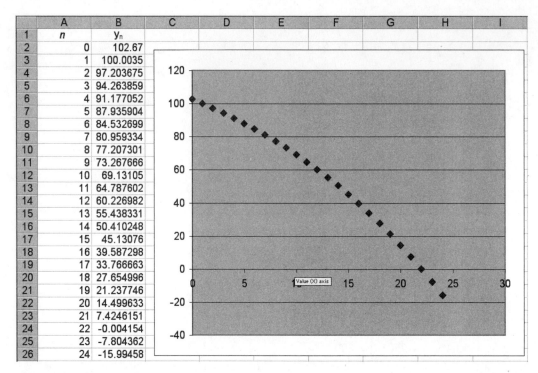

c. $y_7 = 80.96$; $y_n \approx 0$ for $n = 22$

31. a.–b.

n	y_n
0	4
1	7.7
2	4.555
3	7.22825
4	4.9559875
5	6.8874106
6	5.245701
7	6.6411542
8	5.4550189
9	6.4632339
10	5.6062512
11	6.3346865
12	5.7155165
13	6.241811
14	5.7944607
15	6.1747084
16	5.8514978
17	6.1262268
18	5.8927072
19	6.0911989
20	5.9224809
21	6.0658912
22	5.9439925
23	6.0476064
24	5.9595346

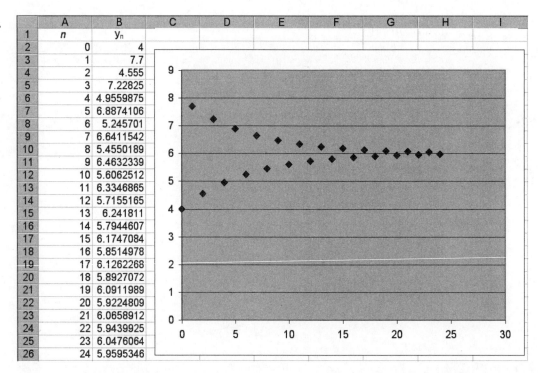

c. $y_{19} = 6.09120$; $|6 - y_n| < .2$ for $n \geq 15$

c. $y_4 = 5.8672$; $y_n < 4$ for $n \geq 9$

33. a.–b.

	A	B
1	n	y_n
2	0	2
3	1	2.75
4	2	3.3125
5	3	3.734375
6	4	4.0507813
7	5	4.2880859
8	6	4.4660645
9	7	4.5995483
10	8	4.6996613
11	9	4.7747459
12	10	4.8310595
13	11	4.8732946
14	12	4.9049709
15	13	4.9287282
16	14	4.9465462
17	15	4.9599096
18	16	4.9699322
19	17	4.9774492
20	18	4.9830869
21	19	4.9873152
22	20	4.9904864
23	21	4.9928648
24	22	4.9946486
25	23	4.9959864
26	24	4.9969898

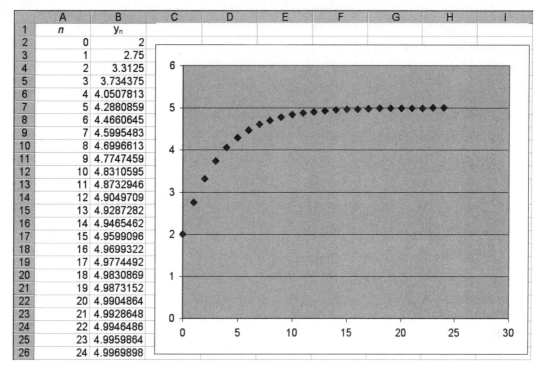

c. $y_9 = 4.77475$; $|5 - y_n| < .5$ for $n \geq 7$

35. a.–b.

	A	B
1	n	y_n
2	0	1
3	1	-0.8
4	2	1
5	3	-0.8
6	4	1
7	5	-0.8
8	6	1
9	7	-0.8
10	8	1
11	9	-0.8
12	10	1
13	11	-0.8
14	12	1
15	13	-0.8
16	14	1
17	15	-0.8
18	16	1
19	17	-0.8
20	18	1
21	19	-0.8
22	20	1
23	21	-0.8
24	22	1
25	23	-0.8
26	24	1

c. The terms alternate between 1 and $-.8$.

Exercises 11.2

1. $b = 5$, $y_0 = 1$
   From formula (2), $y_n = 1 + 5n$.

3. $y_0 = 80$, $i = \dfrac{.09}{12} = .0075$, $n = 5 \times 12 = 60$
   From formula (4), $y_n = 80(1.0075)^{60}$

5. $y_0 = 80$, $i = \dfrac{1}{365}$, $n = 5 \times 365 = 1825$
   From formula (4), $y_n = 80\left(1 + \dfrac{1}{365}\right)^{1825}$.

7. $y_0 = 80$, $i = .07$, $n = 5$
   From formula (3),
   $y_n = 80 + .07 \times 80 \times 5 = 108$

9. $y_0 = 1$

   a. $i = .40$, $n = 1$, $y_n = 1(1.40)^1 = \$1.40$

   b. $i = \dfrac{.40}{2} = .20$, $n = 2 \times 1 = 2$,
   $y_n = 1(1.20)^2 = \$1.44$

   c. $i = \dfrac{.40}{4} = .10$, $n = 4 \times 1 = 4$,
   $y_n = 1(1.10)^4 \approx \$1.46$

11. a. $y_0 = 10$; $y_1 = 2(10) - 10 = 10$;
    $y_2 = 10$; $y_3 = 10$; $y_4 = 10$
    points lie on a horizontal line

    b. $y_0 = 11$; $y_1 = 2(11) - 10 = 12$;
    $y_2 = 2(12) - 10 = 14$;
    $y_3 = 2(14) - 10 = 18$; $y_4 = 2(18) - 10 = 26$;

    points curve upward

c. $y_0 = 9$; $y_1 = 2(9) - 10 = 8$;
   $y_2 = 2(8) - 10 = 6$; $y_3 = 2(6) - 10 = 2$;
   $y_4 = 2(2) - 10 = -6$; points curve downward

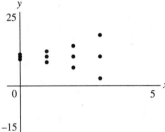

13. $a = .4$, $b = 3$; $y_n = \dfrac{3}{1 - .4} + \left(7 - \dfrac{3}{1 - .4}\right)(.4)^n$
    $= 5 + 2(.4)^n$;
    as n gets large, y_n approaches 5.

15. $a = -5$, $b = 0$;
    $y_n = \dfrac{0}{1 + 5} + \left(2 - \dfrac{0}{1 + 5}\right)(-5)^n = 2(-5)^n$;
    as n gets large, y_n gets arbitrarily large, alternating between being positive and negative.

17. a. $a = 1 + \dfrac{.066}{12} = 1.0055$, $b = -1600$;
    $y_n = 1.0055 y_{n-1} - 1600$, $y_0 = 250{,}525$

    b. $n = 30 \times 12 = 360$; $y_{360} = 0$

19. a. $y_n = 1000(1.03)^n$; $y_0 = 1000$

    b. $y_{20} = 1000(1.03)^{20}$
       $y_{20} = \$1806.11$

21. $a = 1$, $b = \dfrac{50{,}000}{25} = 2000$;
    $y_n = y_{n-1} - 2000$, $y_0 = 50{,}000$;
    $y_n = 50{,}000 - 2000n$

23. $y_0 = 1000$, $i = \dfrac{.06}{4} = .015$; $y_n = 1.015 y_{n-1}$

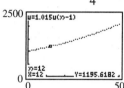

For $n = 3 \times 4 = 12$, $y_{12} = \$1195.62$;

$y_n = 1659$ for $n = 34$, or $8\dfrac{1}{2}$ years;

$y_n \geq 2000$ for $n \geq 47$, so it will double in 47 quarters.

25. $y_0 = 200$, $i = .045$; $y_n = y_{n-1} + 9$

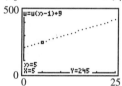

For $n = 5$, $y_5 = \$245$;

$y_n \geq 284$ for $n \geq 10$ years;

$y_n \geq 400$ for $n \geq 23$ years

27.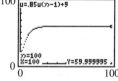

y_n approaches 60.

Exercises 11.3

1. (a), (b), (d), (f), (h)

3. (b), (d), (e), (f)

5. (b), (d), (e), (f)

7. (a), (c), (h), and possibly (g)

9. Possible answer:

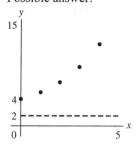

11. Possible answer:

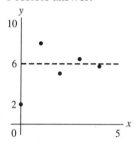

13. Possible answer:

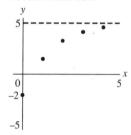

15. Draw $y = \dfrac{b}{1-a} = -2$ as a dashed line.

$a = 3 > 0$, so the graph is monotonic.

$|a| = 3 > 1$, so the graph is repelled from $y = -2$.

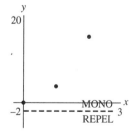

17. Draw $y = \dfrac{b}{1-a} = 6$ as a dashed line.

   $y_0 = 6$, so the graph is constant.

   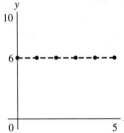

19. Draw $y = \dfrac{b}{1-a} = 4$ as a dashed line.

   $a = -2 < 0$, so the graph is oscillating.

   $|a| = 2 > 1$, so the graph is repelled from $y = 4$.

   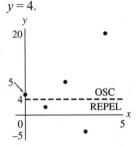

21. Draw $y = \dfrac{b}{1-a} = 10{,}000$ as a dashed line.

   $a = .7 > 0$, so the graph is monotonic.

   $|a| = .7 < 1$, so the graph is attracted to $y = 10{,}000$.

   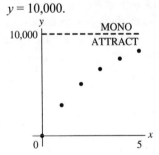

23. Draw $y = \dfrac{b}{1-a} = 1$ as a dashed line.

   $a = -.6 < 0$, so the graph is oscillating.

   $|a| = .6 < 1$, so the graph is attracted to $y = 1$.

   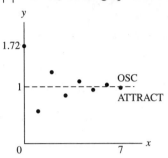

25. $a = 1 + i = 1.0035$, $b = -455$

   The loan, y_0, must be less than $\dfrac{b}{1-a} = \$130{,}000$.

27. $a = 1 + i = 1.015$, $b = -120$

   a. $y_n = 1.015 y_{n-1} - 120$

   b. The deposit, y_0, must be at least $\dfrac{b}{1-a} = \$8000$.

29. $a = 1 + i = 1.028$, $b = -1400$

   $y_n = 1.028 y_{n-1} - 1400$

   The loan, y_0, must be less than $\dfrac{b}{1-a} = \$50{,}000$. If it is greater than or equal to $\$50{,}000$, the loan will never be paid off.

For Exercises 31–36, choose y_0 as the beginning value; choose $a > 0$ or < 0 depending on whether the graph is monotonic or oscillating, also $|a| > 1$ or < 1 depending on whether the graph is unbounded or approaches a value; and if it approaches or is repelled from a value, choose b so that $\dfrac{b}{1-a}$ equals that value.

31. Possible answer: $y_n = .5 y_{n-1} + 4$, $y_0 = 1$

   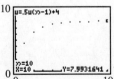

33. Possible answer: $y_n = 2y_{n-1}$, $y_0 = 1$

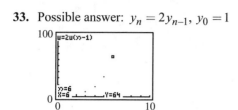

35. Possible answer: $y_n = -2y_{n-1}$, $y_0 = 5$

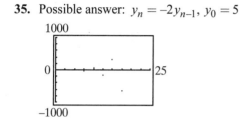

Exercises 11.4

1. $a = 1 + i = 1.0075$, $b = -261.50$;
   $y_n = 1.0075 y_{n-1} - 261.50$, $y_0 = 32{,}500$

3. $a = 1 + i = 1.015$, $b = 200$;
   $y_n = 1.015 y_{n-1} + 200$, $y_0 = 4000$

5. $a = 1 + i = 1.01$, $b = -660$, $\dfrac{b}{1-a} = 66{,}000$

   $y_{120} = 0 = 66{,}000 + (y_0 - 66{,}000)(1.01)^{120}$

   $y_0 = \dfrac{-66{,}000}{(1.01)^{120}} + 66{,}000 \approx \$46{,}002.34$

7. $a = 1 + i = 1.06$, $b = 300$,
   $\dfrac{b}{1-a} = -5000$, $y_0 = 0$

   $y_{20} = -5000 + [0 - (-5000)](1.06)^{20}$
   $\approx \$11{,}035.68$

9. $a = 1 + i = 1.005$, $y_0 = 0$

   $y_{144} = 6000 = \dfrac{b}{-.005} + \left(0 - \dfrac{b}{-.005}\right)(1.005)^{144}$

   $= \dfrac{(1.005)^{144} - 1}{.005} b$

   $b = \dfrac{6000 \times .005}{(1.005)^{144} - 1} \approx \$28.55$

11. $a = 1 + i = 1.005$, $y_0 = 4000$

    $y_{36} = 0 = \dfrac{b}{-.005} + \left(4000 - \dfrac{b}{-.005}\right)(1.005)^{36}$

    $= 4000(1.005)^{36} + 200b[(1.005)^{36} - 1]$

    $b = \dfrac{-4000(1.005)^{36}}{200[(1.005)^{36} - 1]} \approx -121.69$

    \$121.69

13. $a = 1 + i = 1.08$, $b = -4$
    Use $n\text{Min} = 0$, $u(n) = 1.08u(n-1) - 4$,
    $u(n\text{Min}) = \{45\}$
    $u(10) \approx \$39.205$ million
    $u(n) \leq 0$ for $n = 30$ years

15. $a = 1 + i = 1.00225$, $b = 100$
    Use $n\text{Min} = 0$, $u(n) = 1.00225u(n-1) + 100$,
    $u(n\text{Min}) = \{0\}$
    $u(5) \approx \$502.26$
    $u(10) \approx \$1010.19$
    $u(15) \approx \$1523.86$
    $u(n) \geq \$1938.97$ for $n = 19$
    $u(n) > \$3000$ for $n = 30$

Exercises 11.5

1. $a = 1 + .03 - .01 = 1.02$, $b = 0$, $\dfrac{b}{1-a} = 0$

   $y_n = 1.02 y_{n-1}$, $y_0 = 100$ million

   $a > 0$: monotonic; $|a| > 1$: repelled from $y = 0$

3. $a = 1 - .25 = .75$, $b = 0$, $\dfrac{b}{1-a} = 0$

   $y_n = .75 y_{n-1}$

   $a > 0$: monotonic; $|a| < 1$: attracted to $y = 0$

   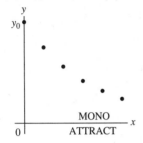

5. $y_n = y_{n-1} + .08(100 - y_{n-1}) = .92 y_{n-1} + 8$,
   $y_0 = 0$

   $a = .92$, $b = 8$, $\dfrac{b}{1-a} = 100$

   $a > 0$: monotonic; $|a| < 1$: attracted to $y = 100$

   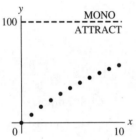

7. $y_n = y_{n-1} + .30(12 - y_{n-1})$
   $= .7 y_{n-1} + 3.6$, $y_0 = 0$

   $a = .7$, $b = 3.6$, $\dfrac{b}{1-a} = 12$

   $a > 0$: monotonic; $|a| < 1$: attracted to $y = 12$

   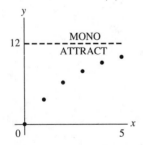

9. $a = 1 + i = 1.05$, $b = -1000$, $\dfrac{b}{1-a} = 20{,}000$

   $y_n = 1.05 y_{n-1} - 1000$, $y_0 = 30{,}000$

   $a > 0$: monotonic; $|a| > 1$: repelled from $y = 20{,}000$

   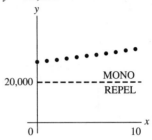

11. $y_n = y_{n-1} + .20(70 - y_{n-1}) = .8 y_{n-1} + 14$,
    $y_0 = 40$

    $a = .8$, $b = 14$, $\dfrac{b}{1-a} = 70$

    $a > 0$: monotonic; $|a| < 1$: attracted to $y = 70$

    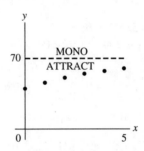

13. $p_n = 20 - .1 q_n = 20 - .1(5 p_{n-1} - 10)$
    $= -.5 p_{n-1} + 21$, $p_0 = 4.54$

    $a = -.5$, $b = 21$, $\dfrac{b}{1-a} = 14$

    $a < 0$: oscillating, $|a| < 1$: attracted to $y = 14$

15. After 2 hours, .5 is left. After four hours, (.5)(.5) = .25 is left. Answer is (b).

17. Use $n\text{Min} = 0$,
    $u(n) = (1+.035-.02)u(n-1)-.0003$,
    $u(n\text{Min}) = \{5\}$
    $u(5) \approx 5.38$ million
    $u(n) > 6$ for $n = 13$, or in the year 2022
    $u(n) \geq 10$ for $n = 47$, or in the year 2056

19. $y_n = y_{n-1} - .13 y_{n-1}$
    $y_n = (.87) y_{n-1},\ y_0 = 130$
    $a = .87,\ b = 0,\ \dfrac{b}{1-a} = 0$
    $y_n = 130(.87)^n$

    After 5 hours, $y_n = 130(0.87)^5 = 64.79$ (This problem can be solved through trial and error or by the use of logarithms)
    After 24 hours, $y_n = 130(0.87)^{24} = 4.6$ milligrams will remain.

Chapter 11 Fundamental Concept Check

1. After the values of $a, b,$ and y_0 are specified, successive terms are generated one at a time by the difference equation.

2. The first value of the sequence generated by the difference equation.

3. For $n = 0, 1, 2, \ldots$, the nth point of the graph is the point whose x–coordinate is n and whose y–coordinate is the nth value of the sequence.

4. For $a \neq 1$, the formula is
   $y_n = \dfrac{b}{1-a} + \left(y_0 - \dfrac{b}{1-a}\right) a^n$. For $a = 1$, the formula is $y_n = y_0 + bn$.

5. Interest is a periodic fee paid by a bank for deposited funds or a fee paid by a borrower of funds. With simple interest, the fee is constant and is based only on the initial amount deposited or borrowed. With compound interest, the fee varies and is based on successive balances on the amount deposited or owed.

6. Increasing: successive values increase; decreasing: successive values decrease; monotonic: either increasing or decreasing; oscillating: successive values alternate between increasing and decreasing; constant: successive values remain the same.

7. constant, increasing without bound, increasing asymptotically to a horizontal line, decreasing without bound, decreasing asymptotically to a horizontal line, oscillating without bound, oscillating asymptotically to a horizontal line.

8. See the tinted box on page 495.

9. A loan taken out to purchase real estate. It has a specified interest rate per period (i), and a payment (R) paid at the end of each period, for a specified number of periods. Successive balances are calculated with the difference equation
   $B_n = (1+i) B_{n-1} - R$,
   $B_0 = [\text{amount of the loan}]$.

10. A sequence of periodic payment of \$D deposited into a savings account earning a specified rate of interest per period (i). Successive balances are calculated with the difference equation
    $B_n = (1+i) B_{n-1} + D,\ B_0 = 0$.

Chapter 11 Review Exercises

1. a. $y_1 = -3(1) + 8 = 5;\ y_2 = -3(5) + 8 = -7;$
      $y_3 = -3(-7) + 8 = 29$

   b. $a = -3,\ b = 8$
      $y_n = \dfrac{b}{1-a} + \left(y_0 - \dfrac{b}{1-a}\right) a^n$
      $= 2 + (1-2)(-3)^n$
      $= 2 - (-3)^n$

   c. $2 - (-3)^4 = -79$

2. a. $y_1 = 10 - \frac{3}{2} = \frac{17}{2}$; $y_2 = \frac{17}{2} - \frac{3}{2} = 7$;

   $y_3 = 7 - \frac{3}{2} = \frac{11}{2}$

   b. $a = 1$, $b = -\frac{3}{2}$

   $y_n = 10 - \frac{3}{2}n$

   c. $10 - \frac{3}{2}(6) = 1$

3. $a = 1 + i = 1 + \frac{.0305}{4} = 1.007625$, $b = 0$,

   $\frac{b}{1-a} = 0$

   $y_n = y_0(1.007625)^n$
   Solve.
   $2474 = y_0(1.007625)^{28}$

   $y_0 = \frac{2474}{1.2370} \approx \$2000$

4. $a = 1 + i = 1.001$, $b = 0$, $\frac{b}{1-a} = 0$, $y_0 = 1000$

   $y_{104} = 0 + (1000 - 0)(1.001)^{104}$
   $= \$1109.54$

5. Draw $y = \frac{b}{1-a} = 6$ as a dashed line.

   $a = -\frac{1}{3} < 0$, so the graph is oscillating.

   $|a| = \frac{1}{3} < 1$, so the graph is attracted to $y = 6$.

   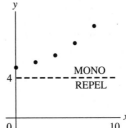

6. Draw $y = \frac{b}{1-a} = 4$ as a dashed line.

   $a = 1.5 > 0$, so the graph is monotonic.

   $|a| = 1.5 > 1$, so the graph is repelled from $y = 4$.

   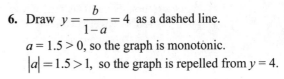

7. a. $y_n = 1.03 y_{n-1} - 600$, $y_0 = 120,000$

   b. $\frac{b}{1-a} = 20,000$;

   $y_{20} = 20,000 + (120,000 - 20,000)(1.03)^{20}$
   $\approx 200,611$

8. a. $y_n = 1.0035 y_{n-1} - 978.03$, $y_0 = 200,000$

   b. $\frac{b}{1-a} = 279,437.14$;

   $y_{84} = 279,437.14$
   $\quad + (200,000 - 279,437.14)(1.0035)^{84}$
   $\approx \$172,904.36$

9. $a = 1 + i = 1.0005$, $y_0 = 0$, $21 \times 52 = 1092$

   $y_{1092} = 36,000$

   $= \frac{b}{-.0005} + \left(0 - \frac{b}{-.0005}\right)(1.0005)^{1092}$

   $= 2000[(1.0005)^{1092} - 1]b$

   $b = \frac{36,000}{2000[(1.0005)^{1092} - 1]} \approx \$24.79$

10. $a = 1 + i = 1.005$, $y_0 = 33,100$

    $y_{240} = 0$

    $= \frac{b}{-.005} + \left(33,100 - \frac{b}{-.005}\right)(1.005)^{240}$

    $= 33,100(1.005)^{240} + 200[(1.005)^{240} - 1]b$

    $b = \frac{-33,100(1.005)^{240}}{200[(1.005)^{240} - 1]} \approx -237.14$

    $\$237.14$

11. $a = 1+i = 1.08$, $b = -2400$, $\dfrac{b}{1-a} = \$30{,}000$

$y_{18} = 0 = 30{,}000 + (y_0 - 30{,}000)(1.08)^{18}$

$y_0 = \dfrac{-30{,}000}{(1.08)^{18}} + 30{,}000 \approx \$22{,}492.53$

12. $y_n = 1.00175 y_{n-1} + 50$, $y_0 = 0$

$a = 1.00175$, $b = 50$, $\dfrac{b}{1-a} = -28{,}571.43$

$y_{48} = -28{,}571.43$
$\quad + [0 - (-28{,}571.43)](1.00175)^{48}$
$\approx \$2501.40$

13. $y_n = y_{n-1} + .10(1{,}000{,}000 - y_{n-1})$
$\quad = .9 y_{n-1} + 100{,}000$, $y_0 = 0$

$a = .9$, $b = 100{,}000$, $\dfrac{b}{1-a} = 1{,}000{,}000$

$a > 0$: monotonic, $|a| < 1$: attracted to
$y = 1{,}000{,}000$

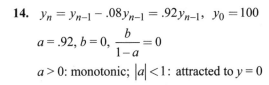

14. $y_n = y_{n-1} - .08 y_{n-1} = .92 y_{n-1}$, $y_0 = 100$

$a = .92$, $b = 0$, $\dfrac{b}{1-a} = 0$

$a > 0$: monotonic; $|a| < 1$: attracted to $y = 0$

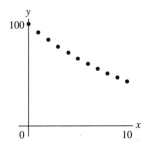

Chapter 12

Exercises 12.1

1. Statement

3. Statement

5. Not a statement—not a declarative sentence.

7. Not a statement—not a declarative sentence.

9. Statement

11. Not a statement—x is not specified.

13. Not a statement—not a declarative sentence.

15. Statement

17. p: The Phelps library is in New York.
    q: The Phelps library is in Dallas.
    Then we have $p \vee q$

19. p: The Smithsonian Museum of Natural History has displays of rocks.
    q: The Smithsonian Museum of Natural History has displays of bugs.
    Then we have $p \wedge q$

21. p: Amtrak trains go to Chicago.
    q: Amtrak trains go to Cincinnati.
    Then we have $\sim p \wedge \sim q$ or $\sim(p \vee q)$

23. a. Ozone is opaque to ultraviolet light, and life on earth requires ozone.

    b. Ozone is not opaque to ultraviolet light, or life on earth requires ozone.

    c. Ozone is not opaque to ultraviolet light, or life on earth does not require ozone.

    d. It is not the case that life on earth does not require ozone.

25. a. Florida borders Alabama or Florida borders Mississippi: $p \vee q$

    b. Florida borders Alabama and Florida does not border Mississippi $p \wedge \sim q$

    c. Florida borders Mississippi and Florida does not border Alabama $q \wedge \sim p$

    d. Florida does not border Alabama and Florida does not border Mississippi $\sim p \wedge \sim q$

Exercises 12.2

1. Since r is a statement form, so is $\sim r$. Then $p \wedge \sim r$ is a statement form, and so is $\sim(p \wedge \sim r)$. Since q is a statement form, $\sim(p \wedge \sim r) \vee q$ is a statement form.

3. Since p is a statement form, so is $\sim p$. Since q and r are statement forms, so are $\sim p \vee r$ and $q \wedge r$, and hence also $(\sim p \vee r) \rightarrow (q \wedge r)$.

5. | p | q | p | $\wedge$ | $\sim$ | q |
    |---|---|---|---|---|---|
    | T | T | T | F | F | T |
    | T | F | T | T | T | F |
    | F | T | F | F | F | T |
    | F | F | F | F | T | F |
    | (1) | (2) | | (4) | (3) | |

7.

p	q	(p	∨	~q)	∨	~p)
T	T		T	F	**T**	F
T	F		T	T	**T**	F
F	T		F	F	**T**	T
F	F		T	T	**T**	T
(1)	(2)		(5)	(3)	(6)	(4)

9.

p	q	~	((p ∨ q)	∧	(p ∧ q))
T	T	**F**	T	T	T
T	F	**T**	T	F	F
F	T	**T**	T	F	F
F	F	**T**	F	F	F
(1)	(2)	(6)	(3)	(5)	(4)

11.

p	q		p	⊕	(~p	∨	q)
T	T		T	**F**	F	T	T
T	F		T	**T**	F	F	F
F	T		F	**T**	T	T	T
F	F		F	**T**	T	T	F
(1)	(2)			(5)	(3)	(4)	

13.

p	q	r	(p	∧	~	r)	∨	q
T	T	T	T	F	F	T	**T**	T
T	T	F	T	T	T	F	**T**	T
T	F	T	T	F	F	T	**F**	F
T	F	F	T	T	T	F	**T**	F
F	T	T	F	F	F	T	**T**	T
F	T	F	F	F	T	F	**T**	T
F	F	T	F	F	F	T	**F**	F
F	F	F	F	F	T	F	**F**	F
(1)	(2)	(3)		(5)	(4)		(6)	

15.

p	q	r	~	[(p	∧	r)	∨	q]
T	T	T	**F**	T	T	T	T	T
T	T	F	**F**	T	F	F	T	T
T	F	T	**F**	T	T	T	T	F
T	F	F	**T**	T	F	F	F	F
F	T	T	**F**	F	F	T	T	T
F	T	F	**F**	F	F	F	T	T
F	F	T	**T**	F	F	T	F	F
F	F	F	**T**	F	F	F	F	F
(1)	(2)	(3)	(6)		(4)		(5)	

17.

p	p	$\vee$	$\sim$	p
T	T	**T**	F	T
F	F	**T**	T	F
(1)		(3)	(2)	

19.

p	q	r	p	$\oplus$	$(q$	$\vee$	$r)$
T	T	T	T	**F**	T	T	T
T	T	F	T	**F**	T	T	F
T	F	T	T	**F**	F	T	T
T	F	F	T	**T**	F	F	F
F	T	T	F	**T**	T	T	T
F	T	F	F	**T**	T	T	F
F	F	T	F	**T**	F	T	T
F	F	F	F	**F**	F	F	F
(1)	(2)	(3)		(5)		(4)	

21.

p	q	r	$(p$	$\vee$	$q)$	$\wedge$	$(p$	$\vee$	$r)$
T	T	T	T	T	T	**T**	T	T	T
T	T	F	T	T	T	**T**	T	T	F
T	F	T	T	T	F	**T**	T	T	T
T	F	F	T	T	F	**T**	T	T	F
F	T	T	F	T	T	**T**	F	T	T
F	T	F	F	T	T	**F**	F	F	F
F	F	T	F	F	F	**F**	F	T	T
F	F	F	F	F	F	**F**	F	F	F
(1)	(2)	(3)		(4)		(6)		(5)	

23.

p	q	$(p$	$\vee$	$q)$	$\wedge$	$\sim$	$(p$	$\vee$	$q)$
T	T	T	T	T	**F**	F	T	T	T
T	F	T	T	F	**F**	F	T	T	F
F	T	F	T	T	**F**	F	F	T	T
F	F	F	F	F	**F**	T	F	F	F
(1)	(2)		(3)		(6)	(5)		(4)	

25.

p	q	r	$\sim$	$(p$	$\vee$	$q)$	$\wedge$	r
T	T	T	F	T	T	T	**F**	T
T	T	F	F	T	T	T	**F**	F
T	F	T	F	T	T	F	**F**	T
T	F	F	F	T	T	F	**F**	F
F	T	T	F	F	T	T	**F**	T
F	T	F	F	F	T	T	**F**	F
F	F	T	T	F	F	F	**T**	T
F	F	F	T	F	F	F	**F**	F
(1)	(2)	(3)	(5)		(4)		(6)	

Chapter 12: Logic

27.

p	q	~	p	∨	~	q	~	(p	∧	q)
T	T	F	T	**F**	F	T	**F**	T	T	T
T	F	F	T	**T**	T	F	**T**	T	F	F
F	T	T	F	**T**	F	T	**T**	F	F	T
F	F	T	F	**T**	T	F	**T**	F	F	F
(1)	(2)	(3)		(5)	(4)		(4)		(3)	

They are identical.

29.

p	q	~	p	⊕	q	(p	∧	q)	∨	~	(p	∨	q)
T	T	**T**	T	F	T	T	T	T	**T**	F	T	T	T
T	F	**F**	T	T	F	T	F	F	**F**	F	T	T	F
F	T	**F**	F	T	T	F	F	T	**F**	F	F	T	T
F	F	**T**	F	F	F	F	F	F	**T**	T	F	F	F
(1)	(2)	(4)		(3)			(4)		(6)	(5)		(3)	

They are identical.

31. For each of the four possible pairs of values for p and q, there are two possibilities, T or F. Thus there are $2^4 = 16$ possible truth tables.

33. a.

p	p	|	p
T	T	**F**	T
F	F	**T**	F
(1)		(2)	

b.

p	q	(p	|	p)	|	(q	|	q)
T	T	T	F	T	**T**	T	F	T
T	F	T	F	T	**T**	F	T	F
F	T	F	T	F	**T**	T	F	T
F	F	F	T	F	**F**	F	T	F
(1)	(2)		(3)		(5)		(4)	

c.

p	q	(p	|	q)	|	(p	|	q)
T	T	T	F	T	**T**	T	F	T
T	F	T	T	F	**F**	T	T	F
F	T	F	T	T	**F**	F	T	T
F	F	F	T	F	**F**	F	T	F
(1)	(2)		(3)		(5)		(4)	

d.

p	q	p	|	((p	|	q)	|	q)
T	T	T	**F**	T	F	T	T	T
T	F	T	**F**	T	T	F	T	F
F	T	F	**T**	F	T	T	F	T
F	F	F	**T**	F	T	F	T	F
(1)	(2)		(5)		(3)		(4)	

35. *p* has truth value T and *q* has truth value F.

    a.   *p* ∨ ∼ *q*
         T **T** T F

    b.   ∼ *p* ∧ *q*
         F T **F** F

    c.   *p* ⊕ *q*
         T **T** F

    d.   ∼ *p* ⊕ *q*
         F T **F** F

    e.   ∼ (*p* ⊕ *q*)
         F T T F

    f.   (*p* ∨ *q*) ⊕ ∼ *q*
         T T F **F** T F

37. (*p* ⊕ *q*) ∧ *r*
       F
    p ⊕ *q*   *r*
     T   F
  p   *q*
  T   F

39. (*p* ∨ *q*) ∧ (*p* ∨ ∼ *r*)
        T
  p ∨ *q*   *p* ∨ ∼ *r*
    T   T
 p *q* *p* ∼ *r*
 T F T T
          r
          F

41. (*p* ⊕ *q*) ∧ ∼ *q*
    T T F **T** T F
    1 will be displayed

43. a.   The calculator will display F.

    b.   The calculator will display F.

45. a.   The calculator will display F.

    b.   The calculator will display F.

47.

p	*q*	*p* ∧ ∼ *q*
T	T	F
T	F	T
F	T	F
F	F	F

49.

p	*q*	*r*	*p* ∧ (∼ *q* ∨ *r*)
T	T	T	T
T	T	F	F
T	F	T	T
T	F	F	T
F	T	T	F
F	T	F	F
F	F	T	F
F	F	F	F

Exercises 12.3

1.

p	q	p	→	~	q
T	T	T	**F**	F	T
T	F	T	**T**	T	F
F	T	F	**T**	F	T
F	F	F	**T**	T	F
(1)	(2)		(4)	(3)	

3.

p	q	(p	⊕	q)	→	q
T	T	T	F	T	**T**	T
T	F	T	T	F	**F**	F
F	T	F	T	T	**T**	T
F	F	F	F	F	**T**	F
(1)	(2)		(3)		(4)	

5.

p	q	r	(~	p	∧	q)	→	r
T	T	T	F	T	F	T	**T**	T
T	T	F	F	T	F	T	**T**	F
T	F	T	F	T	F	F	**T**	T
T	F	F	F	T	F	F	**T**	F
F	T	T	T	F	T	T	**T**	T
F	T	F	T	F	T	T	**F**	F
F	F	T	T	F	F	F	**T**	T
F	F	F	T	F	F	F	**T**	F
(1)	(2)	(3)	(4)		(5)		(6)	

7.

p	q	(p	→	q)	↔	(p	∨	q
T	T	T	T	T	**T**	F	T	T
T	F	T	F	F	**T**	F	T	F
F	T	F	T	T	**T**	F	T	T
F	F	F	T	F	**T**	T	T	F
(1	(2)		(4)		(6)	(3)		(5)

9.

p	q	r	(p	→	q)	→	r
T	T	T	T	T	T	**T**	T
T	T	F	T	T	T	**F**	F
T	F	T	T	F	F	**T**	T
T	F	F	T	F	F	**T**	F
F	T	T	F	T	T	**T**	T
F	T	F	F	T	T	**F**	F
F	F	T	F	T	F	**T**	T
F	F	F	F	T	F	**F**	F
(1)	(2)	(3)		(4)		(5)	

11.

p	q	(p	∨	q)	↔	(p	∧	q)
T	T	T	T	T	**T**	T	T	T
T	F	T	T	F	**F**	T	F	F
F	T	F	T	T	**F**	F	F	T
F	F	F	F	F	**T**	F	F	F
(1)	(2)		(3)		(5)		(4)	

13. $((\sim p) \wedge (\sim q)) \to ((\sim p) \wedge q)$

15. $(((\sim p) \wedge (\sim q)) \vee r) \to ((\sim q) \wedge r)$

17.

~	p	→	q
F	T	**T**	F

19.

q	→	p
F	**T**	T

21.

(p	⊕	q)	→	p
T	T	F	**T**	T

23.

(p	∧	~	q)	→	(~	p	⊕	q)
T	T	T	F	**F**	F	T	F	F

25. $p \to [p \wedge (p \oplus q)]$

T	**T**	T	T	T	T	F

27. $p \leftrightarrow q$

29. $q \to p$; hypothesis: q; conclusion: p

31. $q \to p$; hypothesis: q; conclusion: p

33. $\sim p \to \sim q$; hypothesis: $\sim p$; conclusion: $\sim q$

35. $\sim p \to \sim q$; hypothesis: $\sim p$; conclusion: $\sim q$;
F T **T** F T   TRUE

37. $\sim q \to \sim p$; hypothesis: $\sim q$; conclusion: $\sim p$;
             TRUE

39. a. hyp: A person is healthy.
con: The person lives a long life.

b. hyp: The train stops at the station.
con: A passenger requests the stop.

c. hyp: The azalea grows.
con: The azalea is exposed to sunlight.

d. hyp: I will go to the store.
con: Jane goes to the store.

41. a. If City Sanitation collects the garbage, then the mayor calls.

  b. The price of beans goes down if there is no drought.

  c. Goldfish swim in Lake Erie if Lake Erie is fresh water.

  d. Tap water is not salted if it boils slowly.

43. a. $Z = 0 + 0 = 0$, so the condition fails and $A = 4$.

  b. $Z = 8 + (-8) = 0$, so the condition fails and $A = 4$.

  c. $Z = -3 + 3 = 0$, so the condition fails and $A = 4$.

  d. $X = -3 \leq 0$, so the condition fails and $A = 4$.

  e. $X = 8 > 0$ and $Z = 8 + (-3) = 5 \neq 0$, so the condition is met and $A = 6$.

  f. $X = 3 > 0$ and $Z = 3 + (-8) = -5 \neq 0$, so the condition is met and $A = 6$.

45. a. $B = -6 < 0$, so the condition is met and $Y = 7$.

  b. $B = -6 < 0$, so the condition is met and $Y = 7$.

  c. $C = (-2)(6) = -12 < 10$ and $B \geq 0$, so the condition fails and $Y = 0$.

  d. $B = -1 < 0$, so the condition is met and $Y = 7$.

  e. $A = 4 \geq 0$ and $B = 3 \geq 0$, so the condition fails and $Y = 0$.

  f. $A = 3 \geq 0$ and $B = 1 \geq 0$, so the condition fails and $Y = 0$.

47. a. $A = -1 < 0$ and $B = -2 < 0$, so the condition is met and $C = (-1)(-2) + 4 = 6$.

  b. $B = 8 \geq 6$, so the condition is met and $C = (-2)(8) + 4 = -12$.

  c. $B = 3 \geq 0$ and $B = 3 < 6$, so the condition fails and $C = 0$.

  d. $A = 3 \geq 0$ and $B = -2 < 6$, so the condition fails and $C = 0$.

  e. $B = 8 \geq 6$, so the condition is met and $C = (3)(8) + 4 = 28$.

  f. $A = 3 \geq 0$ and $B = -3 < 6$, so the condition fails and $C = 0$.

Chapter 12: Logic SSM: Finite Math

Exercises 12.4

1. [(p → q) ∧ q] → p
     F T T T T **F** F
   When p is false and q is true, the statement is FALSE.

3. Show that the corresponding bi-conditional is a tautology.

p	q	(p	→	q)	↔	(~	(p	∧	~	q))
T	T	T	T	T	**T**	T	T	F	F	T
T	F	T	F	F	**T**	F	T	T	T	F
F	T	F	T	T	**T**	T	F	F	F	T
F	F	F	T	F	**T**	T	F	F	T	F
(1)	(2)		(3)		(7)	(6)		(5)	(4)	

5.

p	q	c	(p	→	q)	↔	[(p	∧	~	q)	→	c)]
T	T	F	T	T	T	**T**	T	F	F	T	T	F
T	F	F	T	F	F	**T**	T	T	T	F	F	F
F	T	F	F	T	T	**T**	F	F	F	T	T	F
F	F	F	F	T	F	**T**	F	F	T	F	T	F
(1)	(2)	(3)			(6)		(8)		(5)	(4)		(7)

7. False: consider p FALSE and q TRUE.

9. a.

p	~	p	↔	p	|	p
T	F	T	**T**	T	F	T
F	T	F	**T**	F	T	F
(1)	(2)		(4)		(3)	

   b.

p	q	(p	∨	q)	↔	(p	|	p)	|	(q	|	q)
T	T	T	T	T	**T**	T	F	T	T	T	F	T
T	F	T	T	F	**T**	T	F	T	T	F	T	F
F	T	F	T	T	**T**	F	T	F	T	T	F	T
F	F	F	F	F	**T**	F	T	F	F	F	T	F
(1)	(2)		(5)		(7)		(3)		(6)		(4)	

   c.

p	q	(p	∧	q)	↔	(p	|	q)	|	(p	|	q)
T	T	T	T	T	**T**	T	F	T	T	T	F	T
T	F	T	F	F	**T**	T	T	F	F	T	T	F
F	T	F	F	T	**T**	F	T	T	T	F	T	T
F	F	F	F	F	**T**	F	T	F	F	F	T	F
(1)	(2)		(5)		(7)		(3)		(6)		(4)	

   d. $p \to q$
       ⇔ ~p ∨ q   Implication (10a)

$$\Leftrightarrow (\sim p \mid \sim p) \mid (q \mid q) \quad \text{Part (b) above}$$
$$\Leftrightarrow p \mid (q \mid q) \quad \text{Part (a) above}$$

    e. $p \mid q \Leftrightarrow \sim(p \wedge q)$
(Compare truth table for $\sim(p \wedge q)$ with truth table for $p \mid q$.)

11. $p \oplus q \Leftrightarrow (p \vee q) \wedge \sim(p \wedge q) \Leftrightarrow \sim[\sim(p \vee q) \vee \sim(\sim p \vee \sim q)]$

13. $(p \vee q) \rightarrow (q \wedge \sim r)$
        $\Leftrightarrow \sim(p \vee q) \vee (q \wedge \sim r)$ Implication (10a)

15. $\sim(p \wedge \sim q) \rightarrow (p \vee \sim r)$
        $\Leftrightarrow (p \wedge \sim q) \vee (p \vee \sim r)$ Implication (10a) and double negation (1)

17. a. $\sim(p \wedge q) \Leftrightarrow \sim p \vee \sim q$
Arizona does not border California, or Arizona does not border Nevada.

    b. $\sim(p \vee q) \Leftrightarrow \sim p \wedge \sim q$
There are no tickets available, and the agency cannot get tickets.

    c. $\sim(p \vee q) \Leftrightarrow \sim p \wedge \sim q$
The killer's hat was neither white nor gray.

19. $\sim(p \vee \sim q \vee r) \Leftrightarrow \sim p \wedge q \wedge \sim r$

21. a. Jeremy does not take 12 credits and Jeremy does not take 15 credits.

    b. Sandra does not receive a gift from Sally or Sandra does not receive a gift from Sacha.

23. a. I have a ticket to the theater, and I did not spend a lot of money.

    b. Basketball is played on an indoor court, and the players do not wear sneakers.

    c. The stock market is going up, and interest rates are not going down.

    d. Humans have enough water, and humans are not staying healthy.

25. a. If the number of odd numbers in a sum is even, then the sum is even.
True
For example: the number of odd numbers in the sum is even: $1 + 3 + 5 + 7$.
The sum of the odd numbers is 16 (even)

    b. If K is next to W, then the computer keyboard is not standard.
True.

27. a. Contrapositive: If a bird is not a hummingbird, then it is not small (F).
Converse: If a bird is a hummingbird, then it is small (T).

    b. Contrapositive: If two nonvertical lines are not parallel, they do not have the same slope (T).
Converse: If two nonvertical lines are parallel, they have the same slope (T).

    c. Contrapositive: If we are not in France, then we are not in Paris (T).
Converse: If we are in France, we must be in Paris (F).

d. Contrapositive: If you can legally make a U-turn, then the road is not one-way (T).
   Converse: If you cannot legally make a U-turn, then the road is one-way (F).

29. Ask either guard, "If I asked you whether your door was the door to freedom, would you say yes?" Phrasing the question this way forces the guard that always lies to tell the truth as to whether his door is the door to freedom.

31.

p	q	r	(p	$\to$	q)	$\wedge$	(q	$\to$	r)	$\to$	(p	$\to$	r)
T	T	T	T	T	T	T	T	T	T	T	T	T	T
T	T	F	T	T	T	F	T	F	F	T	T	F	F
T	F	T	T	F	F	F	F	T	T	T	T	T	T
T	F	F	T	F	F	F	F	T	F	T	T	F	F
F	T	T	F	T	T	T	T	T	T	T	F	T	T
F	T	F	F	T	T	F	T	F	F	T	F	T	F
F	F	T	F	T	F	T	F	T	T	T	F	T	T
F	F	F	F	T	F	T	F	T	F	T	F	T	F
(1)	(2)	(3)		(4)		(7)		(5)		(8)		(6)	

Exercises 12.5

1. m = "Sue goes to the movies."
   r = "Sue reads."
   1. $m \vee r$ hyp.
   2. $\sim m$ hyp.
   3. r disj. syll. (1, 2)

3. a = "My allowance comes this week."
   p = "I pay the rent."
   b = "My bank account will be in the black."
   e = "I will be evicted."
   1. $(a \wedge p) \to b$ hyp.
   2. $\sim p \to e$ hyp.
   3. $\sim e \wedge a$ hyp.
   4. $\sim e$ subtr. (3)
   5. p mod. tollens (2, 4)
   6. a subtr. (3)
   7. b mod. ponens (5, 6, 1)

5. p = "The price of oil increases.
   a = "The OPEC countries are in agreement."
   d = "There is a U.N. debate."
   1. $p \to a$ hyp.
   2. $\sim d \to p$ hyp.
   3. $\sim a$ hyp.
   4. $\sim p$ mod. tollens (1, 3)
   5. d mod. tollens (2, 4)

7. g = "The germ is present."
   r = "The rash is present."
   f = "The fever is present."
   1. $g \to (r \wedge f)$ hyp.
   2. f hyp.
   3. $\sim r$ hyp.
   4. $\sim r \vee \sim f$ addition (3)
   5. $\sim (r \wedge f)$ DeMorgan (4)
   6. $\sim g$ mod tollens (1, 5)

9. c = "The material is cotton."
   r = "The material is rayon."
   d = "The material can be made into a dress."
   1. $(c \vee r) \to d$ hyp.
   2. $\sim d$ hyp.
   3. $\sim (c \vee r)$ mod. tollens (1, 2)
   4. $\sim c \wedge \sim r$ DeMorgan (3)
   5. $\sim r$ subtraction (4)

11. s = "Salaries go up."
    m = "More people apply."
    If s is false and m is true, then
    $(s \to m) \wedge (m \vee s)$ is true but s is false.
    $(s \to m) \wedge (m \vee s) \Rightarrow s$.
    The argument is invalid.

13. y = "The balloon is yellow."
    p = "The ribbon is pink."
    h = "The balloon is filled with helium."
    1. $y \vee p$ hyp.
    2. $h \to \sim y$ hyp.
    3. h hyp.
    4. $\sim y$ mod. ponens (2, 3)
    5. p disj. syllogism (1, 4)
    The argument is valid.

15. p = "The papa bear sits."
    m = "The mama bear stands."
    b = "The baby bear crawls on the floor."
    1. $p \to m$ hyp.
    2. $m \to b$ hyp.
    3. $\sim b$ hyp.
    4. $\sim m$ mod. tollens (2, 3)
    5. $\sim p$ mod. tollens (1, 4)
    The argument is valid.

17. w = "Wheat prices are steady."
    e = "Exports will increase."
    s = "The GNP will be steady."
    If w and s are true and e is false, then
    $[w \to (e \vee s)] \wedge (w \wedge s)$ is true but e is false.
    $[w \to (e \vee s)] \wedge (w \wedge s) \Rightarrow e$.
    The argument is invalid.

19. i = "Tim is industrious."
    p = "Tim is in line for a promotion."
    l = "Tim is thinking of leaving."
    If l and p are true and i is false, then
    $(i \to p) \wedge (p \vee l)$ is true but $l \to i$ is false.
    $(i \to p) \wedge (p \vee l) \Rightarrow l \to i$
    The argument is invalid.

21. s = "Sam goes to the store."
    m = "Sam needs milk."
    $H_1 = s \to m$
    $H_2 = \sim m$
    $C = \sim s$
    1. s $\sim C$
    2. $s \to m$ H_1
    3. m $\sim H_2$; mod. ponens (1, 2)

23. n = "The newspaper reports the crime."
    t = "Television reports the crime."
    s = "The crime is serious."
    k = "A person is killed."
    $H_1 = (n \wedge t) \to s$
    $H_2 = k \to n$
    $H_3 = k$
    $H_4 = t$
    $C = s$
    1. $\sim s$ $\sim C$
    2. $(n \wedge t) \to s$ H_1
    3. $\sim(n \wedge t)$ mod. tollens (1, 2)
    4. $\sim n \vee \sim t$ DeMorgan (3)
    5. t H_4
    6. $\sim n$ disj. syllogism (4, 5)
    7. $k \to n$ H_2
    8. $\sim k$ $\sim H_3$; mod. tollens (6, 7)

25. j = "Jimmy finds his keys."
    h = "He does his homework."
    $H_1 = \sim j \to h$
    $H_2 = \sim h$
    $C = j$
    Direct proof
    1. $\sim j \to h$ H_1
    2. $\sim h$ H_2
    3. j mod. tollens (1, 2)
    Indirect proof
    1. $\sim j$ $\sim C$
    2. $\sim j \to h$ H_1
    3. h mod. ponens (1, 2) ⎫ contradiction
    4. $\sim h$ H_2 ⎭

27. m = "Marissa goes to the movies."
    k = "Marissa is in a knitting class."
    i = "Marissa is idle."
    $H_1 = \sim m \to \sim i$
    $H_2 = \sim k \to i$
    $H_3 = \sim k$
    $C = m$
    Direct proof
    1. $\sim m \to \sim i$ H_1
    2. $\sim k \to i$ H_2
    3. $\sim k$ H_3
    4. i mod. ponens (2, 3)
    5. m mod. tollens (1, 4)

    Indirect proof
    1. $\sim m$ $\sim C$
    2. $\sim m \to \sim i$ H_1
    3. $\sim i$ mod. ponens (1, 2)
    4. $\sim k \to i$ H_2

5. k mod. tollens (3, 4) ⎫
6. $\sim k$ H_3 ⎭ contradiction

Exercises 12.6

1. **a.** "1 is even or 1 is divisible by 3" is FALSE.

   b. "4 is even or 4 is divisible by 3" is TRUE.

   c. "3 is even or 3 is divisible by 3" is TRUE.

   d. "6 is even or 6 is divisible by 3" is TRUE.

   e. "5 is even or 5 is divisible by 3" is FALSE.

3. Abby's statement: $\forall x \sim p(x)$, or $\sim[\exists x\, p(x)]$. This is surely false. Abby meant to say, "not all men cheat on their wives" $\sim[\forall x\, p(x)]$, or $\exists x \sim p(x)$.

5. **a.** $\forall x\, p(x)$

   b. $\exists x \sim p(x)$

   c. $\exists x\, p(x)$

   d. $\sim[\forall x\, p(x)]$

   e. $\forall x \sim p(x)$

   f. $\sim[\exists x\, p(x)]$

   g. (b) and (d); (e) and (f) · (b) and (d) both say that there are some university professors who don't like poetry; (e) and (f) both say that no university professors like it.

7. **a.** TRUE; $p(4) = $ (4 is prime) $\to$ ($4^2 + 1$ is even) is TRUE, because the hypothesis is FALSE.

   b. FALSE; $p(2) = $ (2 is prime) $\to$ ($2^2 + 1$ is even) is FALSE.

9. **a.** T; every x is either even or odd.

   b. $[\forall x\, p(x)] \lor [\forall x\, q(x)]$
       F   F   F

   c. T; 5 is odd, hence even or odd, for instance.

   d. F; no x is both even and odd.

   e. F; 5 is not both even and odd, for instance.

   f. $[\exists x\, p(x)] \land [\exists x\, q(x)]$
      T   T   T

   g. F; (4 is even) $\to$ (4 is odd) is false, for instance.

   h. $[\forall x\, p(x)] \to [\forall x\, g(x)]$
      F   T   F

11. **a.** Not every dog has iss day.

    b. No men fight wars.

    c. Some mothers are unmarried.

    d. There exists a pot without a cover.

    e. All children have pets.

    f. Every month has 30 days.

13. **a.** "The sum of any two nonnegative integers is greater than 12." FALSE: consider $x = 1$, $y = 2$. "There exist two nonnegative integers whose sum is not greater than 12."

    b. "For any nonnegative integer, there is a nonnegative integer that, added to the first, makes a sum greater than 12." TRUE

    c. "There is a nonnegative integer that, added to any other nonnegative integer, makes a sum greater than 12." TRUE (Try $x = 13$.)

    d. "There are two nonnegative integers, the sum of which is greater than 12." TRUE (Try $x = 6$, $y = 7$.)

15. **a.** FALSE: let $x = 2$, $y = 3$.

    b. TRUE: for any x, let $y = x$.

    c. TRUE: let $x = 1$.

    d. FALSE: no y is divisible by every x.

    e. TRUE: for any y, let $x = y$.

f. TRUE: any x divides itself.

17. a. $S \subseteq T$ translates as $\forall x[x \geq 8 \rightarrow x \leq 10]$.

   b. No; consider $x = 11$.

19. $S = \{2, 4, 6, 8\}$, $T = \{1, 2, 3, 4, 6, 8\}$.
    So, $\forall x[x \in S \rightarrow x \in T)$.

Exercises 12.7

1. $(p \vee q) \wedge (p \vee \sim q)$

3. $((p \wedge q) \wedge \sim r) \wedge (\sim q \vee r)$

5.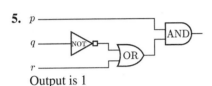
   Output is 1

7.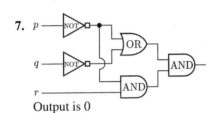
   Output is 0

9. The circuit represents the logic statement
   $(p \wedge q) \vee (p \wedge \sim q)$
   $\Leftrightarrow p \wedge (q \vee \sim q)$ Distributive law (4b)
   $\Leftrightarrow p \wedge t$ (7a)
   $\Leftrightarrow p$ (6d)

11. The circuit represents the logic statement
    $((p \wedge q) \wedge r) \vee \sim ((p \vee q) \vee \sim r)$
    $\Leftrightarrow (r \wedge (p \wedge q)) \vee (r \wedge \sim (p \vee q))$ Commutative law (2b) and DeMorgan's law (8a)
    $\Leftrightarrow r \wedge [(p \wedge q) \vee \sim (p \vee q)]$ Distributive law (4b)

13.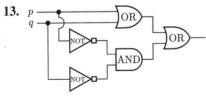

15.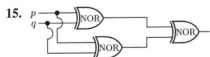

17.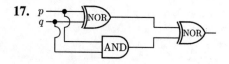

19. A ──┐
    B ──┤XOR├──▷NOT├──

Chapter 12 Fundamental Concept Exercises

1. A declarative statement that is either true or false

2.

p	q	$p \wedge q$	$p \vee q$	$\sim p$	$p \rightarrow q$	$p \leftrightarrow q$
T	T	T	T	F	T	T
T	F	F	T	F	F	F
F	T	F	T	T	T	F
F	F	F	F	T	T	T

3. when p is TRUE and q is FALSE

4. have the same truth tables; compare their truth values

5. $\sim(p \vee q) \Leftrightarrow (\sim p \wedge \sim q), \sim(p \wedge q) \Leftrightarrow (\sim p \vee \sim q)$; to negate statements having the disjunctive or conjunctive connectives.

6. $p; q$

7. $\sim q \rightarrow \sim p, q \rightarrow p$; TRUE, either TRUE or FALSE

8. statement: If a person is a resident of Maryland, then he or she is a resident of the United States.
   Contrapositive: If a person is not a resident of the United States, then he or she is not a resident of Maryland.
   Converse: If a person is a resident of the United States, then he or she is a resident of Maryland.

9. $p \wedge \sim q$

10. AND, $\wedge$; OR, $\vee$; NOT, $\sim$; IF ... THEN, $\rightarrow$; XOR, $\oplus$

11. a statement form that has truth value TRUE regardless of the truth values of its component statements; see whether the final column of its truth table has all Ts

12. $\sim[\exists x p(x)] \Leftrightarrow [\forall x \sim p(x)]; \sim[\forall x p(x)] \Leftrightarrow [\exists x \sim p(x)]$

Chapter 12 Chapter Review Exercises

1. a. Statement

   b. Not a statement—not a declarative sentence.

   c. Statement

d. Not a statement—"he" is not specified.

e. Statement

2. a. If two lines are perpendicular, then their slopes are negative reciprocals of each other.

   b. If goldfish can live in a fishbowl, then the water is aerated.

   c. If it rains, then Jane uses her umbrella.

   d. If Sally gives Morris a treat, then he ate all his food.

3. a. Contrapositive: If the Yankees are not playing in Yankee Stadium, then they are not in New York City; converse: If the Yankees are playing in Yankee Stadium, then they are in New York City.

   b. Contrapositive: If the quake is not considered major, then the Richter scale does not indicate the earthquake is a 7; converse: If the quake is considered major, then the Richter scale indicates the earthquake is a 7.

   c. Contrapositive: If a coat is not warm then it is not made of fur; converse: If a coat is warm then it is made of fur.

   d. Conrapositive: If Jane is not in Moscow then she is not in Russia; converse: If Jane is in Moscow then she is in Russia.

4. a. p = (two triangles are similar) and
      q = (their sides are equal).
      $p \rightarrow q$ negated becomes $\sim(p \rightarrow q)$ or $p \wedge \sim q$, or "two triangles are similar but their sides are unequal."

   b. U = {real numbers} and $p(x) = (x^2 = 5)$.
      $\exists x \, p(x)$ negated becomes $\forall x \sim q(x)$ or "For every real number x, $x^2 \neq 5$."

   c. U = {positive integers},
      $p(n = (n$ is even), and $q(n) = (n^2$ is even).
      $\forall n[p(n) \rightarrow q(n)]$ negated becomes $\exists n \sim [p(n) \rightarrow q(n)]$ or $\exists n[p(n) \wedge \sim q(n)]$, or "There exists a positive integer n such that n is even but n^2 is not even."

   d. U = {real numbers} and $p(x) = (x^2 + 4 = 0)$.
      $\exists x \, p(x)$ negated becomes $\forall x \sim p(x)$ or "For every real number x, $x^2 + 4 \neq 0$."

5. a.
p	p	$\vee$	$\sim$	p
T	T	T	F	T
F	F	T	T	F
(1)		(3)	(2)	

   Tautology

b.

p	q	(p	→	q)	↔	(~	p	∨	q)
T	T	T	T	T	**T**	F	T	T	T
T	F	T	F	F	**T**	F	T	F	F
F	T	F	T	T	**T**	T	F	T	T
F	F	F	T	F	**T**	T	F	T	F
(1)	(2)		(4)		(6)	(3)		(5)	

Tautology

c. Let p and q be TRUE.
$(p \wedge \sim q) \leftrightarrow \sim(\sim p \wedge q)$
T F F T **F** T F T F T
Not a tautology

d. Let p, q, and r be FALSE.
$[p \rightarrow (q \rightarrow r)] \leftrightarrow [(p \rightarrow q) \rightarrow r]$
F T F T F **F** F T F F F
Not a tautology

6. a.

p	q	r	p	→	(~	q	∨	r)
T	T	T	T	**T**	F	T	T	T
T	T	F	T	**F**	F	T	F	F
T	F	T	T	**T**	T	F	T	T
T	F	F	T	**T**	T	F	T	F
F	T	T	F	**T**	F	T	T	T
F	T	F	F	**T**	F	T	F	F
F	F	T	F	**T**	T	F	T	T
F	F	F	F	**T**	T	F	T	F
(1)	(2)	(3)		(6)	(4)		(5)	

b.

p	q	r	p	∧	(q	↔	(r	∧	p))
T	T	T	T	**T**	T	T	T	T	T
T	T	F	T	**F**	T	F	F	F	T
T	F	T	T	**F**	F	F	T	T	T
T	F	F	T	**T**	F	T	F	F	T
F	T	T	F	**F**	T	F	T	F	F
F	T	F	F	**F**	T	F	F	F	F
F	F	T	F	**F**	F	T	T	F	F
F	F	F	F	**F**	F	T	F	F	F
(1)	(2)	(3)		(6)		(5)		(4)	

7. a. True—a version of disjunctive syllogism

b. False—consider p false and q true.

8. a. True—contrapositive

b. False—consider p, q, and r false.

SSM: Finite Math   Chapter 12: Logic

9. a. False—consider p false and q true.

    b. True—modus tollens

10. a. $C = 3(4) + 5 = 17 > 0$ and $B = 5 > 0$, so the condition is met and $Z = 17$.

    b. $B = 2 \not> 3$, so the condition fails and $Z = 100$.

    c. $C = 3(-4) + 5 = -7 \not>$, so the condition fails and $Z = 100$.

    d. $B = -2 \not> 3$, so the condition fails and $Z = 100$.

11. a. $C = 10$, so the condition is met and $Z = 5 \times 10 = 50$.

    b. $C = 10$, so the condition is met and $Z = (5)(-5) = -25$.

    c. $X = -10 \not> 0$, $C = 2 \neq 10$, so the condition fails and $Z = (-10) + (-5) = -15$.

    d. $X = 2 > 0$, $Y = 5 > 0$, so the condition is met and $Z = 2 \times 5 = 10$.

12. a. Cannot be determined

    b. TRUE, by the contrapositive

    c. TRUE

    d. Cannot be determined

    e. Cannot be determined

13. a. Cannot be determined

    b. Cannot be determined

    c. TRUE, by contraposition and DeMorgan

14. a. TRUE (given the additional assumption that at least two mathematicians exist)

    b. FALSE; let $p(x)$ = "like rap music" and U = {mathematicians}.
        $\forall x\, p(x) \quad \sim \exists x\, p(x)$

    c. TRUE

15. a. Cannot be determined

    b. Cannot be determined

    c. True; $\exists x[\sim r(x)] \Leftrightarrow \sim \forall x[r(x)]$

Chapter 12: Logic   **SSM:** Finite Math

16. t = "Taxes go up."
    s = "I sell the house."
    m = "I move to India."
    1. $t \to (s \wedge m)$ hyp.
    2. $\sim m$ hyp.
    3. $\sim s \vee \sim m$ addition (2)
    4. $\sim(s \wedge m)$ DeMorgan (3)
    5. $\sim t$ mod. tollens (1, 4)

17. m = "I study mathematics."
    b = "I study business."
    p = "I can write poetry."
    1. $m \wedge b$ hyp.
    2. $b \to (\sim p \vee \sim m)$ hyp.
    3. b subtraction (1)
    4. $\sim p \vee \sim m$ mod. ponens (2, 3)
    5. m subtraction (1)
    6. $\sim p$ disj. syllogism (4, 5)

18. d = "I shop for a dress."
    h = "I wear high heels."
    s = "I have a sore foot."
    1. $d \to h$ hyp.
    2. $s \to \sim h$ hyp.
    3. d hyp.
    4. h mod. ponens (1, 3)
    5. $\sim s$ mod. tollens (2, 4)

19. a = "Asters grow in the garden."
    d = "Dahlias grow in the garden."
    s = "It is spring."
    1. $a \vee d$ hyp.
    2. $s \to \sim a$ hyp.
    3. s hyp.
    4. $\sim a$ mod. ponens (2, 3)
    5. d disj. syllogism (1, 4)

20. t = "The professor gives a test."
    h = "Nancy studies hard."
    d = "Nancy has a date."
    s = "Nancy takes a shower."
    $H_1 = t \to h$
    $H_2 = d \to s$
    $H_3 = \sim t \to \sim s$
    $H_4 = d$
    $C = h$
    1. $\sim h$ $\sim C$
    2. $t \to h$ H_1
    3. $\sim t$ mod. tollens (1, 2)

4. $\sim t \rightarrow \sim s$ H_3
5. $\sim s$ mod. ponens (3, 4)
6. $d \rightarrow s$ H_2
7. $\sim d$ $\sim H_4$; mod. tollens (5, 6)

Contradiction

21.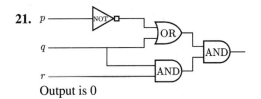

Output is 0

22. $((p \wedge \sim q) \wedge \sim (q \vee \sim r)) \wedge r$
$\Leftrightarrow ((p \wedge \sim q) \wedge (\sim q \wedge r)) \wedge r$ DeMorgan's law (8a)
$\Leftrightarrow ((p \wedge r) \wedge \sim q) \wedge r$ Distributive law (4b) and commutative law (2b)
$\Leftrightarrow (p \wedge (r \wedge r)) \wedge (\sim q \wedge r)$ Distributive law (4b)
$\Leftrightarrow p \wedge (\sim q \wedge r)$ Idempotent law (5b)

Conceptual Exercises

23.

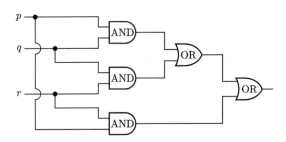

24.

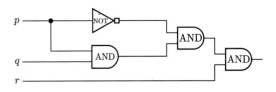

25. (p NOR q) NOR (p AND q)

p	q	(p	NOR	q)	NOR	(p	AND	q)	$\leftrightarrow$	(p	XOR	q)
T	T	T	F	T	F	T	T	T	**T**	T	F	T
T	F	T	F	F	T	T	F	F	**T**	T	T	F
F	T	F	F	T	T	F	F	T	**T**	F	T	T
F	F	F	T	F	F	F	F	F	**T**	F	F	F
(1)	(2)		(3)		(6)		(4)		(7)		(5)